# Reports of General MacArthur

## JAPANESE OPERATIONS IN THE SOUTHWEST PACIFIC AREA

VOLUME II—PART I

COMPILED FROM
JAPANESE DEMOBILIZATION BUREAUX RECORDS

Library of Congress Catalog Card Number: 66–60007

For sale by the Superintendent of Documents, U.S. Government Printing Office, Washington, D.C. 20402

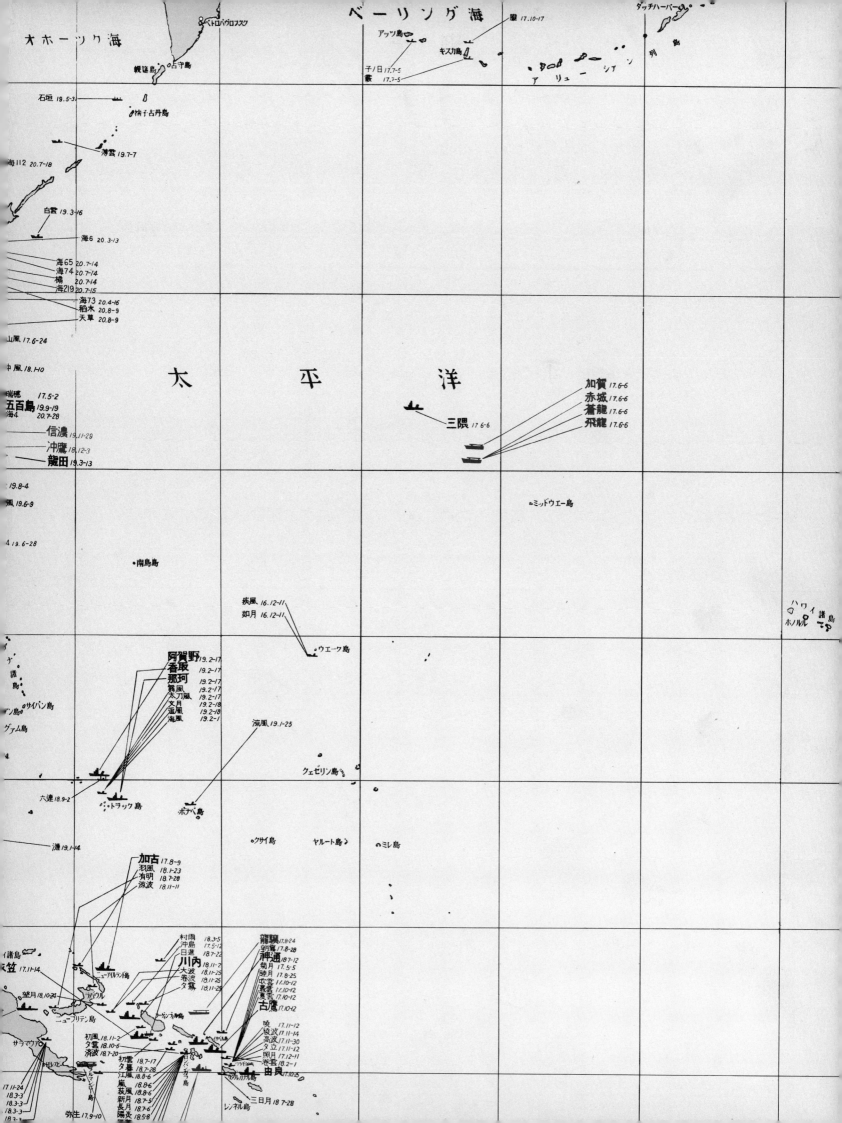

# FOREWORD

The *Reports of General MacArthur* include two volumes being published by the Department of the Army in four books reproduced exactly as they were printed by General MacArthur's Tokyo headquarters in 1950, except for the addition of this foreword and indexes. Since they were Government property, the general turned over to the Department in 1953 these volumes and related source materials. In Army and National Archives custody these materials have been available for research although they have not been easily accessible. While he lived, General MacArthur was unwilling to approve the reproduction and dissemination of the *Reports,* because he believed they needed further editing and correction of some inaccuracies. His passing permits publication but not the correction he deemed desirable. In publishing them, the Department of the Army must therefore disclaim any responsibility for their accuracy. But the Army also recognizes that these volumes have substantial and enduring value, and it believes the American people are entitled to have them made widely available through government publication.

The preliminary work for compiling the MacArthur volumes began in 1943 within the G–3 Section of his General Staff, and was carried forward after the war by members of the G–2 Section, headed by Maj. Gen. Charles A. Willoughby with Professor Gordon W. Prange, on leave from the University of Maryland, as his principal professional assistant. Volume II of the *Reports* represents the contributions of Japanese officers employed to tell their story of operations against MacArthur's forces. The very large number of individuals, American and Japanese, who participated in the compilation and editing of the *Reports* would make a complete listing of contributors relatively meaningless.

Volume I narrates the operations of forces under General MacArthur's command from the Japanese attack on Luzon in 1941 through the surrender in 1945. While service histories have covered much of the same ground in separate volumes, no single detailed narrative of General MacArthur's leadership as commander of the Southwest Pacific Area has yet appeared. Chapters dealing with the reconquest of Borneo, plans for the invasion of Japan, and the Japanese surrender make a distinctly new contribution. Volume I Supplement describes the military phase of the occupation through December 1948, reporting events not treated elsewhere in American publications. Volume II on Japanese operations brings together a mass of information on the enemy now only partially available in many separate works. Collectively, the *Reports* should be of wide interest and value to the American people generally, as well as to students of military affairs. They are an illuminating record of momentous events influenced in large measure by a distinguished American soldier.

Washington, D.C.
January 1966

HAROLD K. JOHNSON
*General, United States Army*
*Chief of Staff*

# PREFACE

This volume parallels the record of Allied operations in the SWPA from the Defense of Luzon, 8 December 1941, to the Surrender Negotiations in Manila, 15 August 1945. It is the Japanese official record, contained in operational monographs furnished by the Japanese Demobilization Bureaux, the successors to the former War and Navy Ministries, developed by Officers of the Japanese Imperial Headquarters, Tokyo, and on the Staffs of major Japanese Commanders in the field. Like Volume I, the material is thus presented by eye witnesses to events, and is supported by official documentary evidence.

It is a record of bitter resistance and tenacious fighting by a first-class Army and Navy, led by Diplomats and Military Politicians through the holocaust of national destruction, an Army that was steeped in medieval cruelty, but fought with the most modern technical skill and savage valor, until superior skill and equal valor broke the spell of the Samurai and the legend of an invincible Empire.

Douglas MacArthur

# TABLE OF CONTENTS—PART I

|  | Page |
|---|---|
| CHAPTER I: Pre-War Japanese Military Preparations 1941 | 1 |
|     Summary | 1 |
|     Pearl Harbor Planning (January-November 1941) | 6 |
|     October 1941 | 8 |
|     November 1941 | 9 |
|     December 1941 | 17 |
| CHAPTER II: Pre-War Japanese Espionage and Intelligence, 1940–1941 | 21 |
|     General | 21 |
|     Philippines | 23 |
|     New Guinea | 24 |
|     East Indies | 29 |
| CHAPTER III: Politico-Military Evolution Toward War | 30 |
|     Historical Background | 30 |
|     Drift Toward Crisis | 31 |
|     War Deliberations | 34 |
| CHAPTER IV: Basic Strategy and Military Organization | 44 |
|     Strategy for a Long War | 44 |
|     Manpower and Materials | 45 |
|     Shipping | 49 |
|     Areas to be Occupied | 50 |
|     Estimate of Allied Strategy | 51 |
|     Timing of the Attack | 52 |
|     Central Command Organization | 52 |
|     Strength and Organization of Forces | 54 |
| CHAPTER V: Initial Offensives | 59 |
|     Planning of Invasion Operations | 59 |
|     Operational Strength | 61 |
|     Operations Orders | 66 |
|     Pearl Harbor Operation | 71 |
|     South Seas and Southern Operations | 74 |
| CHAPTER VI: Conquest of the Philippines | 79 |
|     Preliminary Planning | 79 |
|     Assignment of Forces | 83 |
|     Final Operations Plan | 87 |

|   |   | Page |
|---|---|---|
| Launching of Operations | | 90 |
| The Race for Manila | | 95 |
| The Fall of Manila | | 102 |
| Manila to Bataan | | 103 |
| Bataan, First Phase | | 106 |
| Reinforcement and Preparation | | 110 |
| Bataan, Second Phase | | 114 |
| Fall of Corregidor | | 117 |
| Surrender | | 122 |

CHAPTER VII: Threat to Australia: The Papua Offensive ............ 124

    Invasion of the Bismarcks ............................................. 124
    Advance to New Guinea ............................................... 126
    Plans Against Australia ................................................. 131
    Abortive Sea Advance on Port Moresby ........................ 134
    Plans for a Land Offensive ........................................... 138
    Landing at Buna ......................................................... 142
    Advance to Kokoda ..................................................... 143
    Final Plans Against Moresby ....................................... 143
    Fighting on Guadalcanal .............................................. 146
    Build-up of Forces in New Guinea ............................... 149
    Attack on Milne Bay ................................................... 153
    Owen Stanleys Offensive ............................................. 157
    Retreat from the Owen Stanleys .................................. 164

CHAPTER VIII: Defense of Papua ............................................. 171

    Eighth Area Army Activated ....................................... 171
    Situation in Buna Area ................................................ 173
    First Phase of Fighting ................................................. 176
    Reinforcement Attempts ............................................... 177
    Fall of Buna ................................................................ 180
    Sanananda—Giruwa ..................................................... 181
    Strengthening of Bases in New Guinea ......................... 188
    The Wau Offensive ..................................................... 190
    Evacuation of Guadalcanal .......................................... 193
    Menace of the B-17's .................................................. 197
    Eighteenth Army Reinforcement .................................. 200
    Battle of the Bismarck Sea ........................................... 202
    Shift of Emphasis to Papua ......................................... 205

CHAPTER IX: Fighting Withdrawal to Western New Guinea ......... 208

    Southeast Area Situation, June 1943 ............................. 208

|  | Page |
|---|---|
| Defense of Salamaua | 212 |
| Attack on Lae | 216 |
| Fighting in the Central Solomons | 219 |
| Evacuation of Lae and Ramu Valley Operations | 221 |
| New Strategic Defense Zone | 225 |
| Dampier Strait Defense: Finschhafen | 229 |
| Bougainville | 233 |
| Dampier Strait Defense: New Britain | 236 |
| Saidor | 240 |
| Isolation of Rabaul | 244 |
| Bougainville Counteroffensive | 246 |
| Southeast Area Situation, March 1944 | 248 |

CHAPTER X: Western New Guinea Operations ... 250

| Strategic Planning | 250 |
|---|---|
| Western New Guinea Defenses | 252 |
| Setbacks to Defense Preparations | 257 |
| Hollandia–Aitape | 261 |
| Failure of the Reinforcement Plan | 272 |
| Revision of Defense Plans | 273 |
| Wakde–Sarmi | 276 |
| Biak First Phase | 283 |
| The Kon Operation | 287 |
| Philippines Sea Battle | 292 |
| Biak Final Phase—Noemfoor | 293 |
| Aitape Counterattack | 297 |
| End of the New Guinea Campaign—Sansapor | 303 |

CHAPTER XI: Philippine Defense Plans ... 304

| Strategic Situation, July 1944 | 304 |
|---|---|
| Importance of the Philippines | 305 |
| Local Situation | 309 |
| Southern Army Defense Plans | 312 |
| Battle Preparations No. 11 | 316 |
| Central Planning for Decisive Battle | 319 |
| Army Orders for the *Sho-Go* Operations | 322 |
| Navy Orders for the *Sho-Go* Operations | 328 |
| Preparations for Battle | 330 |
| Final Preparations, Central and Southern Philippines | 338 |

CHAPTER XII: Prelude to Decisive Battle ... 343

| Initial Air Strikes | 343 |

|  | Page |
|---|---|
| Invasion of Palau | 347 |
| Defense of Morotai | 348 |
| Hour of Decision Nears | 353 |
| Formosa Air Battle | 357 |

# ILLUSTRATIONS—PART I

| Plate | | Page |
|---|---|---|
| 1 | Imperial Rescript Declaring War | XIV |
| 2 | Japanese Aerial Photo Coverage, Northern Luzon, Nov–Dec 1940 | 5 |
| 3 | Resting Comrades | 13 |
| 4 | Disposition of Air Forces in South Sea Area | 22 |
| 5 | Disposition of Australian Forces, July 1940 | 26 |
| 6 | Japanese Estimates of Java Troop Strength and Disposition, 1941 | 27 |
| 7 | Japanese Column on the March | 35 |
| 8 | Japan's Basic Strategy, December 1941 | 46, 47 |
| 9 | Central Command Organization, 8 December 1941 | 53 |
| 10 | Army Chain of Command, 8 December 1941 | 55 |
| 11 | Navy Chain of Command, 8 December 1941 | 57 |
| 12 | Disposition of Japan's Military Forces Prior to Outbreak of War | 64, 65 |
| 13 | Pearl Harbor Attack, 8 December 1941 (Tokyo Time) | 69 |
| 14 | Pearl Harbor on 8 December 1941 (Tokyo Time) | 73 |
| 15 | Southern Operations, December 1941–May 1942 | 76, 77 |
| 16 | Japanese Air Operations in Philippines, December 1941 | 84 |
| 17 | Military Topography of Luzon | 85 |
| 18 | Composition and Missions of Landing Forces | 89 |
| 19 | Air Raid on Clark Field | 92 |
| 20 | Invasion of Philippines, 8–25 December 1941 | 96 |
| 21 | Race to Manila, December 1941–January 1942 | 97 |
| 22 | Lingayen—Cabanatuan Operation, 22 December 1941–3 January 1942 | 100 |
| 23 | Attack on Cavite Naval Base | 101 |
| 24 | Bataan Operations, First Phase, Early January–22 February 1942 | 105 |
| 25 | Supply Train Marching Toward the Front | 109 |
| 26 | Fourteenth Army Plan of Attack—Bataan, 22 March 1942 | 113 |
| 27 | Bataan, Second Phase, and Corregidor Operations | 116 |
| 28 | Gun Smoke Road, Corregidor | 120 |
| 29 | Bataan Meeting of Gen Wainwright and Gen Homma | 121 |
| 30 | Conquest of E. New Guinea, Bismarcks, and Solomons, 1942 | 128 |
| 31 | Japanese Landing Operations | 129 |
| 32 | Battle of the Coral Sea, 4–11 May 1942 | 137 |
| 33 | Terrain Along the Kokoda Trail | 140 |
| 34 | Hardships of the Troops in the Owen Stanleys | 144 |
| 35 | First and Second Battles of Solomon Sea, August 1942 | 147 |
| 36 | Army-Navy Cooperation on Guadalcanal | 150 |
| 37 | Operations on Guadalcanal, August–November 1942 | 151 |
| 38 | Landings on Milne Bay, August–October 1942 | 154 |
| 39 | Owen Stanley Penetration, 21 July–26 September 1942 | 158 |

| Plate | | Page |
|---|---|---|
| 40 | Takasago Unit Fighting Through Owen Stanleys | 162 |
| 41 | Looking at Port Moresby from Owen Stanley Mountain Range | 166 |
| 42 | Withdrawal from the Owen Stanleys, September–November 1942 | 167 |
| 43 | Buna–Gona Operation, November–December 1942 | 178 |
| 44 | Fate of Yasuda Force on New Guinea Front | 182 |
| 45 | Buna–Gona Operation, January 1943 | 186 |
| 46 | Withdrawal from Buna and Wau to Salamaua–Lae | 187 |
| 47 | Wau Offensive, January–February 1943 | 191 |
| 48 | Sea Battle in South Pacific | 194 |
| 49 | Suicide Unit Bidding Farewell to Commanding General Sano | 195 |
| 50 | Troops at Work, Southern Area | 199 |
| 51 | Battle of Bismarck Sea, 2–4 March 1942 | 203 |
| 52 | Japanese Dispositions in New Guinea and Solomons, June 1943 | 210 |
| 53 | Salamaua–Lae Operations, June–September 1943 | 214 |
| 54 | Navy Supplying Army Personnel by Submarine | 218 |
| 55 | New Georgia Operation, June–October 1943 | 222 |
| 56 | Ramu Valley Operation, September–November 1943 | 223 |
| 57 | Japan's National Defense Zone, September 1943 | 227 |
| 58 | Operations in Finschhafen Area, September–December 1943 | 231 |
| 59 | Bougainville Operation, November 1943–March 1944 | 235 |
| 60 | Western New Britain Operation, December 1943–February 1944 | 238 |
| 61 | Ramu Valley and Saidor Operations | 242 |
| 62 | Defense of Admiralties, February–March 1944 | 243 |
| 63 | Summary of Japanese Movements in Eastern New Guinea | 247 |
| 64 | Dispositions in New Guinea, 21 April 1944 | 254 |
| 65 | Japanese Engineer Activities in South Pacific | 259 |
| 66 | Army Day Poster: "Develop Asia" | 262 |
| 67 | Hollandia Operation, April–June 1944 | 267 |
| 68 | Deadly Jungle Fighting, New Guinea Front | 271 |
| 69 | Sarmi–Wakde Operation, May–July 1944 | 279 |
| 70 | Biak Operations, May–June 1944 | 286 |
| 71 | Naval Movements During Biak Operation, 2–13 June 1944 | 290 |
| 72 | Fierce Fighting of Otsu Unit in Saipan | 294 |
| 73 | Japanese Staff Conference: West Cave, Biak | 298 |
| 74 | Aitape Counterattack, 10 July–5 August 1944 | 302 |
| 75 | Changes in Shipping Routes, January 1943–August 1944 | 306 |
| 76 | Strategic Position of Philippines, July 1944 | 308 |
| 77 | Unloading Operations, Philippine Area | 313 |
| 78 | Subchaser in Action | 321 |
| 79 | Plans for *Sho* Operation No. 1, August 1944 | 324 |
| 80 | Japanese Air Disposition in Philippines, September 1945 | 332 |

| Plate | | Page |
|---|---|---|
| 81 | Japanese Dispositions in Southern Area, September 1944 | 336, 337 |
| 82 | Japanese Ground Dispositions in Philippines, September 1944 | 339 |
| 83 | Preliminary Operations in the Philippines, Sept–Oct 1944 | 346 |
| 84 | Morotai Operation, September–October 1944 | 351 |
| 85 | Air Force Day: Propaganda Poster | 355 |
| 86 | Transoceanic Air Raid During Typhoon | 359 |

## IMPERIAL RESCRIPT

We, by grace of heaven, Emperor of Japan, seated on the Throne of a line unbroken for ages eternal, enjoin upon ye, Our loyal and brave subjects:

We hereby declare war on the United States of America and the British Empire. The men and officers of Our army and navy shall do their utmost in prosecuting the war, Our public servants of various departments shall perform faithfully and diligently their appointed tasks, and all other subjects of Ours shall pursue their respective duties; the entire nation with a united will shall mobilize their total strength so that nothing will miscarry in the attainment of our war aims.

To insure the stability of East Asia and to contribute to world peace is the far-sighted policy which was formulated by Our Great Illustrious Imperial Grandsire and our Great Imperial Sire succeeding Him, and which We lay constantly to heart. To cultivate friendship among nation and to enjoy prosperity in common with all nations has always been the guiding principle of Our Empire's foreign policy. It has been truly unavoidable and far from Our wishes that Our Empire has now been brought to cross swords with America and Britain. More than four years have passed since China, failing to comprehend the true intentions of Our Empire, and recklessly courting trouble, disturbed the peace of East Asia and compelled Our Empire to take up arms. Although there has been re-established the National Government of China, with which Japan has effected neighbourly intercourse and co-operation, the regime which has survived at Chungking, relying upon American and British protection, still continues its fratricidal opposition. Eager for the realization of their inordinate ambition to dominate the Orient, both America and Britain, giving support to the Chungking regime in the name of peace, have aggravated the disturbances in East Asia. Moreover, these two Powers, inducing other countries to follow suit, increased military preparations on all sides of Our Empire to challenge us. They have obstructed by every means our peaceful commerce, and finally resorted to a direct severance of economic relations, menacing gravely the existence of Our Empire. Patiently have We waited and long have We endured, in the hope that Our Government might retrieve the situation in peace. But our adversaries, showing not the least spirit of conciliation, have unduly delayed a settlement; and in the meantime, they have intensified economic and military pressure to compel thereby Our Empire to submission. This trend of affairs would, if left unchecked, not only nullify Our Empire's efforts of many years for the sake of the stabilization of East Asia, but also endanger the very existence of Our nation. The situation being such as it is, Our Empire for its existence and self-defense has no other recourse but to appeal to arms and to crush every obstacle in its path.

The hallowed spirits of Our Imperial Ancestors guarding Us from above, We rely upon the loyalty and courage of Our subjects in Our confident expectation that the task bequeathed by Our Forefathers will be carried forward, and that the sources of evil will be speedily eradicated and an enduring peace immutably established in East Asia, preserving thereby the glory of Our Empire.

The 8th day of the 12th month of the 16th year of Showa

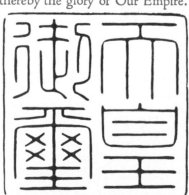

| Signature | Title |
|---|---|
| TOJO, Hideki | Prime Minister, Minister of War |
| HASHIDA, Kunihiko | Minister of Education |
| SUZUKI, Teiichi | Minister of State |
| INO, Sekiya | Minister of Agriculture and Forestry |
| KOIZUMI, Chikahiko | Minister of Health and Social Affairs |
| IWAMURA, Michiyo | Minister of Justice |
| SHIMADA, Shigetaro | Minister of Navy |
| TOGO, Shigenori | Minister of Foreign Affairs |
| TERASHIMA, Takeshi | Minister of Communications |
| KAYA, Okinobu | Minister of Finance |
| KISHI, Nobusuke | Minister of Commerce and Industry |
| HATTA, Yoshiaki | Minister of Railway |

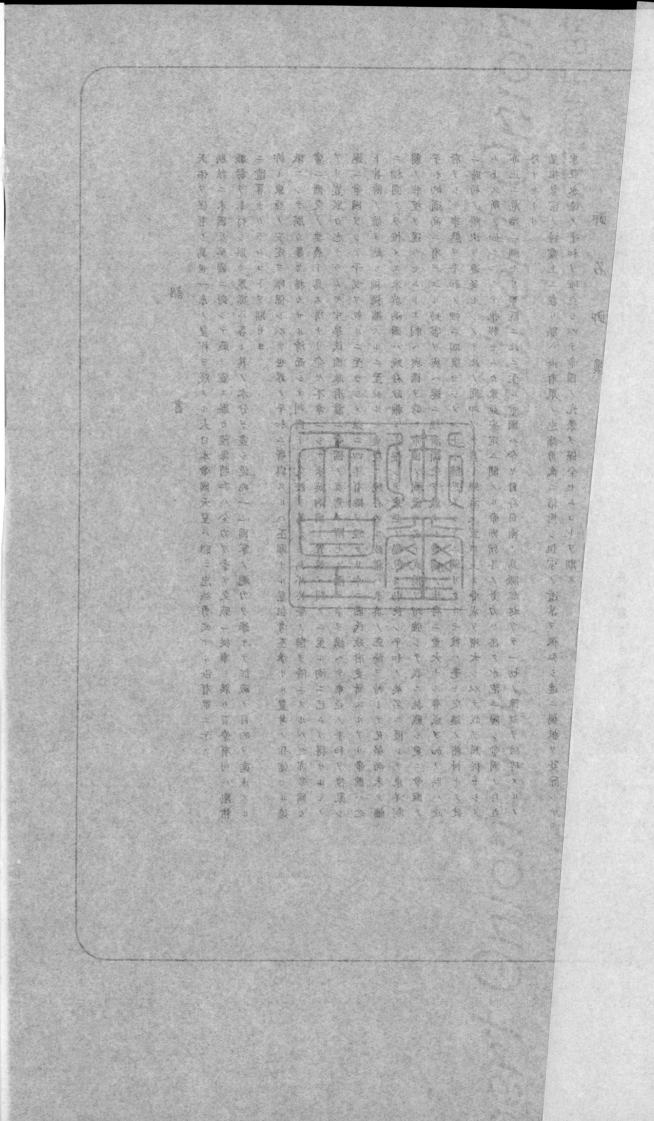

PLATE NO. 1
Imperial Rescript Declaring War

# CHAPTER I

# PRE-WAR JAPANESE MILITARY PREPARATIONS 1941

**Summary**

By 5 November 1941 the Imperial Japanese Government had positively committed itself to taking up arms against the United States, Great Britain, and the Netherlands if, by the first part of December, no diplomatic solution of the Pacific crisis appeared attainable.[1] This formal decision was made at the Imperial conference of 5 November, 17 days after the formation of the Tojo Cabinet.[2]

At this date Japan's military preparations for war were already far advanced. The Imperial conference of 6 September, in view of the unpromising outlook of negotiations with the United States, had decided that such preparations be rushed to completion by the end of October. Under that decision steps were taken to mobilize shipping for Army and Navy use, and the Army began assembling its invasion forces in Japan Proper, Formosa, and South China.[3]

The idea of a surprise attack on Pearl Harbor in the event of war was initially conceived by Admiral Isoroku Yamamoto, Commander-in-Chief of the Combined Fleet, in January 1941.[4] From June of the same year, as Navy leaders became more convinced that their strategy must be based on the hypothesis of fighting the United States and Britain simultaneously, Admiral Yamamoto actively pressed his plan against the opposition of some members of the Navy General Staff, and it was finally accepted in principle on 20 October.[5]

At the annual Navy war-games, held in Tokyo from 10 to 13 September, the general problem under study was fleet operations to establish Japanese control of the Western Pacific, assuming the United States, Britain, and the Netherlands as enemies. The war-games were conducted on the hypothesis of a sortie by the main body of the American fleet into the Western Pacific to block Japanese invasion operations against the Philippines, Malaya, and the Dutch East Indies. However, a special, restricted group of high-ranking staff officers simultaneously studied problems involved in a surprise task force strike at Pearl Harbor.[6]

---

1   Basic material for this chapter and Chapter II is contained in Research Report No. 131, *Japan's Decision to Fight*, 1 Dec 45, published by ATIS (Allied Translator and Interpreter Service), an operating agency of G-2 GHQ SWPA, handling the translation and dissemination of captured enemy documents and the interrogation of prisoners of war. This material has been revised in the light of additional research and Japanese source documents not previously available. All source materials cited in this chapter are located in G-2 Historical Section Files, GHQ FEC.

2   Imperial Conference decisions leading up to the final declaration of war are covered in detail in Chapter III.

3   Cf. Chapter III.

4   ATIS Research Report No. 131, op. cit., p. 66.

5   Cf. n. 2, Chapter V. The final detailed plan of the Pearl Harbor attack was approved by Admiral Osami Nagano, Chief of Navy General Staff, Imperial General Headquarters, Navy Section, on 3 November, two days before the issue of Combined Fleet Top Secret Operations Order No. 1 laying down the general outline of fleet operations.

6   *Showa Juroku Nen no Nihon Kaigun Zujo Enshu no Sogo Hokoku* 昭和十六年の日本海軍圖上演習の綜合報告 (Summary Report on Japanese Naval War Games, September 1941). Compiled by Rear Adm. Sadatoshi Tomioka, Chief, First Bureau (Operations), Imperial General Headquarters, Navy Section.

By 3 November the basic plan for all naval operations attendant upon the commencement of hostilities had been agreed upon and was embodied in Combined Fleet Top Secret Operations Order No. 1, issued on 5 November. Two days later Combined Fleet Top Secret Operations Order No. 2 designated 8 December[7] as the approximate date of the opening of hostilities (Y-Day), and units of the Pearl Harbor Task Force were simultaneously ordered to assemble in Tankan Bay by 22 November. The definitive date for the start of war (X-Day) was not fixed until 2 December, when the Task Force was already well on its way to Hawaii. A Combined Fleet order issued at approximately 5:30 p.m. on that date designated 8 December as X-Day.[8]

Final Army preparations were also completed during November. Imperial General Headquarters on 6 November established the order of battle of the Southern Army under over-all command of General Hisaichi Terauchi, and on 15 November designated the Philippines, British Malaya, the Dutch East Indies, and part of Southern Burma as the areas to be occupied. Invasion assignments were made to the various forces under Southern Army command on 20 November.[9]

Army and Navy operational plans were co-ordinated through an Army-Navy Central Agreement concluded in Tokyo on 10 November between General Terauchi, Commander-in-Chief of the Southern Army, and Admiral Yamamoto, Commander-in-Chief of the Combined Fleet. This basic document, which defined the relative commands, spheres of jurisdiction, missions, and responsibilities of the two services in all areas where joint operations were envisaged, was supplemented by detailed operational agreements concluded in mid-November between the Fleet and Army commanders assigned to each area.

The Japanese military authorities were both far-sighted and thorough in certain of their preparations for the war. Selected units were given specialized training in jungle warfare and amphibious operations; secret agents were dispatched to future zones of operation for purposes of espionage and reconnaissance; maps of crucial areas were prepared far in advance; morale and training literature was written and distributed to units; special striking forces were organized, equipped with tropical issue, and staged to carefully selected assembly areas where a maximum of security was assured; and the necessary transport and convoy facilities were arranged in advance.

Official unit reports refer to the periods 27 July–7 December 1941, 12 October–14 November 1941, and 10 October–8 December 1941 as having been devoted to preparation for the Philippine and Malayan Operations. Units receiving this training were currently in Manchuria, the vicinity of Shanghai, and at Palau.

By 10 November 1941, copies of a pamphlet entitled, "Read This and the War is Won", had been received by 55th Division Infantry Group.[10] The text was clearly premonitory of the imminence of war with the

---

7 7 December West Longitudinal time. Unless otherwise specified, hours and dates throughout this volume are Tokyo time. Japanese Army and Navy operational records employ Tokyo time exclusively regardless of the area under discussion. For purposes of checking against Volume I, Southwest Pacific Area Series: *The Campaigns of MacArthur in the Pacific,* local times are given parenthetically where desirable. In Chapter III, however, the dates of diplomatic notes, official statements, and governmental orders are the dates of the issuing government.

8 ATIS Research Report No. 131, op. cit., pp. 77–8.

9 For details of Army orders covering the Southern operations, consult Chapter V.

10 Textual extracts from *Kore Sae Yomeba Kateru* これさへ讀めば勝てる (Read This and the War is Won) are given later in this chapter.

United States, Great Britain, and the Netherlands. Copies of this were issued to each Japanese soldier before embarkation for overseas.

The significantly named South Seas Detachment was already organized, on paper at least, by 15 November 1941. It comprised the force which took Guam on 10 December and later moved on to Rabaul and New Guinea.[11]

On 15 November, the Commanding General of the South Seas Detachment, Maj. Gen. Tomitaro Horii, issued a "Message to Warriors in the South Seas," addressed to all personnel serving under his command. This message forecasted with great explicitness the coming of war. No date of outbreak was mentioned, but the tenor of the communication was that of a commander to his troops on the eve of battle.

Certain elements of the newly formed South Seas Detachment were being routed to a staging area in Japan as early as 14 November 1941. A part of the 47th Antiaircraft Battalion, for example, left its station in Manchuria and was transported to the port of Sakaide in Northern Shikoku, via Pusan and Ujina. A major portion of the South Seas Detachment appears to have rendezvoused there. On 22, 23, and 24 November, various units embarked and departed for the Ogasawara (Bonin) Islands. The transports arrived at their destination on 27 November 1941, some touching briefly at Chichi-Jima en route to Haha-Jima. At Haha-Jima the troops rested and trained. On 4 December, the convoy sailed to carry out the attack on Guam.

On 29 November 1941, at 1500 hours, 1st Lt. Sakigawa, Commanding Officer of 2nd Company, 55th Transport Regiment, issued Saki Operation Order No. 2. This read in part, "The Detachment will attack Guam Island."

The mounting of the attack on British Malaya has been partially reconstructed from official documents and diaries deriving from the 41st Infantry Regiment, 106th Land Duty Company, Sasebo 5th Special Naval Landing Party, and 77th Air Regiment, elements of all of which participated in the operation.

On 17 November 1941, 41st Infantry Regiment, which had been stationed in the vicinity of Shanghai since at least early October, training for the Malayan Operations, held a review and ceremony in honor of their "departure for the field". On 18 November, elements of the regiment left Shanghai on the *Ryujo Maru* for the assembly point at Samah, on the southern shore of Hainan Island. The diary of one member contained the following entry under date of 18 November: ".... orders have finally arrived. The time has finally come for us to display activity. Are we going to be at war with A, B and D?" On 21 November the *Ryujo Maru* was anchored at Takao. Its date of arrival at Samah is not known, but other elements of the regiment had reached Haikow in Northern Hainan by 20 November. On the same date further elements of the regiment embarked on the *Aobayama Maru* and on 21–22 November sailed from Woosung for Samah. Part of the 106th Land Duty Company left Saigon on the *Tokokawa Maru* on 23 November and arrived at Samah on 25 November. On 25 November also, elements of Sasebo 5th Special Naval Landing Party, while en route to Palau, were ordered to change course and head for Samah. A second section of the 106th Land Duty Company embarked on the *Taikai Maru* at Saigon on 27 November

---

11 The mission assigned to the South Seas Detachment in the first phase of operations was to capture Guam and Rabaul. Its dispatch to New Guinea in July 1942 for the abortive Japanese drive on Port Moresby was not decided until January 1942. *Nanto Homen Sakusen Kiroku Sono Ichi Nankai Shitai no Sakusen* 南東方面作戰記錄其の一南海支隊の作戰 (Southeast Area Operations Record Part I: South Seas Detachment Operations) 1st Demobilization Bureau, Sep 46, pp. 4, 22.

and arrived at Hainan on 1 December. On 3 December, the 2d Squadron of the 77th Air Regiment was ordered to co-operate with the 70th Airfield Company in the air defense of Samah. By 4 December, the assembly was complete. On 4 December, the advance landing forces sailed in convoy for Malaya.

Preparations for the eventual conquest of the Philippines date farther back. There is evidence of extensive pre-war aerial reconnaissance of northern Luzon during the period 27 November to 15 December 1940.[12] (Plate No. 2 shows the photographic coverage obtained and the dates on which the flights were carried out.)

Other evidence indicates that the training of units earmarked for participation in the Philippines campaign was probably under way by early fall of 1941. An extract from Fourth Air Army Ordnance Order No. 12, dated 26 March 1944, reads:

*Death certificate and service record of Sergeant Takeo Goto:*
*Unit: 25th Water Purification Unit.*
*Year of conscription: 1939*
*Service: 27 July 1941 to 7 December 1941, Manchurian Border Defense and preparation for the Philippines operations.*[13]

The main lines of the operational plan for the Philippines invasion were worked out at the joint Army-Navy staff conference held at Iwakuni, on the Inland Sea, from 14 to 16 November. Orders were issued on 20 November to Fourteenth Army units in Formosa directing them to concentrate at the assembly points in readiness for embarkation.[14]

The Tanaka Detachment, one of the advance forces which landed at Aparri and Vigan, on northern Luzon, embarked at Takao between 23 and 25 November and moved to the naval port of Mako, in the Pescadores, which had been fixed as the starting point of the advance invasion convoy. This force and the Kanno Detachment (Vigan landing) remained at Mako until 1700 on 7 December, when the convoy sailed for northern Luzon.[15]

On 16 November and 26 November respectively, the commanders of the heavy cruiser *Kako* and the light cruiser *Katori* addressed their crews in terms clearly indicative of the imminence of war. On the latter date also, Vice Adm. Chuichi Nagumo informed some of the personnel of the task force assembled at Tankan Bay that they were to attack Pearl Harbor.

Various individuals displayed advance knowledge or suspicion of the imminence of war. It is not certain in some cases whether this was based on information derived from reliable official sources or from rumor and popular gossip. Nevertheless, as early as October 1941, the rumor was current on Truk that war would break out with the United States between 25 December 1941 and 1 February 1942. On 18 November, a member of the 41st Infantry Regiment commented on the probable imminence of war with "A, B and D." On 26 November, a member of the 144th Infantry Regiment, South Seas Detachment, wrote, "Our battle zone will be Guam Island." Two other members of the South Seas Detachment displayed similar knowledge of impending hostilities on 29 November.

Between 2–7 December knowledge of the scheduled outbreak of hostilities on 8 December became quite general among members of stri-

---

12 Cf. discussion in Chapter II.
13 ATIS Bulletin No. 1060, p. 10.
14 *Hito Sakusen Kiroku Dai Ikki* 比島作戰記録第一期 (Philippines Operation Record Phase One) 1st Demobilization Bureau, Jun 46, p. 53.
15 Cf. Chapter VI.

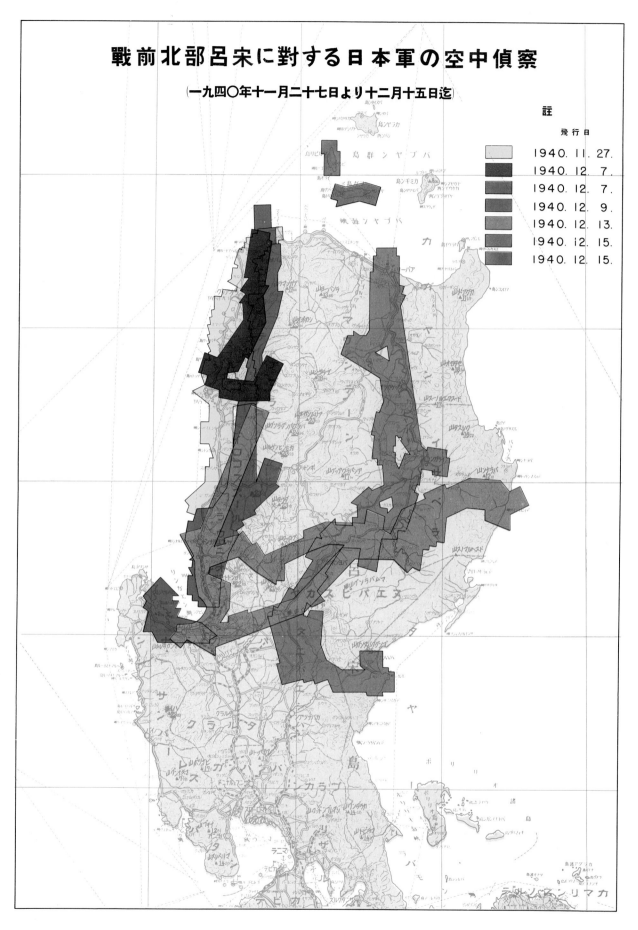

PLATE NO. 2

Japanese Aerial Photo Coverage, Northern Luzon

king forces. On 2 December the captain of the aircraft carrier *Kaga* announced to the crew that war would commence on 8 December. On 4 December Maj. Gen. Horii, Commanding General of the South Seas Detachment, issued a formal statement to the effect that Imperial Japan had, on 2 December 1941, decided to declare war on the United States, Great Britain, and the Netherlands. Thereafter knowledge of their objective appears to have been widespread among members of the South Seas Detachment. Evidence of similarly widespread knowledge among the forces assembled at Samah for the Malayan Operation is more scanty, but there appears to have been some awareness of their goal.

On 1 and 3 December orders for the air defense of Keelung and Samah respectively were issued by the 48th Field Antiaircraft Battalion and 77th Air Regiment. These clearly anticipated the possibility of enemy air attacks.

The evidence upon which the foregoing summary is based is further amplified in the following paragraphs. References have been arranged chronologically in accordance with the date of the most significant entry contained in the diary or other captured document under examination.

### Pearl Harbor Planning (Jan.-Nov. 1941)

When Japanese planes struck immobile United States warships and aircraft at Pearl Harbor on 8 December 1941, they were executing pin-pointed plans conceived months in advance and cloaked in the utmost secrecy. Authoritative Japanese documents obtained since the termination of war and interrogations of the high naval personnel who participated in or had knowledge of this planning make it possible to reconstruct a complete and accurate picture of how the Pearl Harbor attack was conceived and developed over an eight-month period preceding the final outbreak of hostilities.

Prior to 1941 Japanese naval planning for a possible war with the United States had been based upon the assumption that the latter would be Japan's only enemy, and it envisaged awaiting attack by the American fleet in the Western Pacific where Japan's numerically inferior fleet could operate at an advantage.[16] By the end of 1940, however, Japan's entry into the Tripartite Alliance and the United States' aid commitments to Britain had created a new international line-up which made previous Japanese naval planning obsolete. The Japanese Navy began to plan for a simultaneous war against the United States and Britain.

The idea of a surprise attack on Pearl Harbor at the outset of war, with the object of gaining at least temporary naval supremacy in the Western Pacific, was first conceived in early January 1941 by Admiral Yamamoto, Commander-in-Chief of the Japanese Combined Fleet. Admiral Yamamoto at that time ordered Rear Adm. Takijiro Onishi, chief of staff of the Eleventh Air Fleet, to study the feasibility of such an attack.[17]

On the basis of this preliminary study, Admiral Yamamoto in June 1941 began actively pressing for the adoption of his plan by the Navy General Staff as part of Japan's naval strategy in the event of war. Crippling the United States fleet at Hawaii at the start of hostilities, he argued, was absolutely essential to place the Western Pacific under Japanese control for the period necessary to complete the occupation of the strategic areas and economic resources of the South. Were Ame-

---

16 Cf. Chapter IV.
17 ATIS Research Report No. 131, op. cit., p. 66.

rican fleet strength at Hawaii left intact, it could immediately make an incursion into the Western Pacific in the midst of the Southern operations, catching the Japanese fleet dispersed in different areas and unable to deploy for a decisive battle. Under these conditions, he warned, the United States would probably seize Japan's island bases in the Marshalls and transform them into advance bases of operation against Japan.[18]

Despite Admiral Yamamoto's arguments, his plan was vigorously opposed by a section of the Navy General Staff on the ground that swift occupation of the Southern areas was the prime necessity, and that this might fail if Japanese naval strength were divided between operations against Hawaii and support of the Southern invasions. It was further pointed out that detection of the Japanese force en route to Hawaii might result in its complete destruction, and that, even if this did not occur, the attack would be ineffectual if the bulk of the United States fleet was not caught in Pearl Harbor.[19]

This disagreement in the Navy High Command had not been resolved by 10 September, when staff officers of all fleet units assembled at the Naval War College in Tokyo for the annual Navy war games. Just four days earlier the Imperial conference of 6 September had debated the issue of war or peace in a dramatic session and had decided that Japanese military preparations must be speedily brought to completion.[20] The games therefore took place amidst an atmosphere of unusual tension, further heightened by the fact that the central problem of study assumed an American fleet attack into the Western Pacific as a result of Japanese invasion operations in the Southern area.[21]

Admiral Yamamoto himself planned and exercised over-all supervision of the games. A general study session, including chart maneuvers participated in by all officers in tactical command of fleet units, occupied the first three days—10, 11, and 12 September. The last day, 13 September, was devoted to a special study session. Thirteen umpires headed by Rear Adm. Seiichi Ito, Vice-Chief of Navy General Staff, ruled on the execution of maneuvers. The Japanese (Blue) Forces were under command of Rear Adm. Matomi Ugaki, Chief of Staff of the Combined Fleet, and the British-American (Red) Forces under command of Vice Adm. Shiro Takasu, First Fleet Commander.

While the principal games were conducted on the old hypothesis of meeting an American fleet attack in the Western Pacific, a restricted group of staff officers of the Combined Fleet and commanders of those fleet units which eventually made up the Pearl Harbor Task Force[22] met in a separate and top-secret session, the purpose of which was to study problems connected with a possible surprise attack on Pearl Harbor. These problems included:

1. Feasibility of an attack if (as estimated)

---

18 Statement by Rear Adm. Tomioka, previously cited.
19 Ibid.
20 Cf. Chapter III.
21 Summary Report on Japanese Naval War Games. Compiled by Rear Adm. Tomioka, previously cited.
22 In addition to Admiral Yamamoto and other umpires, participants in the special Pearl Harbor study session were: Rear Adm. Ugaki, (Chief of Staff) and nine staff officers of the Combined Fleet: Vice Adm. Chuichi Nagumo, (C-in-C), Rear Adm. Ryunosuke Kusaka (Chief of Staff), and two staff officers of the First Air Fleet; Rear Adm. Tamon Yamaguchi (Commander) and two staff officers of the 2d Carrier Division; Rear Adm. Tadaichi Hara (Commander) and two staff officers of the 5th Carrier Division; the commanders and one staff officer each of the 3d Squadron, 8th Squadron, and 1st Destroyer Squadron. Ibid.

only 50 per cent of American Pacific Fleet strength were in harbor.

2. The possibility of detection by American search planes before the attack could be executed.

3. The refueling at sea of Task Force units with inadequate cruising range.

The conclusions reached with regard to the solution of these problems were those later embodied in the actual operational plan and carried out in the Task Force attack.[23] However, it was not until 20 October, after Admiral Yamamoto had threatened to resign over the issue, that Admiral Osami Nagano, Chief of Navy General Staff, approved the Pearl Harbor plan in principle over General Staff opposition. Preparation of the detailed attack plan was completed during October and finally sanctioned by Admiral Nagano on 3 November.[24] In order to preserve secrecy, knowledge of the plan in its entirety was limited to the Chief and Vice-Chief of the Navy General Staff, the Chief and members of the Operations Section, Navy General Staff, Commander-in-Chief, Chief of Staff, and most staff officers of the Combined Fleet, First Air Fleet and Sixth Fleet.[25] Evidence indicates that Army leaders were not informed until sometime in November, following the issue of Combined Fleet Top Secret Operations Order No. 1.[26]

Even in this order, issued on 5 November, the missions of the Advance (Submarine) Force and the Task Force which were to participate in the Pearl Habor attack were left blank in the printed text, and the missing portions were communicated verbally only to those listed in the preceding paragraph.[27] The commanders of the Task Force units, which assembled in Tankan Bay between 15 and 22 November, were not informed of the attack plan until Vice Adm. Nagumo, commanding the Force, issued Task Force Top Secret Operations Order No. 1 on 23 November, three days before departure for Hawaiian waters.[28] Crew members were told that Pearl Harbor was the target only after receipt of the Combined Fleet X-Day order on 2 December.[29]

## October 1941

Prisoner of war Iwataro Fusei, (JA 145118), a civilian laborer in naval employ present at Truk throughout October 1941, stated that:

*When he was at Truk in October 1941, there were rumors that a war with the United States would start*

---

23   Cf. Chapter V.

24   (1) ATIS Research Report No. 131, op. cit., p. 66.   (2) Statement by Rear Adm. Tomioka, previously cited.

25   (1) ATIS Research Report No. 131, op. cit., p. 67.   (2) Statement by Vice Adm. Mitsumi Shimizu, C-in-C, Sixth Fleet.

26   "I did not know at the time of the October conference (liaison conference between the Government and Imperial General Headquarters following formation of the Tojo Cabinet) that the Navy already had well-laid plans for the Pearl Harbor attack....At a later conference, I believe in November, I was informed of this plan."   Interrogation of General Hideki Tojo, Premier and War Minister, Oct 41–Jul 44.

27   Cf. Chapter V. (extracts from Combined Fleet Top Secret Operations Order No. 1).

28   ATIS Research Report No. 131, op. cit., pp. 78–9.

29   (1) Seaman 3d Class Shigeki Yokota (JA 100037), crew member aboard the aircraft carrier *Kaga* in the Pearl Harbor attack, later taken prisoner of war, stated that on 2 December the Commander of the *Kaga*, Capt. Jisaku Okada, addressed the ship's company and announced that war would be declared against America on 8 December. ATIS Interrogation Report, Ser. No. 230. (2) Another prisoner of war, Seaman Masayuki Furukawa, crew member of the carrier *Shokaku*, stated that the commander "informed the crew of the intended strike on 4 December." JICPOA Preliminary Interrogation Report No. 7, Ser. ADM-101022, 10 Jan 44.

*about 25 December 1941 at the earliest and 1 February 1942 at the latest. When he returned to Japan in November 1941, rumors of war were far less current than at Truk.*[30]

A "Report on Conditions" issued by Lt. Col. Ryuto, Commanding Officer of the 42d Anchorage Group, dated 15 June 1942, states:

*Record of General Situation since Mobilization.*

*The mobilization order was issued on 12 September 1941. Organization from the Hiroshima Western District No. 2 Force was completed by 17 September. We left Ujina on 29 September, sailing to Osaka where 40th Sea Duty Company was attached to us. We left Osaka on 1 October and reached our destination at Palau on 10 October where we established an anchorage headquarters. Then we made preparations for the landing operations which were to accompany the War for Greater East Asia.*[31]

Personal history register of Leading Pvt. Hisazo Kashino of the 41st Infantry Regiment, contains the following entries:

*10 October 1941—Left Ningpo.*

*11 October—Landed at Shanghai.*

*12 October to 14 November—Prepared for Malayan operations in the vicinity of Shanghai.*

*22 November—Left Woosung, Shanghai.*

*8 December—Landed at Singora, Thailand.*[32]

### November 1941

Diary belonging to an unknown member of the 41st Infantry Regiment contains the entries quoted below. This regiment participated in the attack on Malaya, which was mounted from the port of Samah on Hainan Island.

*12 October 1941—Reached Woosung Pier and returned to Kiangwan Barracks.*

*23 October—Okabe Force was assembled and heard an address from the newly appointed brigade commander, Maj. Gen. Saburo Kawamura. There will be a general inspection of the three battalions tomorrow.*

*4 November—Jungle combat training for expected type warfare.*

*13 November—Received rations and other necessary items for tropical combat (medicines and clothing, etc).*

*20 November—Anchored off Haikow on Hainan Island....*

*2 December—Weighed anchor and sailed again for Samah.*

*15 December—Assisted by our guns and tanks, our unit was the first to enter Gubun Street. Capt. Omori, 1st Lt. Nikki, and 2nd Lt. Takahashi were killed. 1st Lts. Okano and Yanagizawa were wounded.*[33]

The pamphlet quoted below, entitled, "Read This and the War is Won", was to be issued to each Japanese soldier before he embarked for overseas. The frontispiece consists of a map of South China, French Indo-China, Thailand, Burma, the Federated Malay States, the Netherlands East Indies, and a small section of the Northwest Australian Coast. The date of publication of the pamphlet is not definitely known. A captured copy, however, was received by 55th Division Infantry Group on 10 November 1941. (Elements of 55th Division figured prominently in the attack on Guam in December 1941 and later in the Burma Campaign). Furthermore, its length and the nature of the contents indicate original preparation at a date considerably prior to this. Pertinent sections of this pamphlet are reproduced below:

*What sort of place is the southern field of operations?*

*(1) It is the treasury of the Orient which has been invaded by the white men of England, America,*

---

30 ATIS Interrogation Report, Ser. No. 97, 14 Hpr 43, p. 16.
31 ATIS Research Report No. 131, op. cit., p. 12.
32 ATIS Current Translations No. 64, 13 Jul 43, pp. 16-7.
33 ATIS Current Translations No. 54, 14 Jun 43, p. 29.

France, and Holland.

(2) One hundred million Orientals are being oppressed by three hundred thousand white men.

It amounts to this—these whites possess scores of Oriental slaves from the moment they are born. Is this the intention of God?

(3) It is a source of world supply of oil, rubber, tin, etc.

Rubber and tin are essential for military supplies, and for these valuable resources the southern countries are the richest in the East. The malevolence of England and America, who have prevented Japan's purchasing these materials by just means, is one of the reasons which necessitates the present military operations.

It is quite clear that the Netherlands East Indies and French Indo-China cannot oppose Japan alone, but with the support and threats of England and America they are showing hostility to Japan. The lack of oil and iron is Japan's weak point, but lack of rubber, tin, and tungsten is the weakest point of America. America's chief sources of supply of these are the South Seas and Southern China. If these could be stopped, it not only would enable Japan to obtain the much-wanted oil and tin but it would stick a knife into America's sorest spot. The essence of America's opposition to Japan's southward advance lies here.

(4) It is a land of perpetual summer.

Bananas and pineapples are plentiful all the year round; at the same time troublesome malaria-mosquitoes are everywhere. In the Java and Singapore areas motor roads have been developed everywhere, but there are many uncivilized places, jungles, and swamps where neither man nor animals pass.

Why must we fight and how must we fight?

(1) By the Imperial will for the peace of the Orient.

The Meiji Restoration saved Japan from invasion by foreign powers. The Showa Restoration, by complying with the Imperial will for the peace of the Orient, must rescue Asiatics from disputes amongst themselves and the invasion of the white race and return Asia to the Asiatics. Peace in Asia will ensue, and this will be followed by peace in the world being firmly established.

Japan is given a great mission to save Manchuria from the design of Soviet Russia, free China from exploitation by the English and Americans, and then aid the independence of Thailand, Annam, and the Philippines, thus to bring about the happiness of the natives of the South Seas and India. This is the spirit of equality and brotherhood.

(2) While destroying the enemy show compassion towards those without crime.

Understanding this war as one between races, we must enforce our just demands on the Europeans, excluding Germans and Italians, without extenuation.

(3) Is the enemy stronger than the Chinese Army?

Comparing the enemy with the Chinese Army, since the officers are Europeans and non-commissioned officers for the greater part natives, the spiritual unity throughout the Army is zero. It must be borne in mind that the number of airplanes, tanks, and guns is far superior to those of the Chinese Army. However, not only are these of old types but their users are weak soldiers, so they are not of much use. Consequently, night attack is what the enemy fears most.

(4) We must be prepared for the war to be a prolonged affair and proceed with every preparation for a drawn-out conflict.

What course will the war follow?

Long voyage followed by landing operations.

All fields of operations are in the South Seas over a thousand miles from Formosa. Some places take a week to ten days to reach. This wide sea is crossed by convoys of several hundred warships and merchantmen. Looking back, our ancestors conquered this rough sea and carried on trade and fought with wooden sailing ships hundreds of years ago. After several days journey in the confines of shipboard, enemy resistance on the shores must be overcome and landings enforced.

What to do aboard ship:

The most important thing in landing operations is the maintenance of secrecy. If the enemy gets to know in advance where we plan to land, it will be very difficult.

There are many instances where a simple thing

*written in a letter has been the cause of the defeat of a whole Army, or where a word dropped over a glass of wine in a cafe just before departure has been the cause of secrets coming to the ears of spies.*

*Remember how the 47 Ronin kept their secret through such trials until they had avenged their Lord; encourage one another to do likewise.*

*There is a timely story of a soldier attached to a certain unit, who landed in Southern China during the present Incident, wrote a letter and dropped it in the sea, sealed in a bottle. The letter was carried by the tide to the coast of Korea. Supposing the letter had reached Vladivostok—what would have been the consequence? Often a clue is caught by aircraft and submarines which are at sea to find out the movements of our transport ships. Care must be observed in the disposal of dirt and rubbish.*

*Battle:*

*(1) Squalls, mist, and night are over all.*

*Europeans are dandies, and delicate and cowardly. Therefore, rain, mist, and night attacks are the things they detest most. They consider night suitable only for dances but not for fighting—we must take advantage of this.*

*(2) Unlike the Chinese soldiers, our present enemy may use gas. If you cast aside your gas mask because of the torment of wearing it in the heat, the consequence may be serious.*

*Action in particular zones:*

*Action in swamps and paddy-fields:*

*French Indo-China and Thailand are, next to Japan, the chief rice-producing countries, and there are paddy-fields everywhere and large swamps here and there. When passing through these places, each soldier must use snowshoes (made of straw and sticks).*

*The present war is a war with Japan's rise or fall at stake. What is at the bottom of America's action of gradually prohibiting the export of oil and iron to Japan, as if to strangle her slowly by "silk-wool"? If they stopped these exports at once, Japan, in her desperation, might march into the South. If the export of the rubber and tin of the South are checked by Japan, America's own sufferings will be far greater than those of Japan, who is harassed for want of oil and ore. It has been the policy of America up to now not to anger Japan, though weakening her.*

*Japan has waited too long—if Japan is patient any longer our aircraft, warships, and motor cars will not move. Five years have passed since the beginning of the China Incident. Over 1,000,000 comrades have exposed their bones on the continent. The arms of Chiang Kai-shek, which killed these comrades, were sold mostly by England and America. Both England and America are prejudiced against the solidarity of the Oriental races as something that stands in the way of their making the Orient their permanent colony and are concentrating every effort on letting Japan and China fight. Our allies, Germany and Italy, are continuing a battle of death in Europe against England, America, and Soviet Russia. America is already assisting England and is essentially participating in the war. For the existence of Japan herself and her obligation to the Tripartite Alliance, not a minute longer must be endured. Japan is confronted with a great mission, bravely to put the last finishing blow, as representatives of the Oriental race, to their invasion of several hundred years. Our incomparable Navy is in full readiness and is infallible: 5–5–3 is the ratio in figures, but if spirit is added, it is 5–5–7. Moreover, half of the British Navy has been smashed by Germany. For the Navy, now is the best time. The Chungking Government's umbilical cord is joined to England and America. Unless this cord is severed soon, the Japan-China Incident will never be permanently settled. The total settlement of the holy war is the present war. The spirits of over a hundred thousand warriors are guarding us. The mass for the dead comrades is to win this war.*

*Whilst showing our heartfelt thanks to the Navy, who, conquering thousands of miles of sea and removing enemy interception, are protecting us without sleep and rest, we must fully repay them for their trouble with good war results. We are privileged with an important and honorable mission to stand as representatives of the Asiatic race and to reverse the history of the world, succeeding our glorious history of 2,600 years and for the trust and reliance in us of His Majesty the Emperor. Both rank and file with one mind must exhibit the real value of Japan's sons in this full-dress display watched by the whole world.*

*The completion of the Showa Restoration to free Asia in realization of the Imperial will, which is for peace in the East, rests on our shoulders.*[34]

The pamphlet entitled, "Message to Warriors in the South Seas", was issued on 15 November 1941 above the signature of Maj. Gen. Horii, Commanding General of the South Seas Detachment, which comprised the main force in the attack on Guam.[35]

> RESTRICTED
> *A Guide for Warriors in*
> *The South Seas*
>
> South Seas Detachment Hqs.
> 15 November 1941
>
> *Horii Force Staff—Educational Pamphlet No. 1*
> *Instructions regarding the attached "Message for Warriors in the South Seas"*
>
> *To all units and militarized civilian personnel under my command:*
>
> *This pamphlet, together with the previously distributed "Collection of Imperial Rescripts," to which are annexed: "Field Service Instructions" and "Read This and the War is Won", is to be used as material for the practical strengthening of morale in the field.*
>
> 15 November 1941
> Tomitaro Horii
> Commanding General,
> South Seas Detachment

*Instructions given to the officers, men, and civilian employees under His Majesty the Emperor and under my command, on the occasion of the formation of the South Seas Detachment and their departure for operations:*

*In obedience to the orders of His Imperial Majesty, I now take command of your honored unit as an independent force, and am about to undertake a vital duty. I cannot repress my deep emotion, and I feel keenly the gravity of my responsibility.*

*I am convinced that the world situation surrounding East Asia faces an unprecedented crisis, and the fate of the Empire hangs in the balance. I believe that all of you, habitually bearing in mind the Imperial Edicts, have obeyed the orders of your superiors and have striven with all your might; however, at this time when your unit has been newly organized and is about to take the field, you are to stress to yourselves these three great principles with fullest courage:—*

*The strict observance of military discipline; the strengthening of esprit de corps; and the determination to fight to the death for certain victory. Whether you be under the higher commands or under the command of subordinate officers, whether you be officers or militarized civilian personnel, true to the spirit of loyalty, you are to have faith in and assist the combined action of the land and sea forces working together as one body; thus you shall do your utmost to utilize the results of your training and to display the combined fighting strength of the detachment.*

*You will take care of yourselves, bear in mind my wishes, and upon the opening of hostilities determine to exalt still more the true worth of the Detachment, swiftly bringing the Holy War to a successful termination, and thereby carrying out the Sacred Imperial Desire.*[36]

The professional notebook of Ensign Toshio Nakamura, contains the following passage:

*Address by our Captain upon the occasion of my boarding ship. Delivered by Capt. Yuji Takahashi (of the heavy cruiser Kako) 16 November 1941:*

*For three years you have studied your duties diligently. And now I believe that as you stand here, at the battlefront, your emotions have been heightened as you sense impending action.*

*When you reflect upon it, this is no training squadron; you have been assigned directly to the front.*

---

34 ATIS Research Report No. 131, op. cit., pp. 13-15.

35 This is the same commander and the identical unit which later operated in the abortive Japanese drive over the Owen Stanley Mountains, in New Guinea, toward the vital Allied base at Port Moresby. The South Seas Detachment had been organized under Maj. Gen. Horii's command by 15 November. ATIS Enemy Publications No. 41, *Miscellaneous Personnel Records of Horii Butai and Sakikawa Butai.* 8 Sep 43, p. 10.

36 ATIS Captured Document No. 89, 17 Oct 42, p. 2.

Original Painting by Sentaro Iwata  Photograph by U.S. Army Signal Corps

## PLATE NO. 3
Resting Comrades

*mightily glad to receive them. At 1900 hours we separated from the special service ship. After anchoring I drank beer and got drunk.*[42]

Personal history register of 1st Class Pvt. Tadatoshi Yamakawa of the 41st Infantry Regiment contains the entry quoted below. The original attack on Malaya was mounted from Samah:

*28 November 1941—During assembly off Samah Harbor on Hainan, was admitted to a hospital ship from Kyushu Maru.*

*1 January 1942—Overtook his own unit at Kampar, Perak.*[43]

Diary, owner and unit unknown, but presumably a member of the South Seas Detachment, contains the following entries:

*18 November 1941—From 1000 hours infantry group held war exercises under General Horii.*

*24 November—Left Marugame at 0630 hours. Boarded Matsue Maru at 1530 hours. Sailed at 1800 hours.*

*28 November—1650 hours stopped over at Haha-Jima, Ogasawara Archipelago.*

*29 November—Went ashore for communication. America has disguised herself till now. We are going to meet the enemy at Guam Island with ever-increasing spirit.*

*3 December—Landed Haha-Jima at 0230 hours to wash clothes. It seems that the Japanese-American talks will finally break down.*

*4 December—Worshipped the Imperial Palace at 0830. Gave 3 Banzais! There was a speech. Japan-America, War! It looks as though the hardships we have borne until now will be rewarded! We have received life for Showa's reign. Men have no greater love than this. Convoy to sail! 0900! Now, prosper, fatherland!*

*4 December—South Haha-Jima at 1422 hours. 'The Empire has decided to go to war against America, Britain, and Holland. The Southern District Army will quickly capture important regions in the Philippines, British Malaya, and the Dutch Indies after beginning attack on 8 December.*

*'For this purpose the first Japanese air attack will be carried out.*

*'The South Seas Detachment will co-operate with Fourth Fleet to capture Guam. If there is no separate order, the landing will take place on 10 December.*

*'Horii Operation Order A, No. 17. Each unit will act according to Order A, No. 7, which has already been issued.'*

*8 December, 1100, war declared!*[44]

Diary, owner and unit unknown, but presumably a member of the South Seas Detachment, contains the entries set forth below. The entry of 29 November 1941 anticipates a Japanese landing north of Talofofo Bay on Guam.

*22 November 1941—0327 hours. Reached Sakaide. 1000 hours. Inspection tour of the Cheribon Maru.*

*23 November—1700 hours. Left Sakaide.*

*27 November—Sighted Bonin Islands. 0800 hours, reached Haha-Jima.*

*28 November—0900 hours. Went to Yokohama Maru for liaison.*

*29 November—Training for boarding motor barges during the morning. It has been decided that battalion will land on the north side of Taro Bay (presumably Talofofo Bay on Guam).*

*2 December—Anchorage point penetration training from 2000 hours.*

*3 December—Battalion officers to meet on Yokohama Maru from 0900 hours. Training in smoke flares and gas. Conference of company commanders, decided to land at Iriya Bay. Two first-class cruisers came to the anchorage point to escort us and we*

---

42 ATIS Current Translations No. 78, 9 Sep 43, pp. 1, 13–5.
43 ATIS Current Translations No. 64, 13 Jul 43, p. 17.
44 Ibid., p. 1.

*feel very safe.*

*4 December—The convoy left at 0900 hours.*

*6 December—Heard the Japanese news broadcast in the salon. Our mission is to attack the United States.*[45]

Diary belonging to Ifuji, a member of Palau No. 3 Defense Unit, contains the following entries:

*29 November—War? All leave was cancelled and I heard that a huge Army unit is out here somewhere. (Written at Palau)*

*5 December—We received a written order from Commanding Officer of No. 3 Base to take up No. 2 Guard Dispositions from today; it is really going to be a serious affair.*

*6 December—It is said that American airplanes are reconnoitering our positions.*

*8 December—Declared war on America and Great Britain.*[46]

## December 1941

Diary, owner and unit unknown, contains the following entries:

*24 November 1941—Embarked on Daifuku Maru (3,523 tons) of N.Y.K. Line at Sakaide.*

*26 November—Destroyer Uzuki is escorting our convoy.*

*2 December—Loaded horses at Haha-Jima.*

*4 December—Order of Tomitaro Horii, South Seas Detachment Commander:*

*On 2 December Imperial Japan decided on war with Great Britain, the United States of America, and Holland. Imperial Japan will, on 8 December, carry out its first air attack against the United States. This detachment will, if there is no special order, land on Guam.*[47]

Diary of Haruichi Nishimura, member of 1st Special Naval Landing Party, Yoshimoto Unit, contains the following entries:

*7 November—Conscripted.*

*30 November—Boarded Kirishima Maru at Ujina. Escorted by Destroyers No. 36 and 37. Headed for Palau.*

*2 December—Heard over radio that American fleet (5 ships) had left harbor. Heard that we are to land on the Philippine Islands after resting at Palau.*

*5 December—Arrived Palau.*

*6 December—Enemy submarine sighted 5000 meters away.*

*7 December—Relations between United States and Japan are getting worse.*

*8 December—War was declared at 0800. Katsuta Maru sunk.*[48]

File of reports, entitled "Thailand Operations," belonging to the 77th Air Regiment, contains the following passage:

*4–7 December 1941—Protection of Twenty-fifth Army transport convoy and preparation for occupation of Thailand.*[49]

Diary belonging to Shigeo Morikami, of Horii Force (South Seas Detachment), Takmori Unit, contains the following entries:

*22 November 1941—Our departure for Sakaide leaving familiar Zentsuji behind. About 1930 hours left Sakaide Harbor behind, bound in oo direction.*

*23 November—About 0500 hours our ship stopped. A mountain can be seen to the east, and a factory zone below it. My friends were saying that it was Senshuji.*

*27 November 1941—We also put in at Chichi-Jima at 0100 hours. We again departed for Haha-Jima at 1100 hours.*

*28 November—Landed the horses at Oki Village Grammar School on Haha-Jima.*

*3 December—Sailing preparations.*

*4 December—Will depart for Guam Island, which is called Omiyajima.*

*5 December—Will depart at 1000 hours. We are cruising safely.*

*6 December—Cruising safely. We will disembark*

---

45 ATIS Current Translations No. 52, 11 Jun 43, p. 31.
46 ATIS Bulletin No. 527, 26 Nov 43, p. 8.
47 ATIS Current Translations No. 23, Mar 43, p. 4c.
48 ATIS Bulletin No. 470, 21 Oct 43, pp. 15–16.
49 ATIS Bulletin No. 1518, 25 Oct 44, pp. 3–4.

in three days.

*8 December 1941—Imperial General Headquarters. War was declared against England and the United States at 1230 hours. In the afternoon, I heard from Captain Takamori that the Hawaiian Islands are being bombed by our Air force. The Philippines and Hong Kong are also being bombed. At 0800 hours of the 8th, our Takamori Unit worshipped the Palace. We will finally begin landing from 1200 hours of the 9th. On the morning of the 8th, some islands could be faintly seen for the first time.*[50]

Diary belonging to Yaichi Takahashi, of South Seas Detachment, Antiaircraft Unit, Takahashi Platoon, contains the following entries:

*14 November 1941—We finally received orders to go to the front. On 28 July we had separated from the friendly 73d Force in Korea and were reorganized as the 47th Antiaircraft Battalion. On 14 November at 0900, we carried out the last ceremony of farewell on the parade ground. When we were leaving for the front, Commander Fuchiyama gave instructions and read a written oath addressed to the Imperial Palace. I have no reluctance in giving my life and being killed in action. We went up to the Goku Shrine to pray for our ultimate victory. We received sacred Sake from the god. Then we shouted "Banzai" three times and dismissed.*

*At 1900, we entrained. We were on a freight car. About 50 troops. All were waiting the time for leaving the friendly Kainei....*

*17 November—At 0600, we eventually arrived at Pusan Station. We stayed in Pusan City today.*

*18 November—Today the Iso Unit is leaving. At 1300, the loading was finished. It was about two years since we were on a ship. The inside of the ship was the same as when we came on her. After a time I noticed that the ship was sailing.*

*19 November—This is Japan. It was two years since I had seen Japan. Ujina—the Iso Unit was divided into two groups here, then we were all embarked on the big ships. I was in the Takahashi Platoon. The ship was the Matsue Maru.*

*On the 23rd at 0600, we arrived at our destination, Sakaide. At 1730, we finally left. We did not know where we were going. On the 28th at 1630 hours, we caught sight of a big island northeast of the ship. Several ships which had come before us were at this island, Haha-Jima. It was four days since we left Sakaide....*

*4 December—At 0930 hours, we eventually left the island. We immediately began to prepare for combat. Approaching enemy position. We were on board 18 days, and every day was the same routine. On 11 December at 0100 hours, we came, at last, face to face with enemy positions. We have a mission on Guam Island.*[51]

Diary and notebook belonging to Yutaka Morita, of 144th Infantry Regiment, contains the following entries:

*22 November 1941—0140 hours. Arrived Sakaide Station in Kagawa Prefecture. Boarded the transport Moji Maru with 9th Company, one company of mountain artillery, 3 guns, 50 horses, cavalry, and part of an engineer unit.*

*1 December—Afternoon. Prepared for landing. Held landing practice. Warships and transports started out of Chichi-Jima at 1800 hours preparatory to departure.*

*2 December—0030 hours. Waited two hours with landing equipment but the motor boats were not ready and the landing was cancelled. The ships and transports returned to Chichi-Jima at 0600 hours, 1330 hours. Four warships, eight airplanes. Loaded some more horses on the ship again.*

*4 December—0930 hours. Warships and transports which were in readiness at Haha-Jima harbor sailed for their destination.*

*5 December—Convoy sailed south.*

*10 December—Landed Guam Island at 0400 hours.*[52]

---

50  ATIS Current Translations No. 49, 9 Jun 43, p. 34.
51  ATIS Current Translations No. 68, 23 Jul 43, pp. 33–4.
52  ATIS Current Translations No. 10, 25 Dec 42, pp. 17–9.

Diary belonging to Leading Pvt. Sagaei Matsuura, of the 144th Infantry Regiment, contains the following entries:

*29 September 1941*—Received induction orders.

*5 October*—Entered service.

*8 October*—Completed mobilization.

*22 November*—Embarked. Sailed in the evening. Arrived off Osaka in the morning. We did not sail during the day. Set sail at night. Headed due south. We sailed southward till the morning of 27th. When I went up on deck in the morning, I saw a little island. It was one of the Bonin Islands.

*27 November*—Reached Chichi-Jima. Departed at 0900 hours the same day. Reached Haha-Jima before noon and anchored. There are not many people living on this island. Ships come here one after another. The bay is filled with large ships. It seems as though there are about seven or eight men-of-war here too. At first there were names on the warships; Uzuki, Yuzuki, and Kikuzuki etc., but the names were taken off. This transport ship had MI written on the smoke stack but it also has been removed. Horses were unloaded on Haha-Jima. Horses and dogs romped around the hills. Those who had previously been here say that the women are not beautiful, but they speak the Tokyo dialect. We fished to pass the time till 4 December. In the meantime horses were loaded. I suppose we are again headed for hot places. We had mosquito nets and lunch boxes made for us.

*4 December*—Today we are really going to set out for our destination. We sailed around 10 o'clock. We started in the morning with a warship as escort. It was the Kurogame. They were practically all carrying airplanes. As soon as we entered this harbor, two airplanes were started as if they had rehearsed going out on reconnaissance. There were many escort ships. As long as the Navy is present, there is nothing to be afraid of.

*6 December*—Tomorrow, we are told, Guam Island will be attacked and occupied. During the voyage all necessary preparation of arms, such as 150 rounds of ammunition, were in readiness. With these we can kill. It is heavy, but I feel like taking more.

*10 December*—At 0200, we will bid farewell to this boat. We got on this boat on the 21st and started to sail on the morning of the 10th. We lived on it for 20 days. At night we made various preparations for tomorrow's landing. I packed food for 3 meals in my haversack along with 150 rounds of ammunition. It is supposed to be packed as light as possible, but it is very heavy. We landed on one portion of the island which was barely visible in the dark. We anticipated enemy fire but did not encounter any. We landed successfully without incident.[53]

Diary belonging to Gumpei Imoto, of French Indo-China Expeditionary Force, 106th Land Duty Company, contains the following entries:

*1 November 1941*—Reached Saigon at 0600 hours.

*27 November*—Left Saigon at about 1400 hours.

*28 November*—En route.

*29 November*—En route.

*30 November*—En route Taikai Maru.

*1 December*—Safely arrived in the morning at Hainan Island.

*2 December*—Still anchored at Hainan.

*3 December*—Remained aboard Taikai Maru until 1600 hours and transhipped to Kashii Maru. Stayed aboard that night.

*4 December*—Departed at 0600 hours for our destination.

*7 December*—Reached Singora safely at 2400 hours.

*8 December*—At 0300 hours, made preparations for opposed landing. Around 0600 hours an unopposed landing was made. Took the enemy completely by surprise.[54]

Diary belonging to Chitoshi Sato, of South Seas Detachment, contains the following entries:

*14 November 1941*—Departed for Pusan.

*15 November*—Travelling south by train.

*16 November*—Still on train.

*17 November*—Approached Keijo.

*19 November*—Loaded guns on ship and sailed from Pusan harbor.

---

53 ATIS Current Translations No. 62, 7 Jul 43, pp. 19–20.
54 ATIS Current Translations No. 57, 26 Jun 43, p. 8.

*20 November*—Entered Moji harbor at 0700 hours, loaded coal at Ujina harbor, was separated from battery commander and 2d Lieutenant Takahashi. Loaded guns on *Matsue*.

*22 November*—Left Ujina harbor for Sakaide. Went through Inland Sea.

*24 November*—Left Sakaide in the evening.

*27 November*—Escorted by warship *Uzuki*.

*28 November*—Arrived at Ogasawara Islands.

*29 November*—Landed at Haha-Jima.

*30 November*—Picked bananas, coconuts, and papayas at Haha-Jima.

*4 December*—0900 hours left Ogasawara....

*10 December*—Infantry made opposed landing at Guam this morning at 0100 hours.[55]

Diary belonging to Susumu Kawano, of 106th Land Duty Company, contains the following entries:

*23 September*—Drilled. Inspection for all mobilized personnel. From 0700 hours visitors were allowed in camp area.

*6 October*—1700 hours arrived Saigon.

*23 November*—Left on transport *Tokokawa Maru*.

*25 November*—Arrived Samah, Hainan Island. Transferred to *Kashii Maru*.

*5 December*—30 transports headed towards the theater of operations with naval escort.

*8 December*—Made opposed landing at Singora, Thailand.[56]

Diary, owner and unit unknown, contains the following entries:

*24 November 1941*—Arrived at Haikow, Hainan Island.

*27 November*—Left Haikow.

*30 November*—Arrived at Humen.

*2 December*—Left Humen.

*4 December*—Arrived Samah harbor.

*5 December*—Sailed from the harbor at 0400 hours for operations.

*8 December*—Arrived at Singora, Malay Peninsula at 0140 hours.[57]

---

[55] ATIS Current Translations No. 74, 18 Aug 43, p. 32.
[56] ATIS Current Translations No. 57, 26 Jun 43, p. 31.
[57] ATIS Bulletin No. 747, 24 Feb 44, p. 6.

# CHAPTER II

# PRE-WAR JAPANESE ESPIONAGE AND INTELLIGENCE
## 1940–1941

### General

Japan's strategic planning of its war operations was based upon intelligence gathered by the armed services and their overseas agents over a considerable period of time preceding the outbreak of hostilities. When the Japanese forces struck on 8 December 1941, they possessed a fairly accurate knowledge of ground, air, and naval strength in the areas attacked, of the locations of airfields and fortifications, and of the terrain and climatic conditions under which they would have to fight. As the operations progressed, gaps inevitably became apparent in Japanese intelligence, but these were not serious in the first phase of hostilities.[1]

Examples of the type of information gathered and available to Imperial General Headquarters for the planning of the initial operations are the original intelligence maps reproduced in Plate Nos. 4 and 5. The sketch map showing the dispositions and strengths of American, British, and Dutch Air forces in the Philippines and Southern area (Plate No. 4) was issued on 6 December 1941 by Army Air Defense Headquarters at Keelung, Formosa. It bears the notation that the map was compiled "before the crisis" and that the air strengths indicated were "estimated currently to be undergoing marked reinforcement." The map reproduced in Plate No. 5 contains detailed order of battle information on Australian ground forces as of July 1940.

Similar maps showing troop dispositions and strengths, airfields, and other military installations on the islands of Java and Sumatra were compiled prior to the war and used in the planning of Southern Army operations. (Plate No. 6)

Augmentations of British troop strength in the Malaya, Singapore, and Burma-Thailand border areas in the months prior to the outbreak of war were noted in a "Simplified Table Showing Changes in the Southern Situation Since August 1941", issued by 20th Division Headquarters. Extracts containing intelligence apparently derived from confidential Japanese sources follow:

*(From Chief of Staff Report, mid-September 1941)*
*Strength on the Burma-Thailand border is approximately 50,000. In Burma there are an additional 2,000 to 3,000 Volunteer Army troops.*

*(From Chief of Staff Report, end October 1941)*
*Increase in strength in Malaya is presumed to be 10,000 Australian troops. Strength at Singapore in mid-August was approximately 5,000 Australians. Although information is lacking on numbers, transports carrying Australian reinforcements had reached Singapore by the end of August. The regular Army strength of 48,000 has reached approximately 60,000. If a rough estimate of the*

---

[1] Until the capture of Manila, intelligence was lacking regarding the existence of strong defense positions on Bataan Peninsula. Cf. Chapter VI. All source materials cited in this chapter are located in G–2 Historical Section Files, GHQ FEC.

of detection by defense installations at Clark Field.[4]

## New Guinea

Detailed reports by military intelligence agents who toured the southern areas prior to the war were also in the hands of the Army planning staffs. One such report, made by Major Tetsuo Toyofuku[5] on the basis of personal observation in March 1941, covered British New Guinea and was used as the basis of an intelligence study on this area compiled by the Army General Staff. The study was reproduced by General Headquarters, Southern Army, in 1942 for use in the New Guinea operations. Text of the study, entitled "Military Data on British New Guinea," follows:

*Part I—Military Value of British New Guinea and Solomon Islands:*

*These possessions, together with the Dutch East Indies Archipelago, form a natural barrier intersecting the Pacific Ocean from north to south. The northern end is within the radius of action of our bombers from most of our South Sea Mandated Islands, and the southern end is within the radius of action of bombers from the northern part of Australia. (It is approximately 1000 kilometers from Truk and Ponape Islands in our South Sea Mandate, to Rabaul, capital of the Australian Mandated Territory; approximately 1250 kilometers from Cooktown, North Australia, to Rabaul, and approximately 600 kilometers to Port Moresby.) They are separated from the Australian Continent by the narrow Torres Strait. Consequently, possession of this territory would make it easy to obtain command of the air and sea in the Southwest Pacific and to acquire "stepping stone" bases for operations against Australia. Control of the southern coast of New Guinea, in particular control of Torres Strait, would cut communications between the South Pacific Ocean and Dutch East Indies as well as the Indian Ocean Area, and would force the enemy fleet to detour to the southern coast of Australia.*

*Part II—Observations on Landing Operations in British New Guinea:*

*The area of the Bismarck Archipelago is approximately 50,000 square kilometers and corresponds to the combined area of our Formosa and Shikoku. However, the population of these territories (New Guinea and Bismarcks) is approximately 850,000 natives, most of whom live in the coastal regions.*

*Nowhere are these territories as yet developed.*

*Since the greater part is uninhabited, the communications facilities naturally are poor, and even the roads are like the government roads of Australian-controlled New Guinea, whose total length is only 136 miles (approximately 218 kilometers) of which 109 miles is in the Central Province, 16 miles in the Eastern Province, and 11 miles in Southeastern Province. It is recognized that these roads only connect the villages in the vicinity of the coast.*

*In regard to present military preparations, it appears that there are small forces and installations in the important political and transportation centers such as Rabaul, Port Moresby, etc. The other sections of the territory are not defended at all.*

*Landing operations on these various islands can, therefore, be carried out easily at any place where it is possible to land. However, advance and occupation from the captured points by land would be extremely difficult and practically impossible in view of the undeveloped road system and the difficulty of supply. Therefore, even if a point is occupied, it will only secure the vicinity of that point, and occupation of the whole territory will be difficult unless the enemy's fighting spirit is completely demoralized.*

*It would be advisable to attempt landing operations at Rabaul, Lae (capital of the Australian Mandated Territory), and Port Moresby (capital of*

---

4 Original aerial photographs, showing date and time of flight, altitude, name and rank of pilot and observer, were obtained from the Imperial Land Survey Bureau, Japanese Government. On file with Engr Intel Div, GHQ FEC.

5 The experience gained by this officer was subsequently utilized through his appointment to the staff of the South Seas Detachment, which landed at Buna in July 1942 and was virtually annihilated in the Owen Stanleys and Buna campaigns.

*Australian controlled New Guinea), which are points of military, political, economic, and communications importance. Considering the weakness of the enemy's present defenses and the strength that will be sent to this area by the Australians in the future, great strength will not be required. There are dwellings in these cities, but commodities, especially food and drinking water, are scarce, and self-sufficiency for a long period would be difficult. Our necessities, especially rice, bean paste (miso), and soy sauce, are not stored at all, so there is no other way but to depend on supply from the rear. The difficulty of supply from the rear must be recognized, and it will be necessary to carry large quantities....*

*Landing Operations at Port Moresby:*

*(1) General Condition of Harbor and City:*

*Port Moresby has a good, wide harbor, and the bay is entered by passing between Hanudamava Island (at the mouth of the harbor) and Bogirohodobi Point, approximately 1.5 miles to the east. At the beginning of 1940 there were approximately 800 Europeans, approximately 20 Chinese, and no Japanese residents. The natives (approximately 2000) have built their village over the water and live apart from the white residents. The city is situated between Tuaguba Hill and Ela Hill on the eastern shore of the harbor, and is the center of the government, military affairs, economics, transportation, communications, etc. of Australian-controlled New Guinea. There are various offices, including government offices and branch offices, a radio station, a government-managed electric power plant, church, school, European and native hospitals, an ice plant, bank, hotels, etc.*

*(2) Value of Port Moresby as a Naval Base:*

*Although the harbor is rather small for a fleet base, it is fairly deep (maximum 10 fathoms), and the bottom is alluvial soil, and one or two squadrons could anchor without difficulty. A space between the coral reefs outside of the harbor offers a very wide anchoring place, large enough for a large fleet to anchor. However, installations for repairs and supply have not been fully established, so it is valuable only as a port of call.*

*(3) Military Preparations:*

*Information obtained by observation of the actual area follows:*

*(a) Garrison Strength:*

*Army:*

*There is a barracks at Granville East (approximately 1 kilometer northeast of the city), which, judging from its size and the amount of equipment, can accommodate approximately 1,000 men. The present garrison appears to be composed wholly of infantry troops, without artillery.*

*Others:*

*A Royal Australian Artillery Detachment (2 officers, 38 non-commissioned officers, and privates, who arrived with 6-inch guns) apparently is stationed on Ela Hill and will be reinforced, judging from the fact that the number of barracks on the hill is being increased.*

*Navy:*

*The strength is not known but appears to be about 30 men. The orderly room is located at the side of the government pier. The station ship has not been identified; only 2 or 3 launches have been identified.*

*(b) Installations:*

*A road for military use has been built to the top of Ela Hill, and two 6-inch guns are placed on top of this hill. The main line of fire of these guns apparently is directed toward Basilisk Passage. The guns are exposed on top of the hill. According to information, they will be increased by two more guns. In addition to the Kila Kila airfield, approximately 4 kilometers east of Port Moresby, an airfield for military use, approximately 11 kilometers from Port Moresby (location unknown), is expected to be constructed. A single road parallel to the coastal highway, and halfway up the hill of Tuaguba, is being constructed.*

*(4) Passage of channels:*

*The greatest difficulty in a landing operation at Port Moresby would be passing through the waterways. There are three channels entering the harbor of Port Moresby. Liljeblad Passage, on the extreme west, has a very strong current and shoals. This passage cannot be used in general because there are shallows before the mouth of the harbor. Therefore, it is difficult to enter this passage. Basilisk Passage, in the center, is*

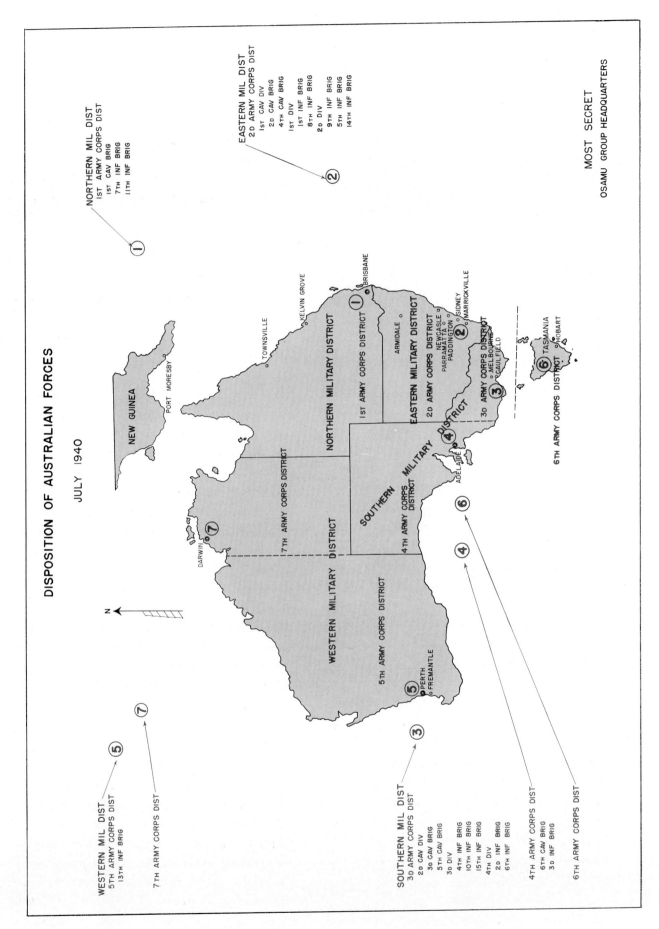

PLATE NO. 5

Disposition of Australian Forces, July 1940

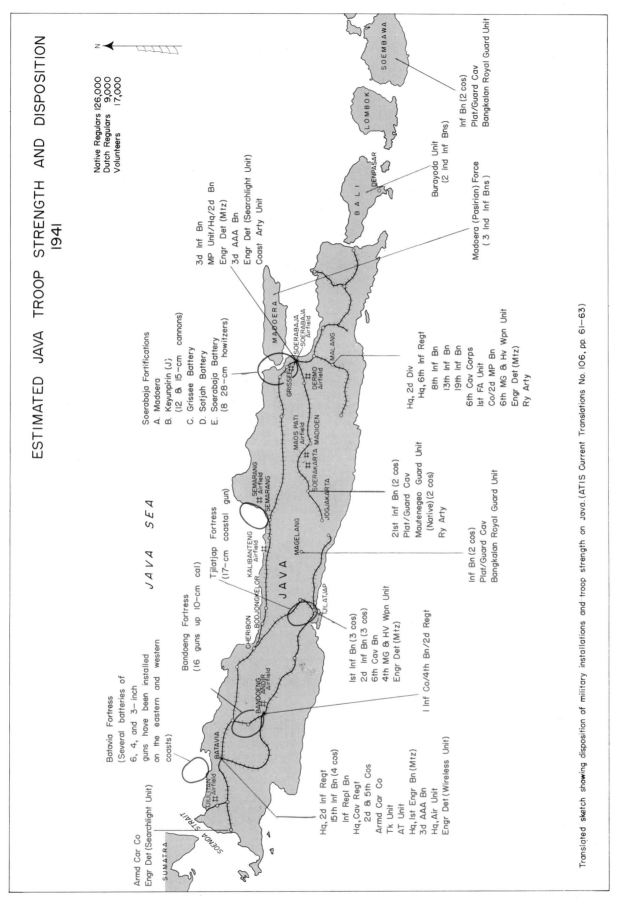

PLATE NO. 6

Japanese Estimates of Java Troop Strength and Disposition 1941

*the channel used by vessels at present, but it is about 6 kilometers from the gun emplacement on Ela Hill and thus is within the guns' effective range. In general, unless the gun emplacement is destroyed, it would be difficult to enter through this channel. Padana Nahua, at the extreme east, is quite wide (about 900 meters) and is outside the effective range of the gun emplacement (about 18,000 meters). This channel should be selected for an entrance. However, all three waterways are neither very deep nor wide, and could easily be covered with mines and other obstacles. These obstacles must be cleared first of all. If a place where the Nateara and Sinavi coral reefs can be passed over with boats could be found, then an approach could be made without risking the danger of passing through the channel. Anchoring outside a coral reef is very difficult, so in such a case the transfer to boats would have to be made while drifting.*

*Fresh water:*

*This area, in general, consists of barren mountains and is dry for the period of eight months between June and January each year. The rainfall is very small. The residents who depend on rainfall sometimes have to transport water from other areas in order to maintain their means of living. The Government has constructed water storage tanks with galvanized iron roofs for use during droughts, at a place 137 meters above sea level southwest of Tuaguba Hill, in back of the city, and this water storage is used in emergencies. Their capacity is said to be several tens of thousands of tons, but this is not definite. The problem of drinking water is most urgent in this area, and the extension of a water supply line is especially important in planning the establishment of a harbor. According to rumor it is planned to bring the water down from the Laloki River, and arrangements for this will be completed about August of this year.*

*Fuel:*

*It is said that 5,000 drums of gasoline, 3,000 drums of heavy oil, a large quantity of motor oil etc., are stored in Carpenter's Warehouse. The nature of the gasoline used for airplanes is not known.*

*Automobiles:*

*For military use—about 50.*

*For civilian use—about 200 (no busses).*

*Part III—Value of Bismarck Islands and British New Guinea as air bases in case of war with Britain and America:*

*1. Essentials:*

*As explained in Part I, the Bismarck Islands are within the radius of action of bombers operating from most of our South Sea Mandated Islands. In turn, Torres Strait and the northern part of Australia are within aerial domination from the Bismarck Islands and British New Guinea, and this territory, in general, is judged to possess an important value from the standpoint of air operations in a war against Britain and America. In particular, if air superiority over Torres Strait is gained, it is thought that it would be advantageous in cutting off enemy communication between the Pacific Ocean and Dutch East Indies, as well as the Indian Ocean.*

*In regard to the possession of airfields in this area, the first principle is to use established airfields, and if necessary to construct new ones on farmland in flat areas.*

*The established airfields are not wholly adequate to fulfill military requirements, but some of them can be utilized immediately and can be extended with a small expenditure of labor.*

*As fuel is difficult to obtain in this area, in general it must be supplied from the rear. Much of the equipment for repairs and construction is inadequate.*

*The condition of the established airfields, landing fields and air network at present in British New Guinea is as shown in Appendix Map No. 10. (Map not attached)*

*2. Value of Bismarck Islands and their vicinity: (Appended Map No. 6, Reference No. 6) [Not attached]*

*(1) Essentials:*

*The established airfields in the Bismarck Islands and their vicinity are two landing fields at Rabaul and Vunakanau, and it is planned to construct others at Kavieng (northeastern end of New Ireland), Namatanai, Buka Passage (the strait separating Bougainville and Buka Islands in the Solomons), and Kieta (the capital of Bougainville). Only one place in the harbor of Rabaul is used for seaplanes, but another is located at Kileg on Galawa Island, across the Lolobo* (northeastern coast of New Britain). While the situation regarding aviation gasoline is not known, apparently almost none at all is stored. Planes operating on a regular schedule apparently refuel at Salamaua or Moresby on the return trip. There are no reports of aviation gasoline being supplied at Rabaul.

3. *Value of Island of New Guinea:*

(1) *Essentials:*

*There are airfields and landing fields on the Island of New Guinea. Most of them are concentrated in Morobe Province because they are used by planes that transport gold from mines developed in the province. The airfields used mainly by land-planes are at Salamaua, Lae, Wau, Madang, Wewak, Port Moresby, etc. Salamaua and Wau are the only two airfields on which we have definite information at present. To discuss the value of the airfields in British New Guinea from the above data is futile. However, since these (the two airfields at Salamaua and Wau) are typical of those used by the regular air lines, I believe they are sound references for use in estimating the others. It seems that the only specially-constructed seaplane base is at Port Moresby.*⁶

## East Indies

The lengthy background of Japanese espionage and subversive activity in the Netherlands East Indies is attested by the following passage from an article entitled, "New Life of the People of Sumatra":

*It was the Achin People who fought against Holland and very bravely defended their independence to the end. The Achin People are naturally fearless. Their native place is on the northern edge of Sumatra; in area it is one and a half times the size of Formosa, and it has the sea on three sides. Again, there are many mountains and geographical factors. Due to these, the natives were able to resist fiercely for 40 years following 1873. Controlled by one family, these warriors hid in the woods and often conducted violent guerrilla warfare, harassing the Dutch troops. The last ten years it was a chronicle of chivalry that the Japanese youth, Motohiko Ban, struggled hard to assist the young tribal chief, Pannamu. However, in 1922, Ban was recalled to Japan by the Foreign Office, and the Achin finally broke up their swords and submitted.*⁷

An insight into more recent phases of possible Japanese espionage in the Netherlands East Indies is provided by the following excerpt from an account of the Japanese operation against Java written by an unnamed staff officer, a lieutenant colonel, and published in the Osaka Mainichi:

*It was dark when we arrived at Bandoeng. Late that night, I went into a room of the old Homan Hotel, where I stayed over two years ago....*

*For the purpose of meeting the Army Commander in the afternoon, I went to the Ifura Hotel, north of Bandoeng. I asked for the old room which I took the year before last.*⁸

---

6 ATIS Research Report No. 131, op. cit., pp. 39–42.
7 Ibid., p. 39.
8 ATIS Enemy Publications No. 32, *Account of the Netherlands East Indies Operation*, 11 Aug 43, p. 11.

# CHAPTER III

# POLITICO-MILITARY EVOLUTION TOWARD WAR

## Historical Background

The sudden, far-flung attacks unleashed by Japan's armed forces against Pearl Harbor and the Asiatic possessions of Great Britain and the United States before dawn on 8 December 1941 rang up the curtain on the Pacific War. It was to be a gigantic struggle, fought over an area covering 38 million square miles of the globe and every kind of terrain from the tundra wastes of the Aleutians to the jungles of Burma and New Guinea.

This desperate act was characterized by the enemy press as "national suicide," but the politico-military clique which gambled Japan's fate in war saw it as the only alternative to a retreat from policies and ambitions to which they stood irrevocably committed.

The Manchurian Incident of 18 September 1931 had evoked a strong reaction in the United States, expressed in repeated diplomatic protests from Washington. Great Britain aligned itself with the United States when hostilities spread to the Shanghai area in March 1932, imperilling British interests, and both nations supported China in an appeal to the League of Nations. The final League report adopted in February 1933 was so adverse that Japan, rather than yield, served notice of withdrawal from League membership.

Anti-Japanese sentiment intensified in Britain and the United States following the outbreak of the Sino-Japanese War on 7 July 1937.[1] On 1 July 1938, six months after the embarrassing sinking of the American gunboat *Panay* by Japanese Navy planes, the United States Government imposed a so-called "moral embargo" on the export of aircraft and aircraft parts to Japan. It was the initial step in a progressively more stringent economic blockade.

On 3 November 1938 Japan proclaimed the establishment of a "New Order for East Asia".[2] The United States and Britain promptly recognized this as a covert threat to China's "Open Door" and countered with loans of 25 million dollars and 50 million pounds,

---

[1] The Japanese Government had endeavored in the early stages to localize the hostilities and achieve a diplomatic settlement. Marquis Koichi Kido, later Lord Privy Seal and closest adviser to the Emperor, recorded in his Diary: "Prince Konoye (then Premier) was deeply concerned over the outbreak of the Sino-Japanese hostilities and exerted every effort to terminate the Incident and prevent its expansion. I did my utmost to support his stand." *Kido Nikki : Kyokuto Kokusai Gunji Saiban ni okeru Kido Hikokunin no Sensei Kokyosho* 木戸日記：極東國際軍事裁判に於ける木戸幸一被告人の宣誓口供書 (Kido Diary: Affidavit of Defendant Koichi Kido in International Military Tribunal for the Far East) p. 34. All source materials cited in this chapter are located in G–2 Historical Section Files, GHQ FEC.

[2] In his speech announcing the New Order (大東亞新秩序), Premier Ayamaro Konoye declared: "Japan does not reject co-operation with other Powers in China, or intend to damage the interests of third Powers. If such nations understand the true intentions of Japan and *adopt policies suitable for the new conditions,* Japan does not hesitate to co-operate with them for the peace of the Orient."

respectively, to the Chungking Government. The League of Nations on 20 January 1939 also proffered aid to Chiang Kai-shek.

Japanese troops occupied Hainan Island, off the South China coast, in February 1939 and at the same time closed the Yangtze to all neutral commercial shipping. On 26 July of the same year, the United States served notice of its intention to abrogate the Japanese-American Treaty of Commerce and Navigation, the trade basis upon which the two countries had operated since 1911. In December 1939 aircraft plans and equipment as well as equipment used in manufacturing high-grade aircraft gasoline were added to the list of items, export of which to Japan was forbidden.

On 30 March 1940 the Wang Ching-wei Government was formally inaugurated at Nanking in opposition to the Chungking Government. The United States promptly refused recognition of the new regime, as a Japanese "puppet," and offered Chiang another loan, this time for 20 million dollars. This was followed on 2 July with enactment of an export control law covering national defense materials, the implied intent of which was to curb the Japanese national potential.

Under this law an export license system was first applied to aircraft materials and machine tools, and was later broadened to include high-grade gasoline, high-grade lubricating oil and first class scrap iron.[3] Thereafter new items were frequently added to the list. Since Japanese domestic production of crude oil supplied but 1,887,000 barrels of the minimum of 34,600,000 barrels annually required to maintain national defense and economic life,[4] the American curb on oil exports alone was regarded in Japanese governing circles as a crippling blow to Japan's basic industry and, indirectly, to her national safety.

## Drift Toward Crisis

On 22 July 1940 the second Konoye Cabinet took office and, five days later, carried out a sweeping revision of basic Japanese policies in the light of changes in the world situation.[5] This revision committed Japan:

1. *To strive for speedy conclusion of the China Incident by cutting off all assistance to Chungking from outside powers.*
2. *To maintain a firm stand toward the United States on one front, while strengthening politicalities with Germany and Italy and ensuring more cordial diplomatic relations with Russia.*
3. *To open negotiations with the Dutch East Indies in order to obtain essential materials.*[6]

Japan's anxiety to end the China stalemate was a paramount consideration. The hostilities on the Continent had bogged down and constituted a severe drain on the nation's resources. Acting under the decisions of 27 July, the Konoye Cabinet therefore concluded a

---

3 "The restrictions of exportation of scrap iron to Japan was paricularly alarmimg to all Japanese in view of the prevailing iron shortage and the production process in Japan." *Kyokuto Kokusai Gunji Saiban ni okeru Hikokunin Tojo Hideki no Sensei Kokyosho* 極東國際軍事裁判に於ける被告人東條英機の宣誓口供書 (Affidavit of Defendant Hideki Tojo in International Military Tribunal for the Far East) Doc. No. 3000.

4 Statement by Maj. Gen. Kikusaburo Okada, Chief of War Plans Section, Economic Mobilization Bureau, War Ministry.

5 The fall of France in mid-July posed the question of the fate of French colonies in the Far East, and it also heightened the belief in high military circles that Germany would successfully overwhelm Britain. The result was to strengthen the hands of those demanding a stronger policy in the South and closer ties with the Axis Powers.

6 Decision taken by a Liaison Conference of the Government and Imperial General Headquarters, 27 July 40. *Juyo Kokusaku Kettei Bunsho* 重要國策決定文書 (File on Important National Policy Decisions).

"Joint Defense Agreement" with the French Vichy Government under which Japanese troops were dispatched to northern French Indo-China, for the purpose of blocking the last remaining supply route to Chungking.[7] Foreign Minister Yosuke Matsuoka explained the limited motive of this act in a special plea to the United States Ambassador in Tokyo, but Washington countered with an added loan of 25 million dollars to Chiang Kai-shek.[8]

In the same month Japan sought relief from the American oil embargo by dispatching a special mission headed by Commerce Minister Ichizo Kobayashi to Batavia to negotiate an agreement with the Dutch East Indies, the major oil-producing country in the Far East. Ambassador Kenkichi Yoshizawa took over the negotiations from December 1940, but the parleys finally ended in failure in June 1941.[9] As a corollary, French Indo-China later failed to deliver to Japan rice and rubber in the amounts fixed by an agreement reached in May 1941.

Four days after the dispatch of troops into northern Indo-China, Japan implemented another decision of the July Liaison Conference by concluding the controversial Tripartite Military Alliance with Germany and Italy on 27 September 1940. The professed object of the alliance was to deter the United States from going to war in either the Atlantic or Pacific,[10] but whatever Japan's real motives, the pact merely increased British and American suspicion of Japanese intentions and brought on new counter-measures.

In October the United States issued a general evacuation order to all Americans within the "East Asia Co-prosperity Sphere". Since early in the year, the bulk of the United States Fleet remained concentrated in Hawaiian waters,[11] and on 13 November Britain established a new Far East Military Command in Singapore. Malaya, Burma, and Hongkong were placed under this command, and military preparations were pushed in close liaison with Australia and New Zealand.

Beginning early in 1941, Japanese fears were heightened by a series of secret staff conferences among high-level Army and Navy representatives of the United States, Britain, China, and the Netherlands. In particular, the Manila conference in April, which was attended by the Commanding General, Philippines Department (Major General George Grunert), the United States High Commissioner to the Philippines (The Hon. Francis B. Sayre), the British Commander-in-Chief for the Far East (Air Marshal Sir Robert Brooke-Popham), the Commander of the United States Asiatic Fleet (Admiral Thomas C. Hart), and the Acting Governor-General of the Netherlands East Indies (The Hon. Hubertus van Mook), was interpreted by Japan as a sign that the so-called ABCD Powers were formulating concrete plans of immediate military collaboration.

Japanese intervention in the border controversy between Thailand and Indo-China in February 1941[12] was followed three months

---

7 Great Britain had closed the Burma Road to supplies for Chungking on 17 July 1940, but in October reopened it in support of American policy.

8 Japanese recognition of the Nanking Government on 30 November 1940 was answered by a further United States loan to Chiang of 100 million dollars.

9 Affidavit of Tojo, op. cit.

10 Japan's diplomatic strategists also entertained the idea that Soviet Russia might be induced to join the Alliance, thereby creating a favorable preponderance of power vis-à-vis the United States and Great Britain.

11 The main body of the United States Fleet moved from San Diego to Hawaii in January 1940. On 7 May the U. S. Navy announced that it would remain at Pearl Harbor indefinitely.

12 The United States considered the "mediation" move a Japanese design to extract new concessions from both disputing parties.

later by new American and British loans of 50 million dollars and ten million pounds, respectively, to the Chungking Government. The United States further bolstered this financial aid by extending the Lend-Lease Act to cover arms shipments to China.

In April 1941 Japan realized one of its major diplomatic objectives with the conclusion of the Japanese-Soviet "Non-Aggression Pact." However, the outbreak of the Soviet-German war only two months later created an entirely new situation. The Konoye Cabinet resigned on 16 July, reassembling two days later under the same Premier but with Matsuoka, the architect of the Axis Pact, replaced as Foreign Minister by Admiral Teijiro Toyoda.[13] The new cabinet was geared to rehabilitate relations with the United States, a course which conservative Navy elements had stoutly advocated.[14]

The American Government refused to take seriously the conciliatory trend of the new government line-up since Japanese troops shortly moved into southern French Indo-China;[15] and the United States retaliated on 26 July by freezing all Japanese assets. London took similar action, also abrogating the British, Indian, and Burmese commercial treaties with Japan, and the Netherlands Government followed suit.

Japan now found its trade cut off with all areas except China, Manchuria, Indo-China, and Thailand. Economic rupture was complete with the United States, Britain, and the Netherlands, who controlled the key materials essential to Japan's national defense and industrial existence. The gradual decline of the nation's power potential was inferentially inevitable.

The stoppage of fuel imports assumed paramount strategic importance. Even if Japan were to suspend all industrial expansion and further military preparations, and to undertake an epochal increase in synthetic petroleum production, it was estimated that approximately seven years would be required before output would reach the annual consumption level of 34,600,000 barrels.[16] Meanwhile, essential industries dependent upon liquid fuels would be paralyzed within a year. In two years the Japanese Navy would be immobilized.

An international impasse was fast approaching, but Japan's leaders in August 1941 hesitated to take the final plunge.

In a war against the material power of Britain and the United States, Japan's inherent economic weakness seemed to make the risk too great. Premier Konoye, who had long

---

13 "I recommended Admiral Toyoda for the foreign portfolio because of my ardent desire to further the Japanese-American negotiations. Admiral Toyoda had served as Navy Vice-Minister, and not only was he versed in Navy affairs....but he was one of those who supported the view that an American-Japanese conflict should be avoided by every means possible." *Konoye Ayamaro Ko Shuki* 近衞文麿公手記 (Memoirs of Prince Ayamaro Konoye) p. 30.

14 In a conversation with Premier Konoye shortly after the conclusion of the Tripartite Alliance in September 1940, Admiral Isoroku Yamamoto, Commander-in-Chief of the Combined Fleet, stated with regard to a Japanese-American war: "If I am told to fight regardless of the consequences, I shall run wild for the first six months or a year, but I have utterly no confidence for the second or third year. The Tripartite Pact has been concluded, and we cannot help it. Now that the situation has come to this pass, I hope you will endeavor to avoid a Japanese-American war." Ibid., p. 3.

15 This move was under a "Joint Defense Agreement" concluded 21 July between Japan and the Vichy Government. The agreement was announced in Tokyo on 26 July simultaneously with the United States freezing order. Japanese troops advanced into Saigon 28 July.

16 Notes of Maj. Gen. Kikusaburo Okada, Chief of War Plans Section, Economic Mobilization Bureau, War Ministry.

subscribed to this view,[17] decided to make a new effort to break the deadlock in the Japanese-American negotiations.[18] His Memoirs record:

*During this period I racked my brains in search of some way to overcome the crisis between Japan and America. Finally I made the firm resolution to attempt a personal meeting with the President. I revealed my intention to the War and Navy Ministers for the first time on the evening of 4 August....*

*The War and Navy Ministers listened tensely to my resolution. They could not reply at that meeting, but later the same day the Navy expressed complete approval and voiced hope for the success of the proposed meeting. The War Minister replied by written memorandum which stated:*

*".... The Army raises no objections, provided however that the Premier firmly adheres to the fundamental principles of the Empire's revised proposal [to the United States] and provided that if, after every effort has been made, the President still fails to understand the Empire's real intentions, and proceeds along the present line of American policy, Japan will firmly resolve to face war with the United States."*[19]

The Konoye proposal was laid before President Roosevelt on 17 August and met with an initially favorable response. However, the State Department's insistence that the meeting be held only after a prior agreement on basic principles resulted in a stalemate.[20] The sands of diplomacy were running out.

## War Deliberations

Amidst this atmosphere of high tension, the Emperor on 6 September summoned the Cabinet and representatives of the Army and Navy High Command to a conference at which, for the first time, the question of peace or war was squarely posed. Deliberation centered upon an "Outline Plan for the Execution of Empire Policies" (*Teikoku Kokusaku Suiko Yoryo*), which provided:

*1. In order to guarantee the existence and defense of the Empire, preparations for an eventual war against the United States, Great Britain and the Netherlands shall be completed approximately by the latter part of October.*

*2. Concurrently with the above, the Empire will exert every effort to secure realization of its demands through diplomatic negotitations with the United States and Great Britain. [The minimum terms which Japan would accept in an agreement with the United States were set forth separately.]*

*3. In the event that these negotiations fail to achieve the Empire's demands by the early part of October, it shall immediately be resolved to go to war with the United States, Great Britain, and the Netherlands.*[21]

Speaking for both Army and Navy High

---

17 "Japan's dependence for materials, particularly war materials, on the United States and Great Britain was her one great weakness. The impossibility of overcoming this was repeatedly confirmed by researches of the Planning Board since the time of the first Konoye Cabinet. The conclusion reported was always: 'Impossible'." Konoye Memoirs, op. cit., p. 4.

18 These negotiations were initiated in April 1941, shortly after the arrival of Admiral Kichisaburo Nomura, newly-appointed Japanese Ambassador, in Washington. The talks virtually came to a standstill following Japan's move into southern Indo-China and the American freezing order.

19 Konoye Memoirs, op. cit., pp. 32, 34–5.

20 At the 17 August interview between the President and the Japanese Ambassador, Admiral Nomura, the President went so far as to mention Juneau, Alaska, and mid-October as the possible time and place for the proposed meeting. However, the formal reply handed to Admiral Nomura on 3 September "evaded a clear-cut expression of his (the President's) stand regarding the meeting and stated that Japanese agreement on fundamental principles was a pre-requisite. Here it became clear that the State Department's views had prevailed." Ibid., pp. 36, 38–9.

21 File on Important National Policy Decisions, op. cit.

Original Painting by Ryohei Koiso

PLATE NO. 7

Japanese Column on the March

Commands, Admiral Osami Nagano, Chief of the Navy General Staff, backed up the plan with a warning that Japan's power to fight was steadily declining due to exhaustion of essential war materials and the increased military preparations of the ABCD Powers. Instead of "letting time slip idly by," he declared, the nation must first push its own war preparations and, if diplomacy fails, "advance bravely into offensive war operations." The statement was especially significant because it reflected the views of the Navy, the role of which would be of paramount importance in war with the United States. Essential extracts follow:[22]

> The High Command sincerely hopes that the Government will exhaust every possible means of settling the present situation diplomatically. However, if Japan should be obliged to resort to war, the High Command, from the standpoint of military operations, is of the opinion that the gradual exhaustion of most of the country's essential materials such as petroleum, is lowering the national defense power, and that, if this continues, Japan in the end will fall irrevocably into a condition of impotency.
>
> Meanwhile the United States, Britain, and other Powers are swiftly reinforcing their military establishments and strategic defenses in the Far East, and war preparations in these countries, especially in the United States, are likewise being greatly accelerated. Consequently, by the latter half of next year, the United States will be far ahead in its preparations, and Japan will be placed in an extremely difficult position.
>
> Under such conditions, it is highly dangerous for Japan to let time slip idly by without attempting to do anything. I think that Japan should, first of all, carry out preparations as best it can; and then, if our minimum demands essential to self-defense and national existence are not accepted in the diplomatic negotiations and war finally becomes inevitable, we should not lose our opportunity but should advance bravely into offensive war operations with firm resolution, thus seeking the salvation of our country.
>
> In regard to the outlook for such operations, it can be considered from the outset that the probability of an extended war is extremely great. Japan, therefore, must have the determination and the preparations to conduct an extended war. It would be just what we are hoping for if the United States, seeking a quick decision, challenged us with its main naval strength.
>
> Considering the present position in the European war, Great Britain can dispatch to the Far East only a very limited portion of its naval strength. Hence, if we could intercept the combined British and American fleets in our own chosen area of decisive battle, we are confident of victory. However, even victory in such a battle would not mean the conclusion of the war. In all probability, the United States will shift its strategy to a long war of attrition, relying upon its invincible position and dominant material and industrial strength.
>
> Japan does not possess the means, by offensive operations, to overcome its enemies and force them to abandon the war. Hence, undesirable as an extended war would be due to our lack of resources, we must be prepared for this contingency. The first requisite is immediate occupation of the enemy's strategic points and of sources of raw materials at the beginning of the war, thus enabling us to secure the necessary resources from our own area of control and to prepare a strong front from an operational viewpoint. If this initial operation succeeds, Japan will be able to establish a firm basis for fighting an extended war even though American military preparations progress according to schedule. For Japan, through the occupation of strategic points in the Southwest Pacific, will be able to maintain an invincible front. Thereafter, much will depend upon the development of our total national strength and the trend of the world situation.
>
> Thus, the outcome of the initial operations will largely determine whether Japan will succeed or fail in an extended war, and to assure the success of the initial operations, the requisites are:
> 1. Immediate decision on whether to go to war, considering prevailing circumstances in regard to relative Japanese and enemy fighting strength;

---

22 Juyo Shorui Tsuzuri 重要書類綴 (File of Important Documents) preserved by Capt. Toshikazu Ohmae, member, Military Affairs Bureau, Navy Ministry.

2. Assumption of the initiative;
3. Consideration of meterological conditions in the zone of operations to facilitate these operations.

It is necessary to repeat that the utmost effort must be made to solve the present crisis and assure Japan's security and development by peaceful means. There is absolutely no reason to wage a war which can be avoided. But to spend our time idly in a temporizing moment of peace, at the price of later being obliged to engage in war under unfavorable circumstances, is definitely not the course to take in view of the Empire's program for lasting prosperity.

Although the conference finally adopted the "Outline Plan," Baron Yoshimichi Hara, President of the Privy Council, pressed for further clarification by the High Command of the apparent subordination of diplomacy to preparations for war.[23] The Emperor himself, in a rare departure from constitutional precedent, intervened to second the demand, voicing regret that the Army and Navy had not made their attitude fully clear.

*With this, His Majesty took from his pocket a sheet of paper on which was written a verse composed by the Emperor Meiji:*

"*When all the earth's oceans are one,*
*Why do the waves seethe and the winds rage?*"

*Reading this aloud, His Majesty said, "I have always endeavored to spread the peace-loving spirit of the late Emperor by reciting this poem."*

*Silence swept the chamber, and none uttered a word....*[24]

After this dramatic moment Admiral Nagano again rose to express "trepidation at the Emperor's censure of the High Command" and to assure His Majesty that "the High Command places major importance upon diplomatic negotiations and will appeal to arms only in the last resort."[25]

Nevertheless, failing diplomatic success within a fixed time limit, Japan now stood committed to war.

In actuality, Japanese military preparations for the "Great East Asia War" far antedated the outbreak of hostilities. Even long before the decision to fight was taken on the highest policy-making level, the Army and Navy had independently begun gathering intelligence, making clandestine aerial surveys, compiling maps, experimenting with new-type weapons

---

23 Hitherto the armed services had, at least outwardly, accepted diplomacy as the primary means of achieving Japan's objectives. In April 1941 Imperial General Headquarters had decided its Basic Policies for the South as follows:

1. The aims of the Empire's immediate policy in the South are to hasten the settlement of the China Incident and to increase the total national defense power. This requires:
    (a) Establishment of close and inseparable military, political and economic co-operation with French Indo-China and Thailand;
    (b) Establishment of close economic relations with the Netherlands East Indies;
    (c) Maintenance of normal trade relations with the other southern countries.
2. Basically, diplomatic measures will be taken to attain the above objectives.
3. In carrying out the above policies, military force will be used for our country's self-defense and existence only if no other solution can be effected when the following situations develop:
    (a) If the Empire's existence is threatened by an American, British, or Dutch embargo;
    (b) If American, British, Dutch, and Chinese encirclement of Japan becomes so serious as to overly endanger national defense. *Juyo Kokusaku Kettei no Keii Gaisetsu* 重要國策決定の經緯概說 (Summary of Circumstances Leading to Decisions on Important National Policies) Foreign Ministry and 1st and 2d Demobilization Bureaus, Jun 46.

24 Konoye Memoirs, op. cit., pp. 43–4.
25 Ibid., p. 44.

and conducting special types of training which were specifically applicable to an eventual war against the United States, Great Britain, and the Netherlands.

As early as July 1940, Japanese Army intelligence possessed detailed information regarding order of battle and troop dispositions in Australia.[26] Between 27 November and 15 December 1940, a year before Pearl Harbor, Japanese aircraft successfully carried out photographic reconnaissance of parts of northern Luzon, including the Lingayen Gulf, Vigan, and Aparri coastal areas[27] where the Philippine invasion forces were to land following the outbreak of war.

Intelligence data regarding troop and air strength, ground force dispositions, airfields, harbors and fortifications were also assembled well in advance of hostilities for Java, Sumatra, Singapore, New Guinea, and the Philippines.[28] To assure the success of the Pearl Harbor attack, special intelligence arrangements were set up to obtain accurate, up-to-date reports on the number and location of American naval units in the harbor.[29]

Midget submarines, the precursors of Japan's *tokko* (special attack) weapons,[30] had been secretly developed by the Navy as early as 1934, but as war with the United States grew imminent during the summer of 1941, experiments were rushed to completion at the Kure Naval Station in attaching these small craft to long-range mother submarines capable of carrying them to a distant zone of operations and then releasing them for attack upon designated targets. Five of these suicide craft were used for the first time in the attack on Pearl Harbor.[31]

During the late summer and fall of 1941 Japanese units destined to take part in the invasions of the Philippines, the Dutch East Indies and Malaya were put through intensive training in amphibious operations and jungle warfare along the South China coast and in special training areas near Canton, on Hainan Island, and in Indo-China. Morale pamphlets, special military manuals, and training guides all based on the assumption of war against Britain and the United States were prepared for advance distribution.[32]

Following the 6 September Imperial conference, the tempo of Japan's war preparations sharply mounted. Steps were taken to mobilize and fit out about 1,500,000 tons of shipping for Army and Navy use. At the same time the assembly of the troops and supplies required for operations against the United States, Britain

---

26 See Plate No. 5.

27 See Plate No. 2 and discussion in Chapter II.

28 Cf. Chapter II on Pre-War Japanese Intelligence.

29 From 15 November 1941 bi-weekly code reports were received in Tokyo from the Japanese Consulate General in Honolulu. Statement by Rear Adm. Kanji Ogawa, Vice-Chief, Third Bureau (Intelligence), Imperial General Headquarters, Navy Section.

30 Japanese suicide weapons and tactics are dealt with more fully in Chapter XVII.

31 Cf. Chapter V, section on Pearl Harbor Operation.

32 A morale pamphlet entitled *Kore Sae Yomeba Kateru* これさへ讀めば勝てる (Read This And War is Won) was distributed to divisional commands in November 1941. Brief extracts follows: "The present war is a war with Japan's rise or fall at stake.... What is at the bottom of America's action in gradually prohibiting the export of oil and iron to Japan, as if to strangle her slowly by silk-wool?.... Japan has waited too long. If we are patient any longer, our aircraft, warships and motor cars will not move.... For the existence of Japan herself and her obligation under the Tripartite Alliance, not a minute longer must be endured. Japan is confronted with a great mission, as representative of the Oriental race, to bravely deliver the finishing blow against Occidental aggression of several hundred years." ATIS Research Report No. 131, *Japan's Decision to Fight*, 1 Dec 45, pp. 13–5.

and the Netherlands, and their concentration in preliminary staging areas in Japan Proper, Formosa, and South China were begun. Actual organization of the various southern invasion forces and the deployment of operational strength in the areas where hostilities were to begin, were to be carried out only after the final decision to go to war had been taken. According to the 6 September plan, this decision had to be made by mid-October.[33]

Only four days after the Imperial conference of 6 September had debated the issue of war or peace, the top-ranking staff officers and fleet commanders of the Navy assembled at the Naval War College in Tokyo to take part in the annual "war games." The problem set for the games was an invasion of the Southern area, but a restricted group of the highest officers of the Combined Fleet simultaneously studied behind barred doors technical problems involved in a surprise attack on Pearl Harbor.[34]

On 10 November, the general terms of Army-Navy co-operation in the Southern operations were agreed upon in Tokyo, and between 14 and 16 November detailed operational plans were elaborated by the Fleet and Army commanders directly concerned in a conference held at the headquarters of the Iwakuni Naval Air Group, on the Inland Sea near Hiroshima.[35]

Meanwhile, the parallel diplomatic efforts to revive the Washington negotiations made no headway. Foreign Minister Toyoda in September pressed for reconsideration by Washington of the proposed Roosevelt-Konoye conference, and the American Ambassador in Tokyo, Mr. Joseph C. Grew, strongly counselled this course in dispatches to the State Department.[36] On 2 October, however, Secretary of State Cordell Hull, in a memorandum handed to Ambassador Nomura in Washington, reiterated that general withdrawal of Japanese troops from both China and Indo-China remained a prerequisite for any Japanese-American agreement.[37]

The Konoye Cabinet, unable to agree on the course that Japan should take in view of these

---

33 Imperial General Headquarters, Army Section estimated that 15 October must be the deadline for the decision if war preparations were to be completed by the end of that month. Statement by Col. Takushiro Hattori, Chief, Operations Section, Imperial General Headquarters, Army Section.

34 The war games lasted from 10 to 13 September.

35 Details of Army-Navy Central Agreement and operational agreements concluded at the Iwakuni conference are given in Chapter V.

36 As paraphrased by the State Department, a dispatch sent by Ambassador Grew on 29 September 1941 stated: "The Ambassador, while admitting that risks will inevitably be involved no matter what course is pursued toward Japan, offers his carefully studied belief that there would be substantial hope at the very least of preventing the Far Eastern situation from becoming worse and perhaps of ensuring definitely constructive results, if an agreement along the lines of the preliminary discussions were brought to a head by the proposed meeting of the heads of the two Governments.... He raises the question whether the United States is not now given the opportunity to halt Japan's program without war, or an immediate risk of war, and further whether, through failure to use the present opportunity, the United States will not face a greatly increased risk of war...." Joseph C. Grew, *Ten Years in Japan*, Simon & Schuster (New York, 1944) pp. 193-4.

37 The American memorandum demanded:
1. That Japan unconditionally accept the following four basic principles:
   (a) Full respect of the territorial integrity and political sovereignty of other nations;
   (b) Non-intervention in the internal affairs of other nations;
   (c) Observance of the principle of equality, including equal opportunity in respect to trade;
   (d) Maintenance of the *status quo* in the Pacific, except where it might be modified by peaceful means.
2. General withdrawal of Japanese troops from China and French Indo-China.
3. Abandonment of exclusive economic arrangements between Japan and China. Summary of Circumstances Leading to Decisions on Important National Policies, op. cit.

conditions, resigned on 16 October, and two days later War Minister General Hideki Tojo formed a new government. Despite the mid-October deadline, Premier Tojo pledged continued efforts for a diplomatic settlement.[38] Then, on 5 November, a newly summoned Imperial conference revamped the 6 September "Outline Plan for the Execution of Empire Policies." Japan's resolution to accept war was reaffirmed; preparations therefor were to be completed by the end of November; however diplomatic negotiations were to be continued in the hope of effecting a compromise.[39]

Explaining the purport of the revised plan before the conference, Premier Tojo declared that eight Liaison conferences of the Government and Imperial General Headquarters, held between 23 October and 2 November, had reached the conclusion that war with the United States, Great Britain, and the Netherlands "was now unavoidable," and had unanimously decided to concentrate effort on war preparations, although still seeking to break the deadlock by diplomatic means.[40]

With the deadline for war now set at the end of November, speed was of the essence. The same day that the Imperial conference took place, the Navy Section, Imperial General Headquarters and Admiral Yamamoto, Commander-in-Chief of the Combined Fleet, issued orders for the fleet to prepare for the outbreak of war.[41] The following day, 6 November, the Army Section, Imperial General Headquarters fixed the order of battle of the Southern Army and directed its commanding general to move his forces to the assembly areas and points of departure for the invasion of the "southern strategic areas."[42]

On the diplomatic front the urgency was no less great. On 6 November Ambassador Extraordinary Saburo Kurusu left by air for Washington to make the final effort for a peaceful solution.[43] Without waiting for his arrival, Japan on 7 November transmitted its Proposal "A" through Ambassador Nomura, and when this was rejected, Proposal "B" for a temporary *modus vivendi* freezing war moves in the Pacific was presented by Ambassador

---

38 "I personally know that on the morning of 18 October, after agreeing to take the portfolio of Navy Minister, Admiral Shimada went to see the new Premier, General Tojo, for the purpose of stipulating a condition for his entry into the Cabinet. This condition was that diplomatic negotiations with the United States must be continued with the avowed objective of reaching a peaceful settlement of the matters in dispute. Admiral Shimada told me and several others at the Navy Ministry that Tojo had expressed complete agreement...." *Kyokuto Kokusai Gunji Saiban ni okeru Shonin Sawamoto Yorio no Sensei Kokyosho* 極東國際軍事裁判に於ける證人澤本賴雄の宣誓口供書 (Affidavit of Witness Yorio Sawamoto, International Military Tribunal for the Far East), Doc. No. 2889.

39 The Imperial conference decided:

"1. In order to bring about a more favorable situation and ensure its defense and national existence, the Empire, with the determination to accept war with the United States and Great Britain, will complete its war preparations by the end of November. At the same time, it will endeavor to effect a compromise through diplomatic negotiations based on Proposals 'A' and 'B', dealt with separately.

"2. In the event that these negotiations fail, decision will be made immediately to go to war against the United States and Great Britain." Affidavit of Tojo, op. cit.

40 File of Important Documents, op. cit.

41 (1) *Daikairei Dai Ichi-go* 大海令第一號 (Imperial General Headqurters Navy Order No. 1) 5 Nov 41. (2) ATIS Limited Distribution Translation No. 39 (Part VIII), 4 Jun 45.

42 *Daihonyei Rikugun Tosui Kiroku* 大本營陸軍統帥記錄 (Imperial General Headquarters Army High Command Record) 1st Demobilization Bureau, Nov 46, pp. 22-4.

43 Ambassador Kurusu, notified only two days previously of his mission, flew to Hongkong where, by arrangement with the United States State Department, departure of a trans-Pacific Clipper was delayed to accommodate him. This haste reflected the new war deadline.

Kurusu on 20 November, three days after his arrival in Washington.[44]

It was at this crucial juncture that the Hull note of 26 November was delivered. Describing the reaction to the note in a statement made after the war, Admiral Shigetaro Shimada, at that time Navy Minister, said:

*It was a stunning blow. It was my prayer that the United States would view whatever concessions we had made as a sincere effort to avoid war and would attempt to meet us half-way, thereby saving the whole situation. But here was a harsh reply from the United States Government, unyielding and unbending. It contained no recognition of the endeavors we had made toward concessions in the negotiations. There were no members of the Cabinet nor responsible officials of the General Staff who advocated acceptance of the Hull note. The view taken was that it was impossible to do so and that this communication was an ultimatum threatening the existence of our country. The general opinion was that acceptance of this note would be tantamount to the surrender of Japan.*[45]

On 21 November, Imperial General Headquarters had ordered the Combined Fleet to move at the appropriate time to positions of readiness for the start of operations.[46] The various naval task forces, though subject to recall in the event of a Japanese-American agreement, left for their designated theaters of operation toward the end of November.

On 29 November a Liaison conference of the Government and Imperial General Headquarters concluded that war must be launched. Instructions were sent to Japan's ambassadors in Germany and Italy to secure commitments whereby:

1. *Germany and Italy would immediately declare war against the United States upon the outbreak of Japanese-American hostilities;*
2. *None of the three Powers would enter into a separate peace with the United States and Great Britain; and*
3. *The three Powers would not make peace with Great Britain alone.*[47]

On 1 December an Imperial conference met to ratify finally the decision to fight. It was a moment of grave solemnity when, in the presence of the Emperor, Premier Tojo rose to announce:

*In accordance with the decision reached at the Imperial conference of 5 November, the Army and Navy have made full preparations for war, while the Government has continued to exert all possible effort to adjust diplomatic relations with the United States. However, the United States has not receded from its original demands. In addition, the United States, Great Britain, the Netherlands and China, in collusion, have demanded a one-sided compromise, adding new conditions such as unconditional withdrawal of our troops from China, repudiation of the Nanking Government and abrogation of the Tripartite Treaty with Germany and Italy.*

*If our country should yield, its prestige would be lost, and the China Incident could not be settled. More than this, the very existence of Japan would be imperilled. It is now clear that our country's claims cannot be realized through diplomatic negotiations.*

*Economic and military pressure by the United States, Great Britain, the Netherlands, and China is increasing. From the standpoint both of national strength and of military operations, the point has finally been reached*

---

44 Proposal "A" offered: 1. Gradual withdrawal of Japanese troops from China, with the exception of garrisons in North China and Inner Mongolia, within two years after the conclusion of peace with China; 2. Withdrawal of troops from French Indo-China as soon as the China war ended. Proposal "B", in addition to calling a halt to fresh war moves in the Pacific, envisaged a limited restoration of commercial relations, including resumption of American oil shipments to Japan.

45 *Kyokuto Kokusai Gunji Saiban ni okeru Hikokunin Shimada Shigetaro no Sensei Kokyosho* 極東國際軍事裁判に於ける被告人嶋田繁太郎の宣誓口供書 (Affidavit of Defendant Shigetaro Shimada, International Military Tribunal for the Far East) Doc. No. 328.

46 *Daikairei Dai Go-go* 大海令第五號 (Imperial General Headquarters Navy Order No. 5) 21 Nov 41.

47 International Military Tribunal for the Far East, Exhibit Doc. No. 1204.

# CHAPTER IV

# BASIC STRATEGY AND MILITARY ORGANIZATION

## Strategy for a Long War

It was obvious to Japan's military strategists that the Pacific War would be a long one. The superior fighting potential of the United States made it improbable that Japan could inflict a crushing defeat on its adversary at the outset. The tremendous distances involved rendered a direct attack on the American mainland impracticable; finally, Japan not only had the United States to contend with, but Great Britain and the Netherlands as well.

Equally obvious was the certainty that possession of the natural resources for war would become a decisive factor. Japan did not have these raw materials within its own territory, and foreign sources of supply were blocked. The supply of liquid fuel, for example, was practically limited to the quantity on hand, and stockpiles were barely adequate for two years of armed conflict.

The first objective of Japan's strategy, therefore, was the conquest of the rich colonial areas in the South, whose vital resources added to those within the Japanese Empire, Manchuria, and Occupied China would provide a firm economic basis for waging an extended war.[1]

The fleet was assigned the vital task of blocking superior enemy naval power and supporting ground-force invasion operations.

In view of the handicap resulting from the pre-war ratio of 7.5 to 10 between the Japanese and American fleets, it was considered that the American fleet must be crippled by a surprise blow at the outbreak of war, giving Japan mastery of the sea long enough to attain its strategic objectives in the Western and Southwest Pacific. With American air and sea power temporarily crushed, and vital American and British bases, as well as the Netherlands Indies, in Japanese hands, it was estimated that Japan could carry on the war successfully for another two years, provided the fleet sustained no serious losses.

Once the initial objectives were taken, Japan would possess an outer defense perimeter extending from Burma through Sumatra, Java, Timor, Western New Guinea, the Caroline and Marshall Islands, and Wake. The vast sea areas within this perimeter, except for the Solomons—New Guinea—Philippines line, were generally favorable to the establishment of a strong strategic inner defense. (Plate No. 8) Within this zone, the Japanese fleet, especially its carrier forces, supplemented by land-based air strength, would be able to operate at great advantage, provided the United States and Britain were unable to build up their air strength sufficiently to swing the balance in their favor.

As regards land operations, it was estimated that the Army, if successful in its initial operations, would be able to secure and maintain its hold on the occupied areas. An anticipated

---

1 *Juyo Shorui Tsuzuri* 重要書類綴 (File of Important Documents) Preserved by Capt. Toshikazu Ohmae, member, Military Affairs Bureau, Navy Ministry. All source materials cited in this chapter are located in G–2 Historical Section Files, GHQ FEC.

British counterattack against Burma could be successfully withstood by utilizing favorable terrain features and furnishing reinforcements when necessary. The southern areas, China, Manchuria, and Japan Proper would be strongly garrisoned, and as long as the destruction of Japanese shipping could be held within reasonable bounds,[2] profitable exploitation of the occupied territories was deemed possible. With all the needed raw materials at its disposal, Japan's economic and military capacity to carry on the war could be guaranteed for about two years.

Beyond that date, however, a number of unpredictable factors made it impossible for Imperial General Headquarters to plan with certainty. The relative position in regard to armaments, fleet strength, and air power three years hence could not be accurately foretold. What changes would Japan's material power and morale undergo? Every shift in the world situation, especially in the European War, would have profound repercussions in the Pacific. These great imponderables rolled up like a stormy wave, making it impossible to see ahead beyond the first two years of war.

## Manpower and Materials

Aside from strategic problems, Japan's war planners devoted special attention to three principal elements which conditioned the nation's over-all fighting strength. These were manpower, raw materials, and transportation, especially shipping.

Paradoxically, as far as manpower was concerned, the very over-population which was one of the pressure factors behind Japanese expansionism now became a factor in Japan's favor. Japan Proper, covering an area of only 381,000 square miles, supported in 1940 a population of 73,114,000.[3] The population, for the past ten years, had been increasing at the rate of 800,000 to one million annually. With this reserve of manpower, immediate mobilization demands could be met easily, and the needs of industry could be filled throughout an extended war.

War-weariness had increased appreciably during the later years of the China Incident. But confronted by the new and graver challenge of a life-or-death struggle against the United States, Great Britain, and the Netherlands, the Japanese people could be expected to rise to the test. Japan's leaders entertained no doubt that traditional loyalty and obedience would keep the people's morale from breaking even under the strain of a long war.

As for raw materials, Japan expected to be practically self-sufficient in coal, iron, and industrial salts within the Japanese-Manchurian-Chinese bloc, if adequate marine transportation could be assured. Kyushu, Hokkaido, Sakhalin, and North China were the main sources of coal for general use. Coal for steel production came chiefly from Sakhalin, Manchuria, and North China. North and Central China could be counted on for iron ore, while industrial salts were available in Korea, Manchuria, North China, and Formosa.

The annual production of steel was approximately 5,000,000 tons. The overall steel plan for 1941 was as follows:

| | |
|---|---|
| Production goal: | 4,760,000 tons |
| Navy Allotment: | 950,000 tons |
| Army Allotment: | 900,000 tons |
| Ordinary consumption: | 2,910,000 tons |

These estimates were revised in November

---

2 Cf. section on Shipping.

3 *Jinko Tokei Soran* 人口統計總覽 (General Compilation of Statistics on Population) Population Branch, Welfare Ministry Research Institute. Sep 43, pp. 2–3.

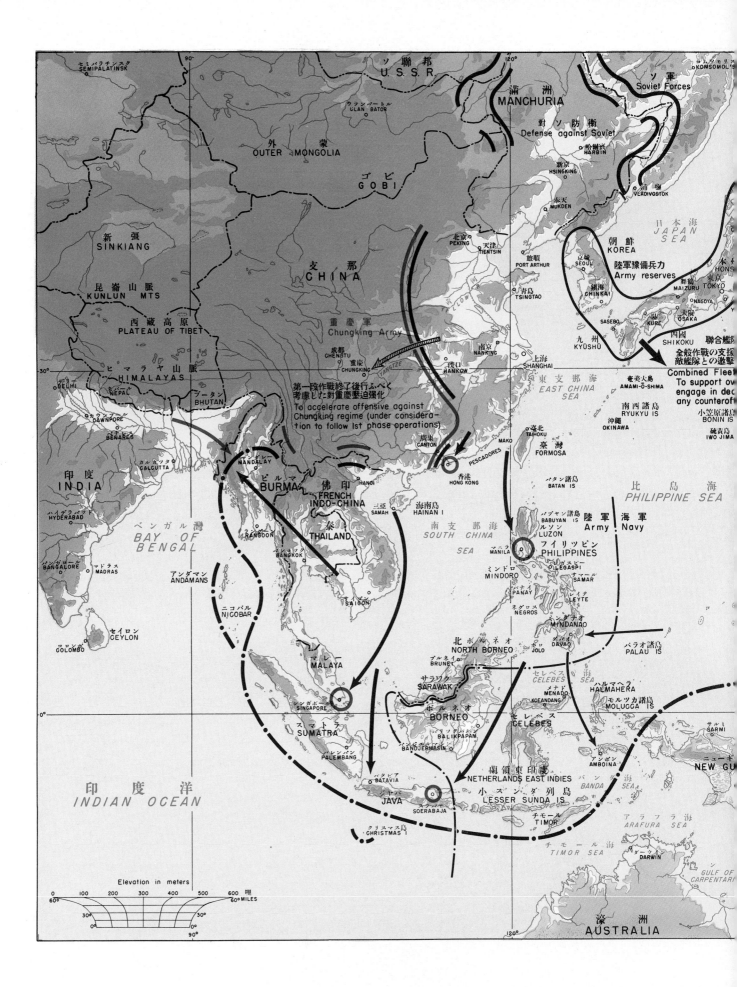

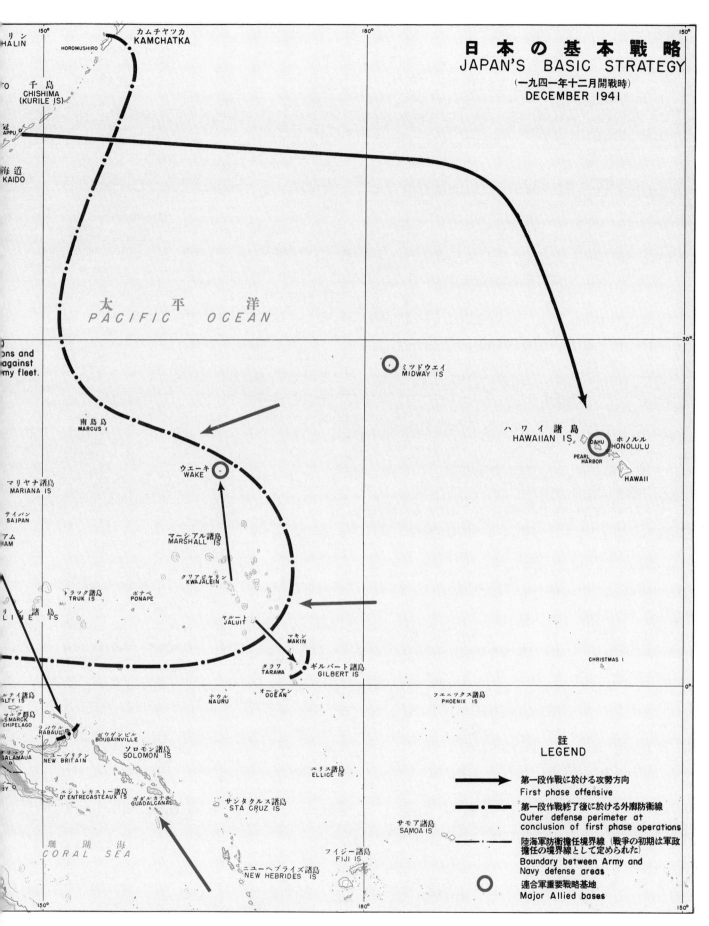

PLATE NO. 8

Japan's Basic Strategy, December 1941

1941 to conform to a decrease in production and an expected increase in Navy requirements in the event of war. The new figures were:

- Production goal: 4,500,000 tons
- Navy Allotment: 1,100,000 tons
- Army Allotment: 790,000 tons
- Ordinary consumption: 2,610,000 tons

Of the 2,610,000 tons allotted for ordinary consumption, it was planned to allocate 300,000 tons to the shipbuilding industry on a priority basis, in order to achieve a ship construction goal of 600,000 tons annually.[4] It was estimated that, if the Southern campaign succeeded, it would be possible to carry on a long war with a steel program of these proportions.

The shortage of liquid fuel was Japan's Achilles heel. The combined output of natural and synthetic oil did not exceed 3,459,000 barrels annually. War needs must be drawn mainly from reserves, at least until the oil-producing territories in the south could be occupied, developed, and fully exploited.

Allowing for the possibility that the oil wells in the southern area might be totally destroyed before they fell into Japanese hands, the Government and Imperial General Headquarters developed a supply plan which would barely meet estimated war requirements:[5]

*Supply:* (in thousand barrels)

Stock pile: 52,836

Domestic production:

|  | 1st year | 2nd year | 3rd year |
|---|---|---|---|
| Crude oil | 1,573 | 1,258 | 1,887 |
| Synthetic oil | 1,887 | 2,516 | 3,145 |
| Total | 3,460 | 3,774 | 5,032 |

Production in the southern areas:

|  | 1st year | 2nd year | 3rd year |
|---|---|---|---|
| Borneo | 1,887 | 6,290 | 15,725 |
| Sumatra | — | 6,290 | 12,580 |
| Total | 1,887 | 12,580 | 28,305 |

*Demand:*

|  | 1st year | 2nd year | 3rd year |
|---|---|---|---|
| Military | 23,902 | 22,644 | 21,072 |
| Non-military | 8,806 | 8,806 | 8,806 |
| Total | 32,708 | 31,450 | 29,878 |

*Balance:* (exclusive of minimum reserve of 9,435,000 bls)

| 1st year | 2nd year | 3rd year |
|---|---|---|
| 16,040 | 944 | 4,403 |

Included in Japan's liquid fuel reserves as of 1 December 1941 were 6,919,000 barrels of aviation gasoline. The production program called for 503,200 barrels during the first year, 2,075,700 barrels during the second year, and 3,396,600 during the third year of the war. Wartime requirements were computed at 4,500,000 to five million per year, and plans were drawn up on the basis of these figures. The margin of safety was so slight, however, that considerable difficulty was anticipated in the second and third years.[6]

To meet Japan's domestic requirements for staple food, 397 million bushels of rice must be available annually. The rice supply plan for 1942 called for domestic production of 298 million bushels, the balance of 99 million bushels to be made up by imports from Korea, Formosa, Thailand, and French Indo-China. Transports returning empty from the zone of operations would be utilized to make up any shortage of non-military shipping. If, due to military operations, imports from Thailand and French Indo-China fell short of the estimated 50 million bushels counted upon from the two countries combined, soy beans,

---

4  Memorandum Report submitted by Lt. Gen. (ret.) Teiichi Suzuki, President of the Planning Board, at the Imperial conference of 5 Nov 41. Preserved in the Notes of Maj. Gen. Kikusaburo Okada, Chief of War Plans Section, Economic Mobilization Bureau, War Ministry.

5  Ibid.

6  Ibid.

sweet potatoes and miscellaneous grains grown in the Homeland, Korea, Formosa and Manchuria would be used to make up the deficit.[7]

Although shortages of a few special materials like cobalt and high quality asbestos were anticipated, control of the southern supply areas and the speedy development of occupied China[8] were expected to produce a steady supply of important materials such as bauxite, raw rubber, raw materials for special steels, metals, non-ferrous metals, leather, cotton, hemp, and oil. This plan, however, depended entirely upon the maintenance of adequate marine transportation, and Imperial General Headquarters realized fully that this factor would prove a decisive one in the Pacific War.

## Shipping

In November 1941, Japan's total shipping amounted to 6,720,000 gross tons, including motor sailboats over 100 tons. Of these, serviceable ships aggregated 5,980,000 gross tons, including 360,000 gross tons of oil tankers.[9]

It was estimated that the level of imports required by the "Materials Mobilization Plan of 1941" could be maintained during hostilities, provided a minimum of three million gross tons of shipping was reserved at all times for non-military use. With this tonnage, approximately five million tons of materials could be transported monthly, even if wartime shipping efficiency dropped by 15 to 20 per cent. Actually, the monthly average of tonnage transported during the first half of 1941 corresponded to this estimate, but since military requirements continued to tie down 2,800,000 tons of shipping long after the Southern operations had entered a relatively inactive phase, the reservation of three million tons of ships for non-military use became a difficult problem.

In view of the vital importance of shipping, Imperial General Headquarters had given careful consideration to probable war losses and replacement construction plans. The Navy estimated that losses would aggregate 800,000 gross tons during the first year, 600,000 the second year, and 700,000 the third year. Imperial General Headquarters, however, estimated that losses during the first year of the war would amount to between 800,000 and one million tons,[10] and that subsequently losses would decline.[11] On this basis, decision was made to build 1,800,000 gross tons of new ships over a three-year period, an average of 600,000 tons annually.[12]

Japan's private shipbuilding capacity at the start of war was approximately 700,000 gross

---

7 Ibid.

8 Extract from report of Chief of Staff of Izeki Force, North China, 25 August 1941: "Due to the present international situation and the increase in national military preparations, the importance of exploiting and utilizing the resources of this area must be considered. The accumulation of these military supplies will be subdivided into procurement by military authorities and purchases by civilian agencies. The items to be acquired by civilians are copper ore, copper wire scrap, copper scrap, brass scrap, cases, melted cases, tin, coin, pewter, and antimony ore.

"Scrap iron in North China will be acquired by the Nippon Iron Industries Company. Other resources to be acquired are nickel, cobalt, tungsten ore, molybdenum ore, copper, lead, zinc, quicksilver, high grade asbestos, high grade mica, nonferrous metals, steel, and other minerals.

"An investigation squad organized by the army has reported the probable presence of iron, sulphur, fluorspar and zinc in the vicinity of Yancheng and of iron in Suehchuanling." ATIS Bulletin No. 1555, 5 Nov 44.

9 Notes of Maj. Gen. Okada, op. cit.
10 Ibid.
11 Statement by Maj. Gen. Okada, previously cited.
12 Notes of Maj. Gen. Okada, op. cit.

*Powers in Europe. Japanese strategists, after carefully weighing the possibilities, estimated that first priority would be given to Europe.*

*6. The United States and Great Britain, already counting Chiang Kai-shek as an ally, would undoubtedly attempt to bring the Soviets into the war.*

## Timing of the Attack

With the decision to fight taken and the areas to be occupied defined, the next vital question facing Imperial General Headquarters was the selection of the most propitious moment for opening the hostilities.[19]

In a war to be waged with inferior forces against three enemy countries, it was deemed absolutely essential that Japan exploit to the fullest the advantage of choosing the moment to strike and seizing the initiative from the start of the operations. Were Japan to wait passively until war finally resulted from a step-by-step process of deterioration, Imperial Headquarters estimated that loss of the initial tactical advantage would make it impossible to attain the basic Japanese strategic objectives.

It was estimated that, if war were not started before March 1942, economic inferiority would be such as to preclude any hope of success.

In order to guard against the remote possibility of an attack by the Soviet Union while Japan would be heavily engaged in the south, it was considered advisable to start hostilities early enough so that the Southern operations would be near completion before the end of the winter, during which a Soviet attack from the north would be unlikely.

In view of the steady tightening of defensive arrangements among the ABCD Powers, particularly joint Anglo-American defense arrangements in the Malaya and Philippines areas, it was deemed advantageous to start hostilities at an early date.

Assuming that the fleet would take the "Great Circle" route to attack Pearl Harbor, navigational and weather conditions would be extremely unfavorable after January. Similarly, navigational conditions off Malaya would become unfavorable in January and February.

To facilitate air and landing operations, it was advisable to select a date during the last-quarter moon.

To achieve a successful surprise attack operations should begin on Saturday or Sunday.[20]

In accordance with the decisions taken by the Imperial conference of 6 September,[21] the High Command first planned to launch hostilities early in November. Then, with the revision of these decisions by the 5 November Imperial conference,[22] the anticipated opening of hostilities was postponed until early December. The final decision on the date of the attack was held in abeyance until the outcome of the Japanese-American negotiations became clear.

Following the Imperial conference of 1 December, which finally ratified the decision to fight, 8 December (Sunday, 7 December in Hawaii and the United States) was fixed by Imperial General Headquarters as the date for the start of the war.

## Central Command Organization[23]

With the outbreak of the China Incident

---

19 Section on "Timing of the Attack" is based on data prepared by Rear Adm. Tomioka and Col. Hattori, previously cited.

20 The Japanese correctly appraised the social and convivial implications of the American "week-end".

21 Cf. Chapter III.

22 Cf. Chapter III.

23 Material in this section is based on statements by Rear Adm. Katsuhei Nakamura, Senior Adjutant of Navy Ministry, and Col. Hattori, previously cited.

# 陸海軍中央指揮組織
# CENTRAL COMMAND ORGANIZATION
(一九四一年十二月八日)
8 DECEMBER 1941

天皇 / EMPEROR

軍事參議院 / BOARD OF MILITARY COUNCILLORS
侍従武官府 / AIDES-DE-CAMP
元帥府 / BOARD OF MARSHALS & ADMIRALS

大本營 / IMPERIAL GENERAL HEADQUARTERS

## ARMY SECTION (陸軍部)

教育総監 / INSPECTORATE GENERAL MILITARY TRAINING
航空総監 / INSPECTORATE GENERAL (ARMY) AVIATION

陸軍大臣 / MINISTER OF WAR
- 陸軍政務次官 / Parliamentary Vice-Minister of War
- 陸軍大臣官房 / War Minister's Secretariat
- 人事局 / Military Affairs Bureau
- 整備局 / Economic Mobilization Bureau
- 興亞局 / Intendance Bureau
- 法務局 / Judicial Bureau
- 陸軍航空本部 / Army Aeronautical Dept
- 整備局 / Air Ordnance Bureau

陸軍大臣 / MINISTER OF WAR
- 属員 / Assistants
- 陸軍次官 / Vice-Minister of War
- 陸軍大臣官房 / War Minister's Secretaries (2)
- 人事局 / Chief, Personnel Bureau — 1 member, Apmt & Asgmt Sec
- 軍務局 / Chief, Mil Affairs Bureau — Chief, Army Affairs Sec, and 1 member
- 軍務局 / Chief, Mil Affairs Bureau — Chief, Mil Affairs Sec, and 2 members
- 整備局 / 1 member, Mil Adm Bureau — 1 member, Economic Mobilization Bureau

参謀総長 / CHIEF OF ARMY GENERAL STAFF
- 兵站総監 / Chief, Line of Comm Bureau
- 運輸通信官部 / Trans & Comm Bureau
- 野戦兵器長官部 / Field Ordnance Bureau
- 野戦航空兵器長官部 / Field Air Ordnance Bureau
- 野戦経理長官部 / Field Intendance Bureau
- 野戦衛生長官部 / Field Medical Bureau
- 副官部 / Adjutant's Office
- 陸軍報道部 / Army Information Bureau

参謀次長 / Deputy-Chief of Army General Staff
- 第二十班幹事部 / 20th Group (Coordinating Group)
- 研究部 / Research Group
- 総務部 / General Affairs Bureau
- 第一部作戦 / 1st Bureau Operations
- 第二部情報 / 2nd Bureau Intelligence
- 第三部運輸通信 / 3rd Bureau Trans & Comm
- 第十八班電波情報 / 18th Group Radio Intelligence
- 第四部戦史 / 4th Bureau Historical

## NAVY SECTION (海軍部)

海軍大臣 / MINISTER OF NAVY
- 海軍政務次官 / Parliamentary Vice-Minister of Navy
- 海軍大臣官房 / Navy Minister's Secretariat
- 成員局 / Mil Preparations Bureau
- 教育局 / Training Bureau
- 医務局 / Medical Bureau
- 法務局 / Judicial Bureau
- 海軍航空本部 / Navy Aeronautical Dept
- 海軍艦政本部 / Navy Construction Dept

海軍大臣 / MINISTER OF NAVY
- 海軍次官 / Vice-Minister of Navy
- 海軍参與官 / Parliamentary Councillor
- 軍務局 / Military Affairs Bureau
- 人事局 / Personnel Bureau
- 軍需局 / Supply Bureau
- 経理局 / Intendance Bureau
- 海軍艦政本部 / Navy Ship & Ordnance Dept

海軍大臣 / MINISTER OF NAVY
- 属員 / Assistants
- 海軍次官 / Vice-Minister of Navy
- 海軍省副官 / Senior Adjutant
- Navy Minister's Secretaries (2)
- 軍務局 / Chief, Mil Affairs Bureau — Chief, 1st Sec, 2nd Sec, and 3 members
- 成員局 / Chief, Mil Preparation Bureau — Chief, 1st Sec, Chief, 2nd Sec, Chief, 3rd Sec
- 人事局 / Chief, Personnel Bureau — Chief, 1st Sec, and 1 member
- 海軍戦備考査部 / War Preparations Analysis Committee

軍令部総長 / CHIEF OF NAVY GENERAL STAFF
- 軍令部次長 / Vice-Chief of Navy General Staff
- 第一部作戦 / 1st Bureau Operations
- 第二部編備 / 2nd Bureau Preparations
- 第三部情報 / 3rd Bureau Intelligence
- 海軍通信部 / Naval Communications
- 特務班 / Special Duty Group (Radio Intelligence)
- 副官部 / Adjutant's Office
- 海軍報道部 / Navy Information Bureau
- 戦史部 / Historical Section

---

In note: In the Army, the following were dual positions:

(a) 参謀次長（兼ねる職務） / Deputy-Chief of Army General Staff; Chief, Line of Communications Bureau
(b) 第三部長（兼ねる通信） / Chief, 3rd Bureau-Transportation & Communications; Chief, Transportation & Communications Bureau
(c) 兵器局長（兼ねる補給局長） / Chief, Ordnance Bureau; Chief, Field Ordnance Bureau
(d) 航空本部総務部長（兼ねる野戦航空兵器長官） / Chief, Air Ordnance Bureau; Chief, Field Air Ordnance Bureau
(e) 経理局長（兼ねる野戦経理長官） / Chief, Intendance Bureau; Chief, Field Intendance Bureau
(f) 医務局長（兼ねる野戦衛生長官） / Chief, Medical Bureau; Chief, Field Medical Bureau

PLATE NO. 9

Central Command Organization

in 1937, Imperial General Headquarters was established as the central directing and coordinating organ of the Army and Navy High Commands. This body was divided into the Army and Navy Sections, in which the Chiefs of General Staff of both services and the chiefs and selected subordinates of the more important bureaus and sections of the War and Navy Ministries and the Army and Navy General Staffs were included. (Plate No. 9)

The Board of Militay Councillors, a special body created in 1887, comprised of selected generals as well as the Board of Field Marshals and Fleet Admirals, composed of all field marshals and fleet admirals, were to advise the Emperor on matters of great military importance, and were also available to the services for consultation.

The Army and Navy Chiefs of General Staff were the Emperor's highest advisers in all matters involving the operational use of the fighting forces. Such matters did not pass through the Premier or the Cabinet. In other words, it was the special characteristic of the Japanese military High Command that it enjoyed complete independence from control by political organs of the Government in military matters.

To unify political and military strategy during war and promote closer co-ordination, the Government and High Command, in November 1937, established the Imperial General Headquarters-Government Liaison Conference. Members ordinarily included the Premier, Ministers of War, Navy, and Foreign Affairs, and the Chiefs of the Army and Navy General Staffs. Decisions arrived at by this body were to be implemented by the responsible military or governmental agency.[24]

## Strength And Organization Of Forces

Originally, Japan's basic policy was to maintain its controlling position in the Far East. Its armament, therefore, was developed mainly for use in its own and neighboring territories. The two potential enemies were the United States, a naval power on the east, and Russia, a military threat on the Continent to the west. For this reason, Japan could not subordinate one service to the other, and maintained an Army and Navy of equal strength.

Although both services made tremendous advances during the China Incident, a great disparity still existed between their strength and that of the United States Navy and the Soviet Army. Armament building and operational plans had been scaled to a conflict with one enemy before the Pacific War. Military operations against more than two nations had hitherto not been considered.

Japan had no independent Air force. Each service possessed its own air arm. After the outbreak of the China Incident in 1937, Japan began expanding its Air forces. Growing realization of the importance of air power as proved

---

24 The Liaison conference convened only when necessary until November 1940, when meetings began to be held twice weekly at the Premier's official residence. In July of the following year, after Germany invaded Russia, the members agreed to make more active use of the Council, and the meeting place was then changed to the Imperial Palace. Following the establishment of the Koiso Cabinet in July 1944, the Council was newly designated as the "Supreme War Direction Council," but its functions remained unchanged.

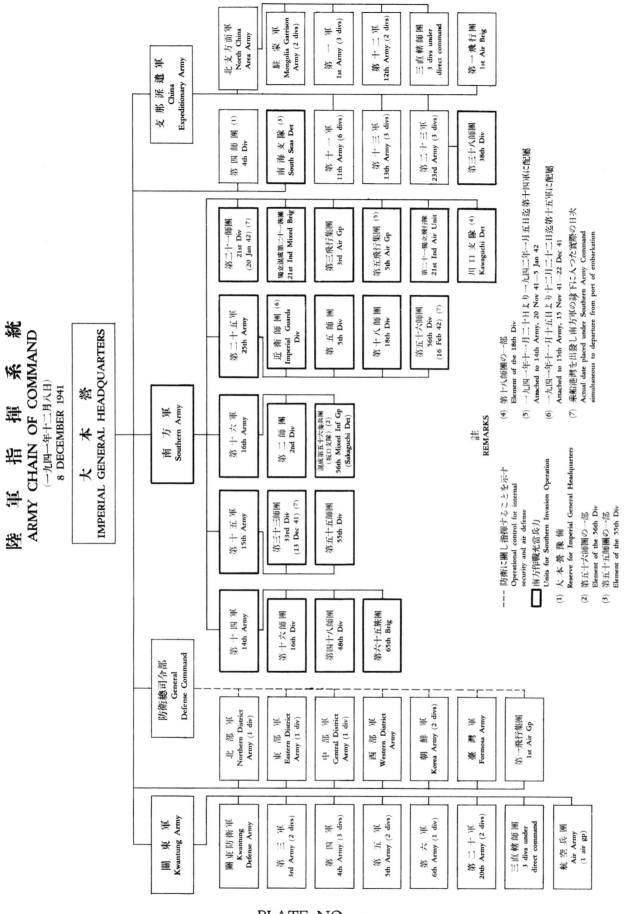

PLATE NO. 10

Army Chain of Command

by active operations stimulated this expansion.[25] The late start of the program, however, made it impossible to build up an air strength of planes of advanced design in the amount considered desirable before the beginning of the Pacific War. The Army Air forces had been operating with planes of short or medium range. Development of long range aircraft had not been emphasized.

At the time of its organization, the Japanese Army had been modeled first on the French and then on the Prussian Armies. From experience gained in the Russo-Japanese War, however, and profiting by the lessons of World War I, Japan adopted a unique system of organization, tactics, and training suitable to the requirements for operations in East Asia.

Following the outbreak of the Manchurian Incident, the potential enemy was obviously the Soviet Union. To oppose superior Soviet forces training emphasized offensive operations, individual courage, and proficiency in all branches of military science. Fire power and mechanization were not, however, stressed.

During the six years' interval between the Manchurian and China Incidents, military personnel strength was enormously increased, and further augmentation between 1937 and 1941 brought the total strength up to 51 divisions at the outbreak of war.[26] However, due to a lack of raw materials and limited budgets, little new equipment of improved design was issued. The Chinese forces, poorly trained and equipped, offered little stimulus to efficiency, and as a result Japanese staff officers grew slack as hostilities dragged on, and the whole standard of training deteriorated.

As the political situation vis-à-vis the

---

25      Growth of Army and Navy Air Forces 1935–1941[1]

| | Army[2] | | | | Navy[3] | | | | | |
| Year | Fighters | Bombers | Rcn | Total | Carrier Based Fighters | Land Based Bombers | Torpedo Planes | Other Types | Total |
|---|---|---|---|---|---|---|---|---|---|
| 1935 | * | * | * | * | 188 | 108 | 24 | 132 | 138 | 590 |
| 1936 | * | * | * | * | 216 | 120 | 144 | 132 | 160 | 772 |
| 1937 | 210 | 210 | 120 | 540 | 216 | 132 | 204 | 108 | 178 | 838 |
| 1938 | 240 | 330 | 130 | 700 | 269 | 132 | 228 | 132 | 200 | 961 |
| 1939 | 280 | 450 | 180 | 910 | 201 | 132 | 288 | 156 | 228 | 1,005 |
| 1940 | 360 | 500 | 200 | 1,060 | 167 | 132 | 264 | 180 | 306 | 1,049 |
| 8 Dec 1941 | 550 | 660 | 290 | 1,500 | 684 | 252 | 443 | 92 | 198 | 1,669 |

1  Statistics include only first-line aircraft.
2  Compiled by 1st Demobilization Bureau, Japanese Government.
3  Compiled by 2d Demobilization Bureau, Japanese Government.
*  Figures not available.

26   From a strength of 17 divisions (not including 13 reserve divisions) during the period 1924–36, the Army expanded as follows:

| Year | No. of Divisions |
|---|---|
| 1937 | 24 (Not including 6 reserve divs.) |
| 1938 | 34 |
| 1939 | 41 |
| 1940 | 50 |
| 1941 | 51 |

(Statistics compiled by 1st Demobilization Bureau, Japanese Government)

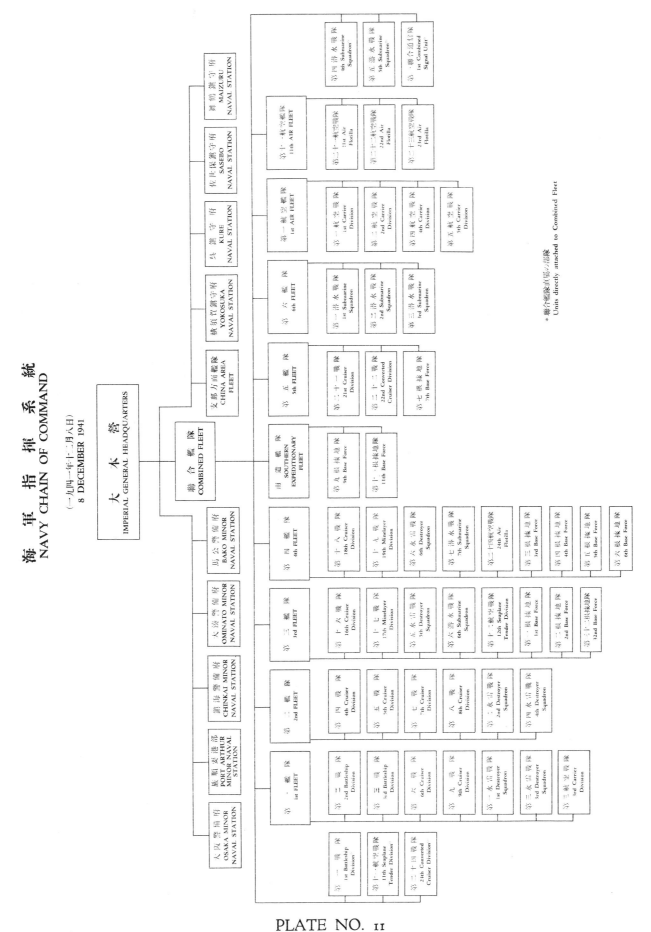

PLATE NO. 11

Navy Chain of Command

success could be assured by maneuvering available Japanese military strength so as to develop a three-to-one local superiority of forces in all invasion sectors.[13]

Achievement of this superiority required adherence to a carefully determined invasion schedule and the double use of troops and shipping in successive operations. Thus, it was decided that forces and shipping assigned to the Philippines, Hongkong, Guam and Malaya operations would be used again in succeeding operations.[14] The Burma operations were to be carried out by troops diverted from other combat zones where they were no longer needed.

To provide the 11 divisions called for by the invasion plans, five divisions were diverted from the China front, and six were taken from the homeland. These were further reinforced by the addition of the bulk of the Kwantung Army's service troops, which were withdrawn from Manchuria.[15] The main strength of the 3d Air Group was detached from the China Expeditionary Forces, and the main strength of the 5th Air Group was taken from Manchuria. Both were reorganized to include the best air units from China, Manchuria, and Japan Proper.

Just before the outbreak of hostilities, the tactical grouping and disposition (Plate No. 12) of Army forces allocated to the Southern Operations were as follows:[16]

**Southern Army:** General Headquarters in Saigon, French Indo-China

14th Army: Philippines Attack Force
- Army Headquarters ⎫
- Army troops (65th Brig. & other elements) ⎬ in Formosa
- 48th Division (main elements) ⎭
- 16th Division in Amami-Oshima
- Remaining elements in Pescadores and Palau

15th Army: Burma Attack Force
- Army Headquarters ⎫
- Elements 55th Division ⎬ in S. Indo-China
- Army troops ⎫
- 55th Division (less elements) ⎬ in N. Indo-China
- 33d Division in Central China

16th Army: East Indies Attack Force
- Army Headquarters ⎫ in Japan
- 2d Division ⎭ Proper
- Elements 56th Division (56th Mixed Inf Group) in Palau

25th Army: Malaya Attack Force
- Army Headquarters ⎫ on Hainan
- 5th Division ⎭ Island
- Imperial Guards Division in S. Indo-China[17]
- 18th Division in Canton
- Army troops in Formosa

Army reserves:
- 21st Division in North China
- 56th Division (main elements) in Japan Proper

3d Air Group: Malaya Attack Force
- 5 fighter groups ⎫
- 4 light bomber groups ⎬ in South China and N. Indo-China
- 4 heavy bomber groups
- 2 reconnaissance groups ⎭

---

13 Data on Imperial General Headquarters planning prepared by Lt. Gen. Tanaka, and Col. Hattori, previously cited.

14 Plans for the Philippines operation called for the transfer of the 48th Division, after the fall of Manila, to the South for employment in the invasion of Java.

15 The 5th, 18th, 21st, 33d, 38th and Imperial Guards Divisions were taken from China, and the 2d, 16th, 48th, 55th and 56th Divisons from Formosa and Japan Proper.

16 Statement by Col. Hattori, previously cited.

17 The Imperial Guards Division was temporarily transferred to the Fifteenth Army for initial operations (pacification of Thailand) but was then restored to the Twenty-fifth Army for participation in the Malaya campaign.

5th Air Group:    2 fighter groups ⎫
Philippines         2 light bomber groups ⎬ in S.
Attack Force       1 heavy bomber group ⎪ Formosa
                           1 reconnaissance Unit ⎭

**China Expeditionary Forces:** General Headquarters in Nanking

23d Army:      Army Headquarters ⎫ in
Hongkong     38th Division        ⎬ Canton
Attack Force                             ⎭

**Guam Occupation Force:** Directly under Imperial General Headquarters

South Seas     Detachment Headquarters ⎫
Detachment:    144 Inf. Regt            ⎬ in Bonin Islands
                 1 artillery battalion      ⎪
                 Other elements              ⎭

The Imperial General Headquarters decided that virtually the whole of the Navy's "outer combat force" (外戰部隊)[18] would be employed in the operations against the United States, Great Britain, and the Netherlands. The tactical grouping of this force and mission assignments in the initial operations were as follows:[19]

**Combined Fleet**

Main Body:     Under direct command C-in-C, Combined Fleet.
                Mission: To support overall operations.
                6 battleships, 2 aircraft carriers, 2 light cruisers, 1 destroyer.

Task Force:     Under C-in-C, 1st Air Fleet.
                Mission: To attack the American Fleet in the Hawaii area and subsequently support operations of the South Seas and Southern Forces.
                6 aircraft carriers, 2 battleships, 2 heavy cruisers, 1 light cruiser, 11 destroyers, 3 submarines.

Advance (Submarine) Force:    Under C-in-C, 6th Fleet.
                Mission: To reconnoiter Hawaiian waters in advance of Pearl Harbor attack, cooperate with Task Force in execution of attack, and attack enemy naval forces along west coast of the United States.
                27 submarines, 1 submarine tender, 1 coastal defense ship.

South Seas Force:    Under C-in-C, 4th Fleet.
                Missions: To occupy Wake; defend and patrol inner South Seas area and protect surface traffic; cooperate with the Army in the successive occupation of Guam and Rabaul.

Southern Forces:    Under over-all command of C-in-C, 2d Fleet.
                Missions: To destroy enemy fleet and air strength in the Philippines, Malaya, and Dutch East Indies areas; act as surface escort and support landings of Army forces in Philippines, Malaya, Borneo, and Thailand; prepare for invasion operations in the Dutch East Indies, Timor, and Burma.

Main Body:     Under direct command C-in-C, 2d Fleet.
                2 battleships, 2 heavy cruisers, 10 destroyers.

Philippines Force:    Under C-in-C, 3d Fleet.
                1 aircraft carrier, 5 heavy cruisers, 5 light cruisers, 29 destroyers, 4 torpedo boats, 4 minesweepers, 3 base forces.

Malaya Force:    Under C-in-C, Southern Expeditionary Fleet.

---

[18] The Japanese Navy employed this term to cover combat forces for employment in operations outside Japanese home waters, as distinguished from the "inner combat force" (內戰部隊) which operated only in home waters.

[19] Statement by Capt. Toshikazu Ohmae, member, Military Affairs Bureau, Navy Ministry.

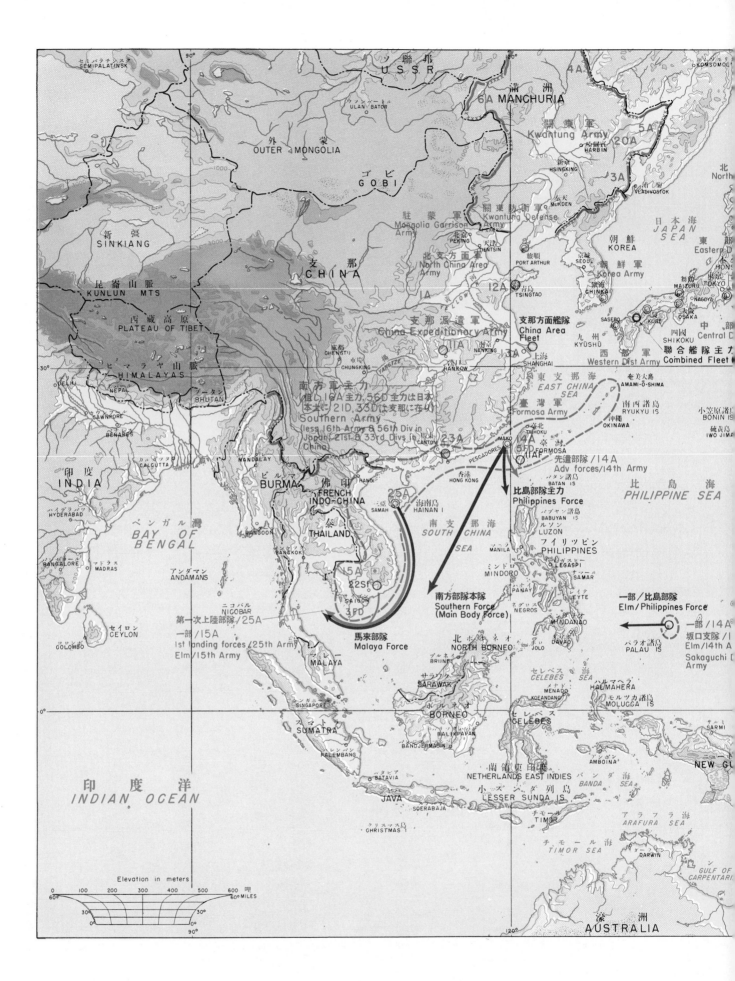

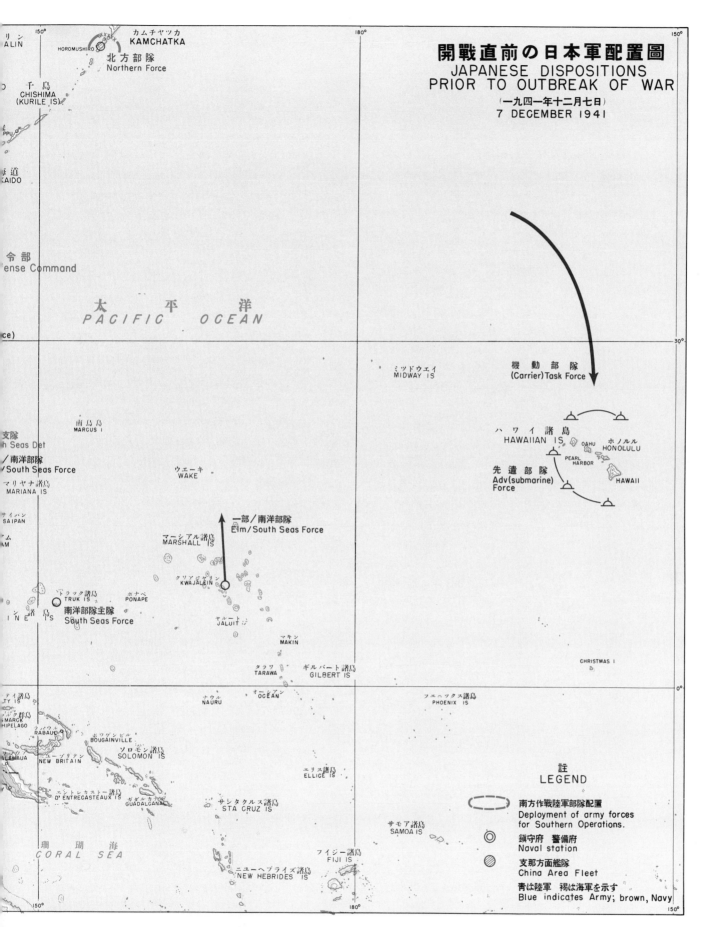

PLATE NO. 12

Disposition of Japan's Military Forces Prior to Outbreak of War

| | |
|---|---|
| | 5 heavy cruisers, 3 light cruisers, 15 destroyers, 16 submarines 1 minelayer, 1 coastal defense ship, 1 naval air flotilla, 2 base forces. |
| Air Force: | Under C-in-C, 11th Air Fleet. 2 naval air flotillas (shore-based), 2 destroyers. |
| Submarine Force: | Under Commander, 5th Submarine Flotilla. 2 submarines, 1 submarine tender. |
| Northern Force: | Under C-in-C, 5th Fleet. Missions: To patrol and defend waters east of Japan Proper; defend the Bonin Islands; guard the route of the Task Force; protect surface traffic. 2 light cruisers, 1 destroyer, 2 torpedo boats, 1 shore-based air group, 1 base force. |
| China Seas Fleet: | Under C-in-C, China Seas Fleet. Missions: To continue operations in China and destroy local enemy forces; cooperate with the Army in taking Hongkong; protect surface traffic in Chinese coastal waters; cooperate with ground forces. 2 coastal defense ships (old-type heavy cruisers), 1 light cruiser, 3 destroyers, 6 torpedo boats, 13 gunboats, 5 base forces, 4 Special Naval Landing Parties. |

## Operations Orders

The basic orders directing the Army and Navy forces to prepare for hostilities in early December were issued by Imperial General Headquarters immediately following the 5 November Imperial conference which fixed the end of November as the final deadline for the decision on war in case of failure to achieve a diplomatic settlement.[20]

The Imperial General Headquarters Navy Directive No. 1, issued on 5 November, ordered the Commander-in-Chief, Combined Fleet, to prepare "for the eventuality that war with the United States, Great Britain, and the Netherlands may become unavoidable in the first part of December."[21]

It directed that the necessary forces be assembled "at the appropriate time at initial staging areas", and laid down the general plan of fleet operations, which was incorporated in Combined Fleet Top Secret Operations Order No. 1, issued on the same date.[22] Essential portions of this order follow:

### Preparations for War and Start of Hostilities

*1. Preparations for War*

*a. The Empire anticipates the outbreak of war with the United States, Great Britain, and the Netherlands. When it has been decided to complete over-all operational preparations, an order will be issued setting the approximate date (Y-Day) for the commencement of operations and announcing "First Preparations for War." The various forces will, upon receipt of this order, act as follows:*

---

20 Cf. Chapter III.

21 ATIS Research Report No. 131, *Japan's Decision to Fight*, 1 Dec 45, p. 75.

22 The original text of Combined Fleet Top Secret Operations Order No. 1 was recovered from the Japanese cruiser Nachi, sunk in Manila Bay, in April 1945. Translated in full in ATIS Limited Distribution Translation No. 39 (Part VIII), 4 Jun 45, pp. 2–54. The contents of Combined Fleet Top Secret Operations Order No. 1 had been worked out during October, but it was not drawn up in final form until Admiral Yamamoto and his staff were summoned to Tokyo early in November in anticipation of the Imperial conference decision. The order numbered approximately 120 pages and was reproduced in 700 copies.

(1) *All fleets and naval units, without special orders, will be organized according to the allocation of forces for the First Period Operations of First Phase Operations, and will complete battle preparations. At the appropriate time, as directed by their commanders, they will proceed to alert (待機) areas prior to the start of operations.*

(2) *All units will be sharply on the alert for possible attacks by American, British and Dutch forces.*

\* \* \* \* \*

*b. When the necessary advance forces for the execution of operations are dispatched to the areas of operations, "Second Preparations for War" will be ordered. The various forces will, upon receipt of this order, act as follows:*

(1) *Submarine forces attached to the Advance Force, Task Force, Commerce Destruction Force,[23] Southern Force, and South Seas Force will, at the appropriate time as directed by their commanders, leave for their respective areas of operations.*

(2) *The remainder of these Forces will, as directed by their commanders, proceed so as to be in designated positions for the start of hostilities.*

\* \* \* \* \*

2. *Start of hostilities*

*a. The date for the start of hostilities (X-Day) will be fixed by Imperial Order (to be issued several days in advance). After 0000 hours on X-Day, a state of war will exist.[24] All forces will commence operations according to plan.*

\* \* \* \* \*

### First Phase Operations

1. *Operational Plan*

*a. The Advance Force, Task Force, South Seas Force, Northern Force, and Main Body will operate against the American Fleet.*

*The Advance Force [will scout and carry out surprise attacks on enemy naval forces in the Hawaii area and on the west coast of the United States.]*

*The Task Force [will attack and destroy enemy naval forces at Hawaii at the start of hostilities.][25]*

*The South Seas Force will occupy or destroy enemy key points in the vicinity of its operational area[26] and prepare to meet enemy naval forces in the Australian area.*

*The Northern Force will be charged with patrolling against the Soviet Union.*

*b. The Southern Forces, while holding local superiority, will annihilate enemy naval forces in the Philippines, British Malaya, and Netherlands Indies areas, and will carry out the following operations in cooperation with the Army.*

(1) *Operations against British Malaya and the Philippines will be launched simultaneously. The initiative will be taken in launching a sustained air offensive against enemy air and naval forces in these areas, and Army advance expeditionary groups will be landed as quickly*

---

23 The "Commerce Destruction Force", a subsidiary unit of the Combined Fleet, consisted of only three converted cruisers.

24 In case of a serious enemy attack before X-Day, the Combined Fleet Top Secret Operations Order No. 1 stipulated that "forces attacked will counterattack immediately." It further directed that military force might be used after "Second Preparations for War" had been ordered, (1) "if American, British or Dutch ships or planes approach the vicinity of our territorial waters and their action is deemed to constitute a danger"; (2) "if our forces operating outside the vicinity of our territorial waters encounter positive actions by American, British or Dutch forces such as endanger our forces." ATIS Limited Distribution Translation No. 39 (Part VIII), op. cit., p. 5.

25 As a special precaution to guard the secrecy of the Pearl Harbor attack plan, the bracketed portions were left blank in the printed text of the order and were communicated verbally only to a restricted number of high Navy General Staff officers and staff officers of Combined Fleet, First Air Fleet and Sixth Fleet Headquarters. Statement by Rear Adm. Tomioka, previously cited.

26 Separate Table 1 of the Combined Fleet Top Secret Operations Order No. 1 specified that the South Seas Force would "invade Wake and Guam" and would also "invade Rabaul if the situation warrants" during the first period of hostilities. ATIS Limited Distribution Translation No. 39 (Part VIII), op. cit., p. 45.

*as possible in strategic areas of Malaya, the Philippines and British Borneo. Air forces will be moved forward and air operations intensified.*

*(2) Following the successful completion of these operations, the main bodies of the Army invasion groups will be landed in the Philippines and Malaya and will quickly occupy these areas.*

*(3) During the first period of operations, strategic points in the Celebes, Dutch Borneo, and southern Sumatra will be occupied. If favorable opportunity arises, strategic points in the Moluccas and Timor will also be taken, and necessary air bases established in these areas.*

*(4) As these air bases are completed, Air forces will gradually be moved forward, and enemy air strength in the Java area will be destroyed. When this has been accomplished, the main body of the Army invasion group will be landed and will occupy Java.*

*(5) After the capture of Singapore, strategic points in northern Sumatra will be taken, and operations will be carried out at the appropriate time against Burma to cut the enemy supply route to China.*[27]

The basic Fleet order quoted above was followed on 7 November by Combined Fleet Top Secret Operations Order No. 2, which fixed Y-Day, the approximate date for the start of hostilities, as 8 December and ordered "First Preparations for War".[28] A further order of the same date ordered the Task Force to assemble at Tankan Bay, in the Kuriles, and take on supplies until 22 November.[29]

On 21 November Imperial General Headquarters Navy Order No. 5 directed the Commander-in-Chief of the Combined Fleet to advance the necessary forces at the appropriate time to positions of readiness for the start of hostilities. At the same time Imperial General Headquarters Navy Directive No. 5 stipulated that these forces should immediately be ordered to return to home bases in the event of a Japanese-American agreement.[30] A Combined Fleet operations order issued on 25 November stated:

*The Task Force will move out of Tankan Bay on 26 November and, taking every precaution to conceal its movements, will advance by late evening of 3 December to a rendezvous point at 42 degrees N. 170 degrees W., where refueling will be speedily carried out.*[31]

Following the Imperial conference decision to go to war, Imperial General Headquarters Navy Section on 1 December issued an order to the Commander-in-Chief, Combined Fleet, which stated:

*1. The Empire of Japan has decided to open hostilities against the United States, Great Britain, and the Netherlands during the first part of December.*

*2. The Combined Fleet will destroy enemy naval and air forces in the Far East and will repulse and destroy any enemy naval forces which may come to the attack.*

*3. The Commander-in-Chief, Combined Fleet, will cooperate with the Commander-in-Chief, Southern Army, in executing swift attacks on American, British, and Dutch strategic bases in East Asia and in oc-*

---

27 Part VIII of the Combined Fleet Top Secret Operations Order No. 1 outlined Second Phase Operations to begin after the capture of the Netherlands East Indies, specifying the following areas "to be occupied or destroyed as speedily as operational conditions permit": (1) Eastern New Guinea, New Britain, Fiji and Samoa; (2) Aleutians and Midway; (3) Andaman Islands; (4) Strategic points in the Australian area. Ibid., pp. 6, 9.

28 Full text of the Combined Fleet Top Secret Operations Order No. 2 read: "First preparations for war. Y-Day 8 December." Ibid., p. 55.

29 ATIS Research Report No. 131, op. cit., p. 77.

30 This directive was implemented by a Combined Fleet Operations Order dated 22 November, which stated: "In the event an agreement is reached in the negotiations with the United States, the Task Force will immediately return to Japan." Ibid.

31 Ibid., p. 78.

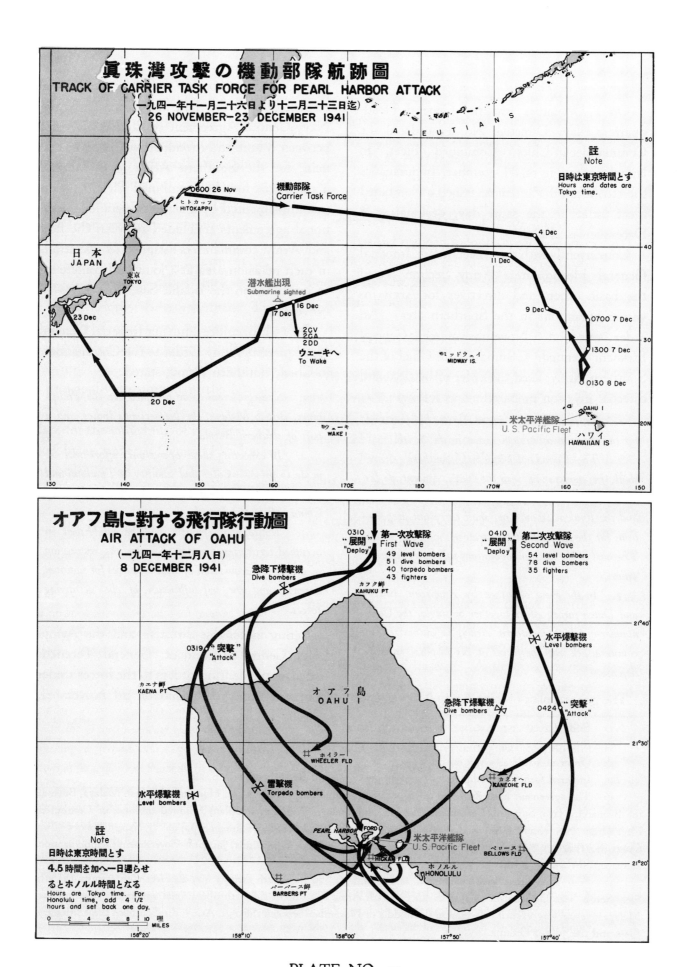

PLATE NO. 13

Pearl Harbor Attack, 8 December 1941

At 0400 4 December, the Task Force altered its course to the southeast and proceeded until 0700 on X-1 Day, 7 December, when it headed due south and began the final run toward Oahu at a speed of 24 knots. (Plate No. 13) At 0130 on 8 December, from a point approximately 200 nautical miles north of Oahu, the First Attack Unit of 183 planes took off from the decks of the six carriers, formed over the Task Force, and at 0145 headed for Pearl Harbor.[44]

Flying at 3,000 feet over dense but broken cloud formations, the first wave sighted the northern shoreline of Oahu at 0310 and immediately deployed, receiving the order to "attack" at 0319 (0749 Hawaii time). Dive bomber groups spearheaded the attack with swift strikes at Wheeler, Hickam and Ford Island airfields, crippling enemy fighter strength before it had a chance to get off the ground. Immediately thereafter torpedo plane and level bomber groups converged on the fleet anchorage at Ford Island and attacked the heavy units lying at berth.

The second wave of 167 planes took off from the carriers at 0245, reaching offshore the eastern coast of Oahu at 0424 (0854 Hawaii time), when the "attack" order was given. Dive and level bombers again swept in on the fleet anchorage, striking at ships not severely hit in the first attack. Fighter groups went in as escorts with both first and second waves, and when enemy air opposition failed to develop, they strafed ground targets. Both attacks continued from thirty minutes to one hour.

By 0830 (1300 Hawaii time) all aircraft, except nine missing from the first wave and 20 from the second, had returned to the carriers, and the Task Force began its withdrawal to the northwest at full speed. On 16 December the carriers *Soryu* and *Hiryu* (2d Carrier Division) and cruisers *Tone* and *Chikuma* (8th Squadron) broke off from the Task Force to take part in softening-up air attacks against Wake. The rest of the force continued toward home bases, arriving in the Inland Sea on 23 December.

On the basis of photographic analysis and reports by flight personnel, the Navy estimated the results of the Pearl Harbor air strike as follows: Sunk—four battleships, one cruiser, two tankers; heavily damaged—four battleships; lightly damaged—one battleship. Approximately 248 planes were estimated destroyed on the ground, 17 shot down in the air, and possibly 230 destroyed in hangars.[45]

Coordinated with the air strike were simultaneous attacks by the Advance (Submarine) Force, under command of Vice Adm. Shimizu. This Force, consisting of 27 of the Navy's best submarines, had left its bases in Japan and Kwajalein, in the Marshalls, between 16 and 24 November, and by X-1 Day had taken up positions controlling the entrance to Pearl Harbor. Its missions were to observe enemy fleet movements prior to the Task Force attack, to carry out torpedo attacks (with A-Target midget submarines)[46] simultaneously with the

---

44 Battle Lessons of the Greater East Asia War, op. cit., p. 49.

45 (1) Ibid., pp. 58–9. (2) An Imperial General Headquarters Navy Information Bureau communique issued at 1300 18 December announced the results as: "Sunk—5 battleships, 2 A or B-Class cruisers; heavily damaged—3 battleships, 2 light cruisers, 2 destroyers; medium or light damage—1 battleship, 4 B-Class cruisers. 450 enemy planes destroyed by bombing and strafing; 14 shot down." ATIS Research Report No. 132, *The Pearl Harbor Operation*, 1 Dec 45, p. 17.

46 These craft were carried aboard long-range "mother" submarines fitted with a mechanism for releasing them in the zone of operations. Mechanical improvements and training of the crews were completed barely in time to permit their employment in the Pearl Harbor operation.

一九四一年十二月八日の眞珠灣 (日本時間) (奇襲攻擊)　　　　藤田嗣治　昭和十七年

Original Painting by Tsuguji Fujita　　　　Photograph by U. S. Army Signal Corps

PLATE NO. 14

Pearl Harbor on 8 December 1941 (Tokyo Time)

air strike, to attack any enemy ships trying to put to sea, and to watch the movements of surviving enemy warcraft after the Task Force withdrawal.[47]

Between 2012 and 2303 on 7 December, several hours in advance of the air strike, five midget submarines were released from their "mother" submarines at positions from five to twelve nautical miles from Pearl Harbor and, aided by moonlight, gradually made their way toward the harbor entrance. Since radio communication was then discontinued, exact knowledge of their actions was lacking, but it was believed on the basis of offshore observation and later radio reports that at least three of the craft had successfully penetrated into the harbor. A heavy explosion witnessed at 1631 8 December was believed to indicate that a large warship had been sunk or severely damaged, presumably by midget submarine action.[48]

Although rescue submarines remained off Oahu for several days to pick up any of the midget craft which might have survived the attack, none returned.[49] Until early January part of the Advance Force continued to operate in the vicinity of Hawaii, largely to observe fleet activity and interfere with the anticipated transport of reinforcements to the Far Eastern zone of operations. Most of these submarines, at different times, proceeded to the west coast of the United States to attack shipping.[50]

## South Seas and Southern Operations[51]

While Vice Admiral Nagumo's Task Force temporarily crippled the United States Pacific Fleet at Pearl Harbor, Japanese forces in the South Seas area and Southeast Asia began operations in execution of other phases of the over-all war plan. (Plate No. 15)

The Navy's South Seas Force, charged with operations in the general area of the Japanese mandated islands, began air attacks on 8 December to knock out American air bases on Guam, Wake, and Howland Islands. On 10 December troops of the Army's South Seas Detachment, with the naval support of the South Seas Force, effected surprise landings on the northwestern and eastern shores of Guam before dawn and occupied the island without serious resistance.[52] This eliminated the isolated enemy base in the heart of the Japanese mandated islands.

At Wake, following repeated attacks by Navy planes based in the Marshalls, 1,000 special naval landing troops attempted a dawn landing on 10 December but were forced to withdraw due to effective air attack by remaining American planes and heavy seas. Following the arrival of the aircraft carriers *Soryu* and *Hiryu*, diverted from the Task Force returning from Hawaii, and 500 additional naval landing troops, a successful landing was accomplished during the night of 22–23 December, and the

---

47 ATIS Limited Distribution Translation No. 39 (Part VIII). op. cit., p. 44.
48 ATIS Research Report No. 131, op. cit., pp. 71–2.
49 One of the midget submarines which failed to penetrate into the harbor attacked small enemy craft on 8 December until it was finally disabled. One of its two crew members, Ensign Sakamaki, was taken prisoner and was the only survivor.
50 ATIS Research Report No. 131, op. cit., p. 74.
51 Excluding the Philippines, covered in Chapter VI.
52 The Guam invasion force (South Seas Detachment) sailed from Haha-Jima, in the Bonins, on 4 December. *Kaigum Nanyo Butai Sakusen no Gaiyo narabini Butai Shisetsu no Ippan Jokyo* 海軍南洋部隊作戦ノ概要並ニ部隊施設ノ一般狀況 (Outline of South Seas Naval Force Operations and General Situation of Facilities) 2d Demobilization Bureau, Jul 49, p. 3.

island was completely occupied the following day.[53]

In the Gilbert Islands, naval landing parties occupied Makin and Tarawa on 10 December and immediately constructed an advance air base on Makin. The capture of these islands and of Wake, enabling their utilization as air bases, strengthened the Navy's strategic outer defense line against American counterattack from the Central Pacific.

In the principal theater of operations in Southeast Asia, the Japanese forces struck swiftly at the strategic center of British strength in Malaya. The advance invasion units of the Twenty-fifth Army (main strength of the 5th Division and elements of the 18th Division) embarked from Hainan Island on 4 December. Early on 8 December[54] these forces, supported by the main strength of the Navy's Malaya Force and under air cover provided by the 3d Army Air Group, began landing operations at Singora and Pattani, in southern Thailand, and Kota Bharu, in northern Malaya. The Kota Bharu force, severely attacked by British planes after it landed on the beach, temporarily withdrew but, with reinforced air cover, succeeded in a second landing later the same day.[55]

Concurrently with the landing operations, land-based bombers of the 22d Naval Air Flotilla flew from Indo-China bases at 0500 on 8 December to bomb enemy military installations at Singapore. Two days later, on 10 December, Navy torpedo planes and bombers crippled the British Far Eastern Fleet by sinking the powerful battleships *Prince of Wales* and *Repulse* and a destroyer in the waters east of Malaya.[56]

With the occupation of Singora, Pattani and Kota Bharu, Army Air units immediately began operating from these advance bases, gained mastery of the air over Malaya and provided direct support for the ground forces advancing on Singapore. The Twenty-fifth Army's drive progressed smoothly despite sporadic enemy resistance, and by late January 1942, all units had reached the Johore Straits at the southern tip of Malaya. Singapore fell on 15 February.[57]

To the north, the Imperial Guards Division (temporarily attached to the Fifteenth Army) moved across the Indo-Chinese border into Thailand on 8 December, while some of its elements landed by sea at points along the Kra Isthmus. These operations were accomplished without resistance. In January, the main strength of the Fifteenth Army (55th and 33d Divisions) concentrated at Rahaeng and Bangkok in preparation for the invasion of Burma.[58]

In the Borneo and Celebes area, Japanese operations likewise proceeded according to plan. Embarking at Camranh Bay, French Indo-China, on 13 December, the Kawaguchi Detachment (three infantry battalions plus Yoko-

---

53 Outline of South Seas Naval Force Operations and General Situation of Facilities op. cit., pp. 5–8.

54 Landings began at the following times: Kota Bharu 0215 (0015 Malay Time); Singora 0410 (0210 M. T.); Pattani 0430 (0230 M. T.) *Marai Sakusen Kiroku Dai Nijugo Gun* 馬來作戰記錄第二十五軍 (Malay Operations Record: Twenty-fifth Army) 1st Demobilization Bureau, Sep 46, pp, 42–3.

55 Ibid.

56 This striking success bolstered Japanese morale and strongly influenced subsequent air operational methods. Two Japanese accounts of the engagement are published in ATIS Enemy Publications No. 6, *The Hawaii-Malaya Naval Operations*, 27 Mar 43, pp. 12–8.

57 Unconditional surrender was signed at 1950 on 15 February at a meeting between General Yamashita, Commander-in-Chief of the Malaya Invasion Forces, and Lt. Gen. Sir A. E. Percival.

58 Imperial General Headquarters on 22 January issued an order to the Commander-in-Chief, Southern Army, to launch operations jointly with the Navy for the occupation of important points in Burma. Imperial General Headquarters Army High Command Record, op. cit., p. 83.

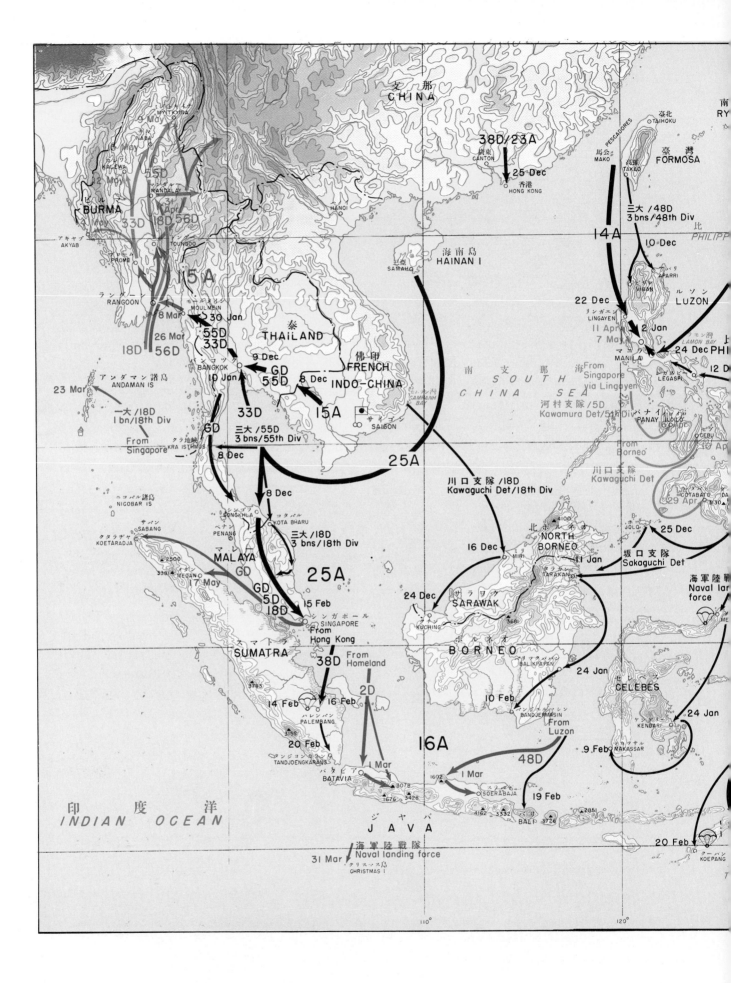

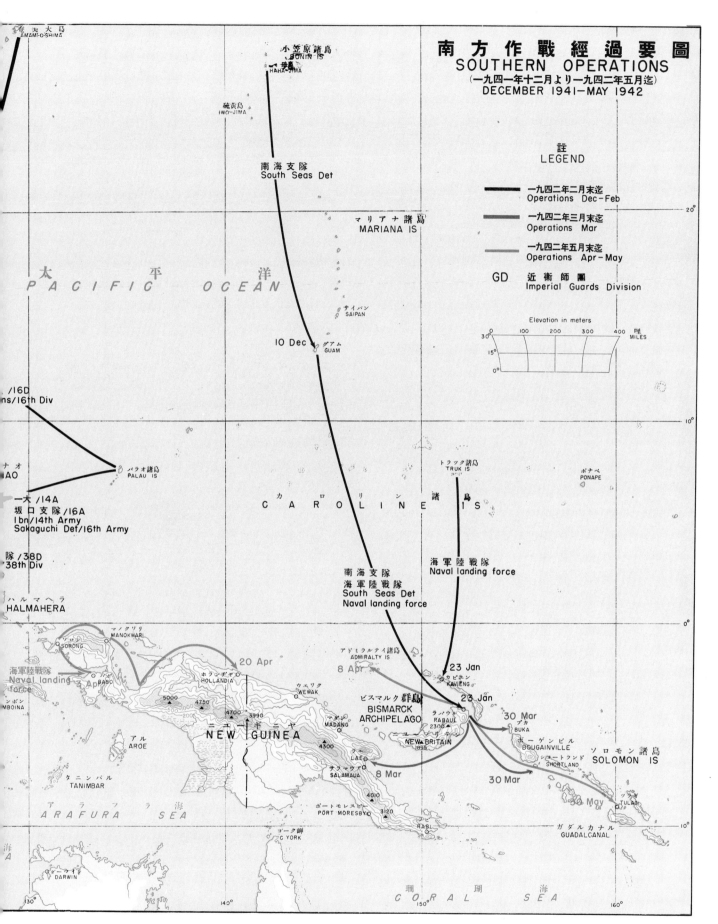

PLATE NO. 15

Southern Operations, December 1941–May 1942

suka 2d Special Naval Landing Force) landed near Miri, on the coast of British Borneo, on 16 December and occupied the oil fields and airfield.[59] The detachment, moving by sea, took Kuching on 23 December. Brunei, Labuan Island, Jesselton, and Tawau were taken in subsequent operations.[60]

Strategic points in Dutch Borneo were occupied by elements of the Sakaguchi Detachment which, after taking Davao in the Southern Philippines, had hopped to Jolo Island, in the Sulu Archipelago. This force occupied Tarakan on 11 January 1942 and Balikpapan on 24 January. Simultaneously with these operations, Navy forces invaded the Celebes, taking Menado on 11 January, Kendari on 24 January and Amboina on 31 January.[61] These operations gave the Japanese forces control over important oil-producing areas and at the same time provided strategic forward bases for continuation of the southward advance on Java.

In the China area, joint Army-Navy plans at the start of hostilities called for the invasion of Hongkong as soon as the Malaya landings had been accomplished. The 38th Division (Twenty-third Army) and the Second China Expeditionary Fleet were assigned to this operation.[62] The 38th Division moved from Canton to the Kowloon Peninsula on 14 December, and joint Army-Navy amphibious operations against Hongkong began on 18 December. On 25 December the British forces surrendered.[63] Meanwhile Japanese Army and Navy units in the Shanghai and Tientsin areas took control of the foreign concessions there.

The initial offensives of the Japanese armed forces on virtually every front thus attained a measure of success that was beyond original expectations. The United States and Great Britain were forced to assume the defensive, while the security of the Japanese homeland against Allied counterattack was greatly strengthened through the seizure of strategic areas. Acquisition of the resources of the southern regions not only cut off the flow of these resources to the United States and Great Britain, but placed Japan in a favorable economic position for the prosecution of an extended war.

Only in the Philippines, despite the early capture of Manila, did the Japanese Army fall sharply behind its invasion timetable as a result of the wholly unexpected and bitter resistance offered by General MacArthur's isolated forces on Bataan.[64] The protracted American defense of Bataan, which was brought into sharp relief by the unexpectedly early conquest of Singapore, a modern fortress with facilities far in excess of the rundown, antiquated installations of Corregidor, required extraordinary measures by Imperial General Headquarters.[65]

---

59 *Boruneo Sakusen Kiroku* ボルネオ作戰記錄 (Borneo Operations Record). 1st Demobilization Bureau, Dec 46. pp. 7–8.

60 Ibid., pp. 11–12.

61 *Ranryo Higashi Indo Koryaku Sakusen* 蘭領東印度攻略作戰 (Netherlands East Indies Naval Invasion Operations) 2d Demobilization Bureau, Oct 49, pp. 23–9.

62 Imperial General Headquarters Army Order to Commander-in-Chief, China Expeditionary Forces, 1 December 1941: "1. Commander-in-Chief, China Expeditionary Forces, in cooperation with the Navy, will capture Hongkong, using as the main body the 38th Division under the command of the Commanding General, Twenty-third Army. 2. Operations will commence immediately after the landings in Malaya or upon confirmation of the air attack." Imperial General Headquarters Army High Command Record, op. cit., p. 75.

63 *Shina Homen Sakusen Kiroku* 支那方面作戰記錄 (China Area Operations Record) 1st Demobilization Bureau, Dec 46, Vol. I, pp. 26–30.

64 Cf. Vol. I: Southwest Pacific Area Series: *The Campaigns of MacArthur in the Pacific*, Chapter I, p. 16, n. 29.

65 Cf. Chapter VI.

# CHAPTER VI

# CONQUEST OF THE PHILIPPINES

## Preliminary Planning

As the westernmost bastion of American military power in the Pacific, the Philippines in December 1941 were clearly marked as one of the first objectives of the Japanese armed forces.[1] The primary purposes which Imperial General Headquarters planned to achieve through their capture were not necessarily political or economic, but essentially strategic in character:[2]

> *To deny to American ground, sea and air forces the use of the Philippines as an advance base of operations.*
>
> *To secure the line of communications between the occupied areas in the south and Japan Proper.*
>
> *To acquire intermediate staging areas and supply bases needed to facilitate operations in the southern area.*[3]

Serious study of the tactical and logistic problems involved in an invasion of the Philippines simultaneously with operations against Malaya and the Dutch East Indies got under way in September 1941, when over-all international developments had convinced Imperial General Headquarters that an eventual Japanese move against British and Dutch possessions in Southeast Asia would almost certainly bring the United States into war. By the early part of October, when special Army war games took place in Tokyo to test the tactical plans being elaborated for the invasion of the southern area, the main lines of the Philippines operation plan had been tentatively worked out and were subjected to study as part of the games.[4]

---

[1] This chapter was originally prepared in Japanese by Col. Ichiji Sugita, Imperial Japanese Army. Duty assignments of this officer were as follows: Staff Officer (Intelligence), Imperial General Headquarters, Army Section, Feb 39–9 Nov 41; Staff Officer (Intelligence), Twenty-fifth Army, 9 Nov 41—23 Mar 42; Staff Officer (Intelligence), Imperial General Headquarters, Army Section, 23 Mar—9 Nov 42; Staff Officer (Intelligence), Eighth Area Army, 15 Nov 42—15 May 43; Staff Officer (Intelligence), Imperial General Headquarters, Army Section, 15 May—15 Oct 43; Chief, Intelligence Section, Imperial General Headquarters, Army Section, 15 Oct 43—31 Mar 44; Staff Officer (Operations), Imperial General Headquarters, Army Section, 1 Apr 43—16 Jul 45; Staff Officer (Operations), Seventeenth Area Army, 16 Jul—23 Aug 45. All source materials cited in this chapter are located in G-2 Historical Section Files, GHQ FEC.

[2] Cf. Chapter IV, section on Areas to be Occupied; n. 17, pp. 48–9.

[3] (1) Interrogation of General Hideki Tojo, Premier and War Minister, 1941–4. (2) Statements by Rear Adm. Sadatoshi Tomioka, Chief, First Bureau (Operations), Imperial General Headquarters, Navy Section, and Col. Takushiro Hattori, Chief, Operations Section, Imperial General Headquarters, Army Section.

[4] Theoretical plans for an invasion of the Philippines in the event of war with the United States had previously been formulated by both the Army and Navy General Staffs as part of normal military preparedness against major potential enemies. However, until the summer of 1941, no concrete plans were seriously considered, and Army strategists saw a possibility of by-passing the Philippines and avoiding war with the United States, even if Japan embarked on operations against Britain and the Netherlands. By September, decision had been reached that such a course would be too risky, and that the Philippines must therefore be included in the overall plan of operations. (Interrogations of Lt. Gen. Shinichi Tanaka, Chief, First Bureau (Operations), Imperial General Headquarters, Army Section, and General Tojo, previously cited.)

the range of Japanese Army planes operating from southern Formosa, it was necessary to obtain the cooperation of naval air strength, including long-range bombers based in southern Formosa, as well as seaplane and carrier forces. The boundary of air operations between the Army and Navy was to be fixed at 16 degrees N. Lat., placing all the enemy's major bases in the Manila area within the Navy's operational sphere. (Plate No. 16)

Imperial General Headquarters estimated that enemy air resistance would be sufficiently neutralized within two to four days to permit execution of the next step in the operational plan: the landing of advance detachments on northern and southern Luzon with the mission of seizing air bases at strategic points and quickly preparing them for operational use by the Japanese forces. The airfields at Aparri, Laoag and Vigan were designated as the initial objectives on northern Luzon, while the southern Luzon force was to seize the airfield at Legaspi. Prior to the advance landings on Luzon Proper, occupation of Batan Island, 150 miles north of Aparri, was planned as a preliminary step to facilitate fighter cover of the north Luzon landings.[14]

Airfield construction and maintenance units, going in with the advance forces, were to prepare the occupied fields for operational use within a few days of their capture, and Army and Navy Air units were then to move immediately forward and resume offensive operations. Allowing a further brief period for these operations to complete the destruction of enemy air power, Imperial General Headquarters initially estimated that the main landings could be carried out on X-Day plus 9 at Lingayen Gulf, and X-Day plus 11 at Lamon Bay.[15] These estimates were revised upward by five days in the final operations plan.

The basic plan of attack against Manila envisaged a two-pronged pincers movement, the main invasion forces landing at Lingayen Gulf and driving toward the capital from the north, while a strong secondary force was to land at Lamon Bay[16] and advance on Manila from the southeast, splitting the enemy defense effort. Since it was the shortest route, it was decided to direct the main effort toward Man a via Tayug and Cabanatuan, skirting the eastern edge of the Luzon plain.[17] (Plate No. 17)

Parallel with the main operations on Luzon, the over-all invasion plan called for the seizure by small forces of Davao, on the southern coast of Mindanao, and Jolo Island, in the Sulu Archipelago. Strategically, occupation of these points was designed to obtain air bases for impeding a possible southward withdrawal of the American forces in the Philippines and were also needed as staging points for the scheduled invasion of Celebes and eastern Borneo.

Imperial General Headquarters, taking into consideration the troop requirements for other phases of the southern operations, tentatively set the basic infantry strength to be employed in the Philippine landings at a total of 21 battalions.[18] The allocation of these forces by landing area was as follows: Northern Luzon

---

14 Philippine Operations Record, Phase One, op. cit., pp. 24–5.
15 Ibid., pp. 37–8.
16 In the initial planning stage, Batangas Province, on the west coast of Luzon, was also considered as a possible site for the secondary landing. Lamon Bay was chosen because it offered a shorter and less dangerous route of sea approach. Ibid., pp. 50–1.
17 Ibid., p. 39.
18 The Fourteenth Army, after its assignment to the Philippines invasion, requested a strength of two and one-half first-line combat divisions for the execution of the Philippines landings, but Imperial General Headquarters refused the request. Transport tonnage allotted to the Philippines was also pared down from an originally estimated 800,000 tons required to 630,000 tons. (1) Philippine Operations Record, Phase One, op. cit., pp. 48–50. (2) Interrogation of Lt. Gen. Masami Maeda, Chief of Staff, Fourteenth Army.

advance landings, three battalions; Legaspi advance landing, two battalions; Lingayen Gulf, nine battalions; Lamon Bay, three battalions; Davao and Jolo, four battalions.[19]

The Navy, in addition to furnishing the bulk of the air strength to be employed in the initial phase of the operations, was assigned the missions of destroying enemy fleet and air strength in the Philippines area, protecting the assembly points of the invasion convoys, providing surface escort and naval support of the landing operations, and guarding against possible counterattacks by Allied naval forces. Imperial General Headquarters, Navy Section anticipated that the major threat of such counterattacks would come from the American Asiatic Fleet, possibly reinforced by Allied fleet units. In the event, however, that the main body of the United States Pacific Fleet sortied into the Western Pacific, plans were made to divert the main strength of the Navy's Southern Forces to counter the attack.

Assembly points of the invasion convoys were selected with special attention to the maintenance of secrecy and safety from enemy submarine and air attack. To avoid overlarge concentrations of ships in southern Formosan harbors, it was decided to stage the main invasion forces from three ports: Keelung and Takao, on Formosa, and the naval base of Mako, in the Pescadores. The Lamon Bay and Mindanao landing forces were to stage respectively from Amami-Oshima, in the Ryukyu Islands, and Palau, in the western Carolines.

Imperial General Headquarters estimated that the occupation of key areas in the Philippines could be accomplished within a period of about fifty days.[20] On the basis of this estimate, it was tentatively decided to withdraw one combat division as soon as the major military objectives had been achieved, and to reassign it to the invasion of Java. Most of the naval forces were to be withdrawn at the same time and reorganized as the Dutch Indies Force. This would leave relatively weak Army and Navy forces to complete the occupation of the islands and secure them against enemy counterattack, but it was anticipated that Filipino cooperation could readily be won through political concessions and that the islands would be safe from counterattack behind the rampart of Japan's defenses in the mandated islands.

## Assignment of Forces

In accordance with the over-all plans elaborated by Imperial General Headquarters, the Southern Army allotted the mission of executing the Philippines invasion to the Fourteenth Army, under command of Lt. Gen. Masaharu Homma, peacetime commander of the Formosa Army. To provide Army air support, the 5th Air Group, under command of Lt. Gen. Hideyoshi Obata, was transferred from Manchuria to Formosa and placed under Fourteenth Army command.[21] Naval missions incident upon the operation were assigned by the Combined Fleet to the Philippines Force under Vice Admiral Ibo Takahashi, Third Fleet Commander, and the Eleventh Air Fleet under Vice Admiral Nishizo Tsukahara.[22]

---

19 Philippine Operations Record, Phase One, op. cit., p. 50.
20 Statement by Col. Hattori, previously cited. Cf. Chapter V, n. 4, p. 58.
21 *Hito Koku Sakusen Kiroku Dai Ikki* 比島航空作戰記錄第一期 (Philippine Air Operations Record, Phase One) 1st Demobilization Bureau, Jun 46, pp. 1, 19.
22 ATIS Limited Distribution Translation No. 39 (Part VIII) 4 Jun 45, p. 45. (2) Operational Situation of the Japanese Navy in the Philippine Invasion, op. cit., pp. 2-3.

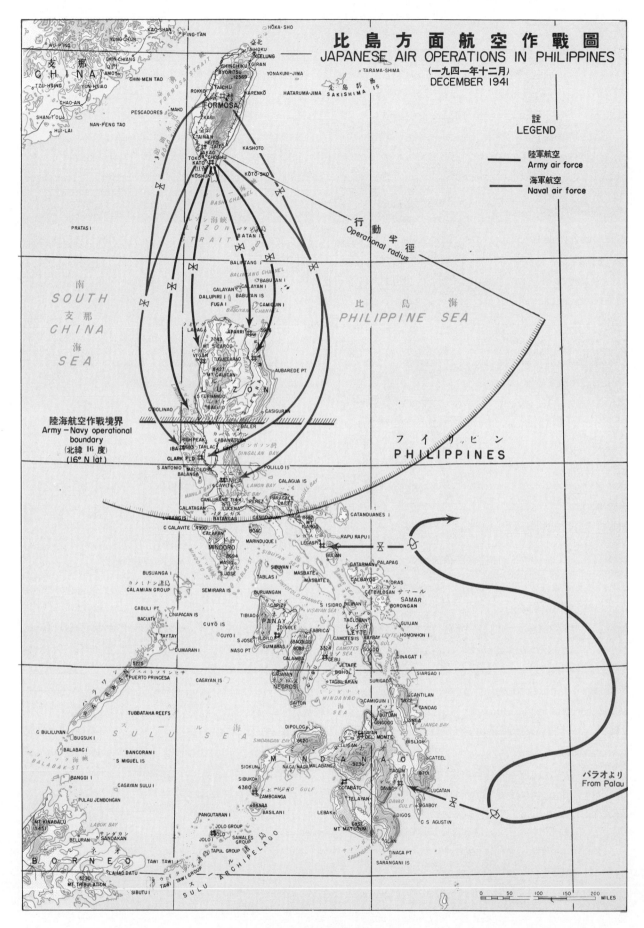

PLATE NO. 16

Japanese Air Operations in Philippines, December 1941

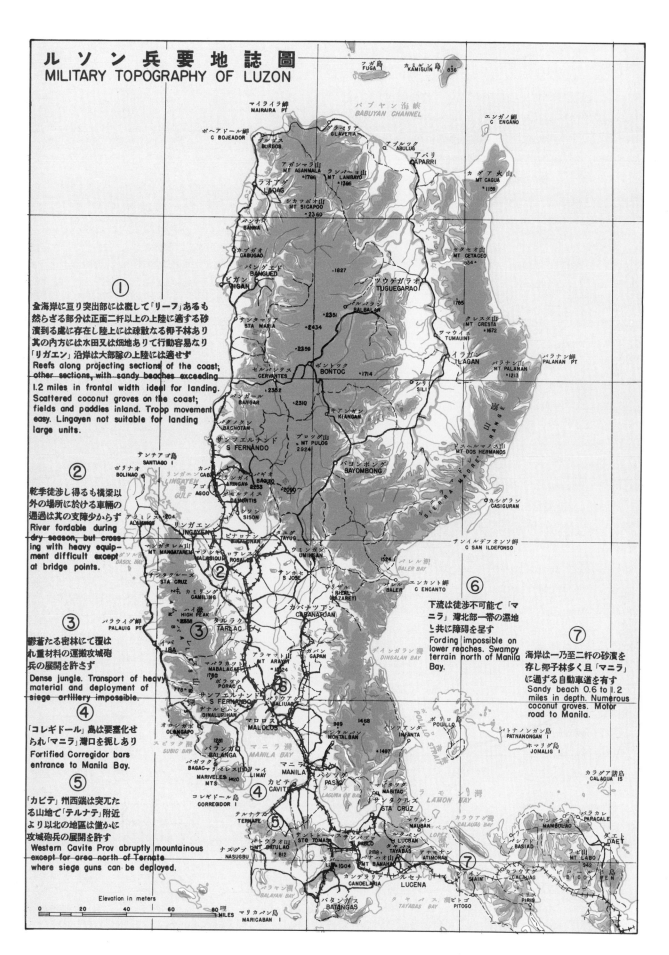

PLATE NO. 17

Military Topography of Luzon

Ground force strength assigned to the Fourteenth Army for the accomplishment of its mission centered around two first-line combat divisions, the 16th and 48th, which were to execute the initial phases of the operations, and the 65th Brigade, which was to move in subsequently as a garrison force.[23] The 48th Division, based in Formosa, was among the Japanese Army's most experienced units and specially trained in amphibious operations. The 16th Division, scheduled to execute the secondary landing at Lamon Bay, was picked as one of the best divisions then available in Japan Proper.

These units were reinforced by two tank regiments, five heavy field artillery battalions (Army artillery), approximately five field antiaircraft artillery battalions, four independent antitank companies, and an independent mortar battalion. To meet the special requirements of the operation, an unusually strong complement of independent engineer units and bridge companies was included in the Army's attached service forces.

Combat strength of the 5th Air Group consisted of two fighter regiments, two light bomber regiments, and one heavy bomber regiment, plus an independent reconnaissance and observation unit. Strength in Army aircraft aggregated 192, including 72 fighters, 81 bombers and 39 reconnaissance and observation planes.[24]

Principal units composing Fourteenth Army order of battle for the first phase of the Philippines operation were as follows:[25]

*Fourteenth Army Headquarters*
*16th Division*
*48th Division*
*65th Brigade*
*3d, 8th, 9th, 10th Ind. Antitank Cos.*
*4th, 7th Tank Regiments*
*1st, 8th Heavy Artillery Regts.*
*9th Ind. Heavy Artillery Bn.*
*40th, 45th, 47th, 48th Field A. A. A. Bns.*
*30th, 31st Ind. Field A. A. A. Cos.*
*15th Ind. Mortar Bn.*
*3d, 21st Ind. Engr. Regts.*
*3d Ind. Engr. Co.*
*26th, 28th Bridge Material Cos.*
*13th, 14th River Crossing Material Cos.*
*Army signal units*
*6th Railway Regt.*
*Shipping units*
*Line of Communications units*
*1st Field M. P. Unit*
*5th Air Group*
  *Headquarters*
  *4th Air Brigade*
    *50th Fighter Regt*
    *8th Light Bomber Regt.*
    *16th Light Bomber Regt.*
    *14th Heavy Bomber Regt.*
  *24th Fighter Regt.*
  *10th Ind. Air Unit*[26]

To permit employment of virtually the Army's full strength in the crucial assault on Luzon, Imperial General Headquarters and the Southern Army decided to transfer the initial mission of occupying Davao and Jolo to the Sixteenth Army, assigned to operations against eastern Borneo and Java. Under this arrangement, the Sixteenth Army's Sakaguchi Detachment (56th Mixed Infantry Group Hqs.; 146th Infantry Regt. reinf.) provided the main strength of the Davao landing force. One infantry battalion of the 16th Division was

---

23 The 65th Brigade consisted of three infantry regiments. Cf. p. 100.
24 Philippine Air Operations, Phase One, op. cit., p. 22.
25 *Hito Sakusen Kiroku Dai Ikki Bessatsu* 比島作戰記錄第一期別冊 (Philippine Operations Phase One, Supplement) 1st Demobilization Bureau, Jul 46.
26 The 10th Independent Air Unit was composed of the 52d, 74th and 76th Independent Air Companies, all equipped only with reconnaissance and observation craft.

temporarily attached for occupation duty, thus effecting early release of the Sakaguchi Detachment for its further missions on Jolo and in eastern Borneo. The 16th Division force remaining at Davao was then to revert to Fourteenth Army command.

Naval surface strength allotted to support the Philippines operation comprised the bulk of the Third Fleet, reinforced by the attachment of two destroyer squadrons (less elements) and one cruiser division from the Second Fleet, the 4th Carrier Division (*Ryujo* and one destroyer) from the First Air Fleet, and the 11th Seaplane Tender Division from the Combined Fleet.[27] This gave the Philippines Force an aggregate strength in combat ships of one aircraft carrier, five heavy cruisers, five light cruisers, three seaplane tenders, 29 destroyers, four torpedo boats, 13 minesweepers and four minelayers.

To carry out its missions, the Eleventh Air Fleet assigned the bulk of its land-based forces, the 21st and 23d Air Flotillas, with a combined strength of 146 bombers, 123 fighters, 24 flying boats, and 15 reconnaissance planes.[28] These were in addition to 16 fighters and 18 torpedo planes composing the complement of the *Ryujo*, and a total of 68 seaplanes operating from surface units.[29] Aggregate naval air strength assigned to the Philippines operation thus reached 412 planes. Combined initial allotment of Army and Navy aircraft totaled 604.

## Final Operations Plan

By early November, the Commander-in-Chief of the Southern Army and Combined Fleet had completed study of the Imperial General Headquarters outline plan of operations and had reached agreement on the general terms of Army-Navy cooperation. The commanders and principal staff officers of the Fourteenth Army, 5th Air Group, Third Fleet and Eleventh Air Fleet were then summoned to participate in the joint Army-Navy conference at Iwakuni from 14 to 16 November,[30] and the final Fourteenth Army operational plan for the Philippines invasion was drawn up. Its essentials were as follows:[31]

*1. The objective of the operations will be to crush the enemy's principal bases and defeat his forces in the Philippines. In cooperation with the Navy, the Army will land its main forces on Luzon, overcome enemy resistance, and quickly occupy Manila. Thereafter it will occupy other vital points in the Philippines.*

*2. Prior to the start of hostilities, the Army will assemble its advance elements at Mako and Palau, and its main forces on Formosa and the Nansei (Ryukyu) Islands. All necessary precautions will be taken to guard the secrecy of the above movements and of the operational preparations.*

*3. The operations will begin with air attacks on the Philippines. The Army Air forces will be responsible for attacking enemy air power north of 16 degrees N. Lat., and the Navy Air forces will be responsible for same mission south of that line.*

*4. At the appropriate time following the start of air attacks, the advance elements of the Army and Navy will execute landings and occupy enemy air bases as follows:*

*"A" Detachment will occupy Aparri.*

*"B" Detachment will occupy Vigan and Laoag.*

*"C" Detachment will occupy Legaspi.*

---

27 (1) *Hito Koryaku Sakusen* 菲島攻略作戰 (Philippine Invasion Operations) Combined Fleet Headquarters, Jun 42, pp. 2-3. (2) ATIS Limited Distribution Translation No. 39 (Part VIII), op. cit., p. 45.

28 Battle Lessons of the Great Asia War, op. cit.

29 *Nihon Kaigun Hensei Suii oyobi Heiryoku Soshitsu Hyo* 日本海軍編制推移及び兵力喪失表 (Table Showing Organizational Changes and Losses of Japanese Naval Forsces) 2d Demobilization Bureau, Oct 49, pp. J-1 and L-1.

30 Cf. Chapter III, p. 37.

31 Philippine Operations Record, Phase One, op. cit., pp. 21-8, 31.

between 23 and 25 November to the nearby naval port of Mako, in the Pescadores, where final landing preparations were completed. The main strength of the Division simultaneously began assembling at Takao, Mako and the northern Formosan port of Keelung. By the end of November, the 65th Brigade had also completed its movement from the Japanese mainland and was assembled in Formosa.[34]

Concurrently with the assembly of the invasion troops, the 5th Air Group and Eleventh Air Fleet rapidly concentrated at southern Formosan bases in readiness for the launching of the initial air offensive. All units of the 5th Air Group were assembled at their bases at Heito, Koshun, Choshu and Kato by 6 December. Land-based bombers and fighters of the 21st and 23d Air Flotillas prepared to operate mainly from bases at Tainan, Takao, Taichu and Palau.

Between 24 November and 5 December, operational orders were issued by the Army commander to all units specifying the composition and missions of the various landing forces (Plate No. 18), air force assignments, and essential points of the landing operations. Details of cooperation were worked out in agreements concluded between the Army and Navy commanders directly assigned to each landing operation.[35]

## Launching of Operations

The land, sea and air forces were now poised for the attack. Organization for combat was completed, and morale high. On 1 December Lt. Gen. Homma transferred his headquarters to Takao, and Vice Admiral Takahashi, Commander of the Third Fleet, raised his flag aboard the *Ashigara* at Mako. On 2 December orders were received from General Hisaichi Terauchi, Southern Army Commander-in-Chief, and Admiral Isoroku Yamamoto, Commander-in-Chief of the Combined Fleet, designating 8 December as X-day.

The Army Air forces began operations according to plan early on 8 December. Taking off before dawn from bases in southern Formosa, 43 Army planes struck the first blows at enemy airfield at Tuguegarao and barracks at Baguio, on northern Luzon. The attacks were made at approximately 0800 (0700 local time), about four hours and a half after the first bombs from Japanese carrier planes struck Pearl Harbor.

Due to heavy fog over their airdromes at Tainan and Takao, naval land-based aircraft scheduled for the initial attacks were late in taking off, finally clearing their fields at about 0930. This force, made up of 108 Navy land-based attack planes and 84 fighters, had as its objectives Clark Field and the American fighter base at Iba, on the west coast of Luzon. The formations arrived over their targets shortly after 1330 (1230 local time) and carried out highly successful attacks. Meanwhile, carrier planes took off from the *Ryujo* at a point 100 miles east of Mindanao during the early morning hours and carried out an effective strike on Davao.

Since radio intelligence showed that Philippines defense installations had been alerted at 0430 on 8 December, it was assumed that the enemy had already received news of the Pearl Harbor attack and that the Japanese air units would encounter energetic resistance from both intercepter aircraft and antiaircraft artillery. Resistance, however, proved much weaker than anticipated, with the result that the attacks achieved a spectacular degree of success, es-

---

34 Ibid. p., 60.
35 Ibid., pp. 58–9.

pecially at Clark Field.[36]

While the first air attacks were being mounted, the advance force convoys were at sea, heading toward the various invasion objectives. Surface and air cover was furnished the convoys by the Third Fleet, and by Army and land-based Navy units operating from airfields in southern Formosa.

At dawn on 8 December the Batan Island landing force made an unopposed landing and seized the airstrip. On 9 December fighters of the Army's 5th Air Group landed on the strip and found it suitable for operational use. Airfield construction units swiftly effected necessary improvements, and fighter units moved forward to support the landing operations at Vigan and Aparri.

On 10 December, while the Navy Air force carried out heavy neutralization strikes against the airfields in the Manila area, the Tanaka (2d Battalion, 2d Formosa Infantry, reinf.) and Kanno (1st and 3d Battalions, 2d Formosa Infantry, reinf.) Detachments effected their dawn landings at Aparri and Vigan against no opposition. The airfields were quickly occupied. The Kanno Detachment immediately pushed north from Vigan along the coast road and took the airfield at Laoag on 12 December. Meanwhile, a small element of the Tanaka Detachment advanced up the Cagayan River and took Tuguegarao. There was no enemy air reaction to these operations except individual sorties against Vigan anchorage by large-type American aircraft.[37]

The captured airdromes were rapidly prepared for use as advance operational bases, and units of the Army's 5th Air Group quickly moved forward according to plan. The 24th Fighter Regiment advanced to Vigan on 11 December, and on 12 and 14 December the 50th Fighter Regiment and one element of the 16th Light Bomber Regiment arrived at Aparri.[38] On the 13th more than 100 navy bombers carried out neutralization strikes on Del Carmen, Clark, Iba and Nichols fields. Also on the 13th, 15 Army heavy bombers and fighters hit Clark Field.

The successful exploitation of advance bases soon gave the Japanese Air forces an overwhelming superiority which was to have a great effect on later operations. On 15 December, it was estimated that the combat strength of the United States Air Forces had been reduced to about ten bombers, ten flying boats and twenty fighters.[39] It was presumed that enemy air strength had been dispersed to the central Philippines and to Iloilo, Del Monte, and Jolo to the south. In less than a week the Japanese had gained control of the skies over the Philippines.

In the interim, amphibious operations continued to progress satisfactorily. The Kimura Detachment landed in the vicinity of Legaspi at 0245 on 12 December without encountering any enemy opposition and quickly overran the nearby airfield.[40] Naval Air units supported the operation by continuing the neutralization of enemy airfields in the Manila area on 12–13 December. Japanese air losses were negligible.

In the north the Tanaka and Kanno Detachments, having accomplished their mission, were regrouped for further operations. General Homma, seeing that the enemy was not conducting an aggressive defense in northern

---

36 Operational Situation of the Japanese Navy in the Philippines Invasion, op. cit., p. 4.
37 Ibid., pp. 4–5.
38 The Group Commander, Lt. Gen. Hideyoshi Obata, advanced to Aparri on 18 December. Philippine Air Operations, Phase One, op. cit., pp. 32, 34, 38, 47.
39 Operational Situation of the Japanese Navy in the Philippines Invasion, op. cit., p. 5.
40 Responsibility for operation of this field was assigned to the Navy. Philippine Operations Record, Phase One, op. cit., pp. 62–3, 69.

ber and headed for Lamon Bay under escort by elements of the Third Fleet. Meanwhile the 48th Division and other elements of the Lingayen landing force embarked at Keelung, Takao, and Mako in three convoys with a total of 76 transports. These convoys with their naval escorts sortied on 17–18 December en route to Lingayen Gulf. (Plate No. 20)

The main attack force entered Lingayen Gulf at 1001 22 December without encountering any opposition. In the darkness an error was made as to the point of anchorage, the lead transports advancing too far south. The frontal spread of this disposition was 15 miles. For this reason long distance surface movements with small craft became necessary.

The plan of assault called for the first echelon to land on the right in the vicinity of Agoo at 0540, the 47th Infantry Regiment in the assault. The second echelon (less the Uejima Detachment) was to land in the center at Caba, near Santa Lucia, at 0550, the 1st Formosa Infantry Regiment in the assault. The Uejima Detachment was to effect landings on the left at Bauang at 0730, the 1st Battalion, 9th Infantry Regiment in the assault. The 3d Battalion of the 9th Infantry Regiment from the third echelon was to be committed at 0730 in the vicinity of Santiago. The third echelon was constituted as a floating reserve.[46]

The initial landings were effected as scheduled on 22 December. Although enemy fire from the beaches was heavy during the approach, causing some casualties, resistance on the beaches was found to be moderate and was quickly dispersed.

Soon after the first landings a sudden deterioration of the weather threatened to impede operations, but it was decided to continue according to plan. During the day the transport area received both air and submarine attacks, but no heavy casualties were sustained.[47] The Navy's 2d Base Force completed the defense installation of the advance base early in the day, while surface units continued to patrol the gulf entrance.

Air support for the operation was fully effective. The Army Air forces, responsible for supporting and protecting the landing, maintained an air umbrella over the anchorage with planes dispatched from northern Luzon bases. Meanwhile bombers attacked Nichols, Camp Murphy, Limay, Clark, Del Carmen, and Batangas airfields in neutralization strikes, while direct air support was afforded in landing forces.

After dispersing the light resistance encountered in the vicinity of the beach, the main force pressed inland. The main body of the 48th Division immediately turned south toward Rosario, taking two routes, the main coastal road and a parallel road slightly to the east. Along the coastal road more than ten U. S. tanks were destroyed. At about 1900 hours on the night of 22 December, the division advance guard reached Damortis and nearby Rosario.

The Uejima Detachment, charged with taking San Fernando and covering the Army's left flank, had meanwhile carried out its scheduled landing near the mouth of the Bauang River at 0730 on the 22nd. Stubborn enemy resistance was met, but, by 1400 hours, the defenders were driven inland, and a junction was effected at San Fernando with the Tanaka Detachment, which had advanced down the coast road from Vigan according to plan. On the right of the Uejima Detachment, the 3d Battalion of the 9th Infantry Regiment, responsible for driving inland and seizing the Naguilian airfield, simultaneously landed at Santiago, and, meeting little enemy resistance, carried out its assigned mission by the evening of 22 December. This battalion then assembled in

---

46 Philippine Operations Record, Phase One, Suppl., op. cit., pp. 1–2.
47 Philippine Operations Record, Phase One, op. cit., pp. 88–9.

the Naguilian area and prepared to advance on Baguio.

On 23 December landing operations continued under improved weather conditions, but progress was slow. The remaining elements of the 48th Division, including Army artillery units and rear echelon units under Army control, were still not ashore. During the morning General Homma landed at Bauang and established the command post of Fourteenth Army.

Meanwhile front-line units pushed ahead. The 48th Division routed a Philippine-American force of approximately 1,700 men near Sison and occupied the town by evening. Elements pushing down the coast road occupied Mabilao. The Tanaka Detachment, moving south, reached the 48th Division area by evening and reverted to 48th Division control.

During the next two days unloading operations progressed smoothly, and on 25 December debarkation of the 48th Division was completed. The debarkation point had been shifted to the south, so that by the 26th unloading was being accomplished over the beach in the vicinity of Damortis. The landing of the greater part of the Army was completed by 28 December. Because of a typhoon, the departure of the second invasion convoy carrying the 65th Brigade from its staging area on Formosa was postponed until 30 December.

The secondary landing on the east coast of Luzon had also been successfully executed. Shortly after midnight on the night of 23–24 December the main body of the 16th Division was landed between Atimonan and Siain on Lamon Bay, the 1st and 3d Battalions, 20th Infantry Regiment, in the assault. The 2d Battalion, 20th Infantry was landed at Mauban.[48] The main force encountered light enemy resistance but soon cleared the area east of the Atimonan isthmus ridge. An element advanced towards Calauag via the coastal road from Siain in order to cut the route of withdrawal of the enemy force retiring before the Kimura Detachment, then pushing west from Daet. On the Atimonan—Siain beach the 1st Naval Base Force took over base construction, and unloading continued until 28 December.

### The Race for Manila

Fourteenth Army operations on all sectors were proceeding with complete success. No large scale counterattack against the Lingayen landing force had materialized, and the lack of resistance encouraged the Army Commander to drive rapidly to the final objective—Manila, with no change in plans. The morale of officers and men was extremely high. The two divisions, the 48th from Lingayen and the 16th from Lamon Bay, began a race for the honor of entering the capital city first.[49] (Plate Nos. 21 & 22)

The 48th Division, not waiting for the landing of its rear echelon, moved rapidly southward. Mountainous terrain restricted forward movement to a narrow front. The initial objective was to seize the Agno River crossings. Advance units of the 1st Formosa Infantry Regiment and the 48th Reconnaissance Regiment crossed the Agno against opposition on 26 December, and thereafter, took Carmen,

---

48 The 2d Battalion, 20th Infantry, which landed at Mauban, first encountered stiff resistance by Philippine-American troops deployed along the coast, and later ran into an enemy force with more than ten tanks at Piis, three miles northwest of Lucban. After beating off enemy counterattacks, the battalion succeeded in rejoining the main force. (Statement by Maj. Shoji Ohta, Staff Officer (Intelligence), 16th Division.)

49 Since the objective of the Imperial General Headquarters was to capture Manila, the main strength of the 48th Division, anxious to beat the 16th Division into Manila, was rushed to the city. (Interrogation of Lt. Gen. Maeda, previously cited.)

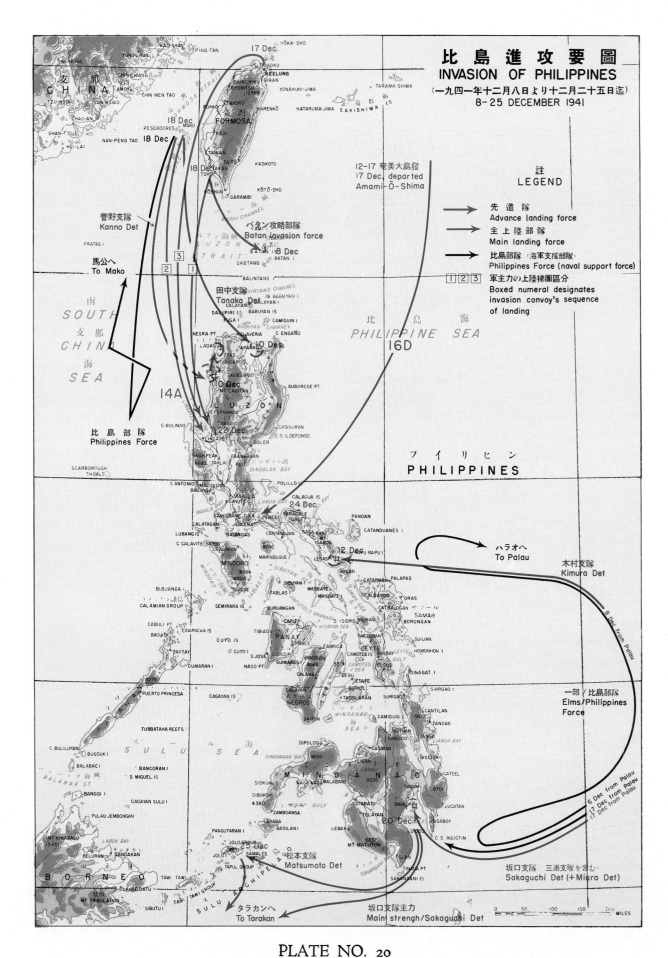

PLATE NO. 20

Invasion of Philippines, 8—25 December 1941

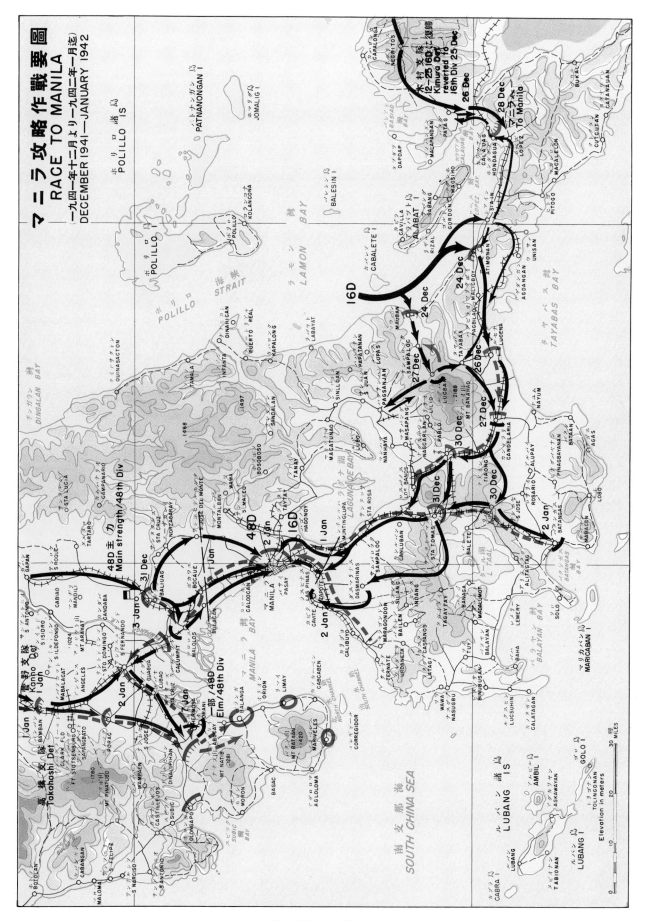

PLATE NO. 21

Race to Manila, December 1941—January 1942

Rosales, and Tayug. In order to protect the Lingayen anchorage and secure the Army's right flank, the Uejima Detachment took Dagupan on the same day. On 27 December, in the center, the 47th Infantry Regiment crossed Agno and occupied Umingan. On the same day Baguio fell to the 3d Battalion, 9th Infantry. The main body of the Army was now disposed north of Agno. Crossings had been secured. Flank guards were out to right and left, and the Army stood poised for the final effort.

The 16th Division advancing from the Atimonan area had not been idle. Destroying armed resistance in its advance, the division pressed on to Candelaria and Lucban on 27 December. In the area northeast of Calauag, the 1st Battalion, 20th Infantry (less one company) linked up with the Kimura Detachment which had driven up the Bicol Peninsula from Legaspi. (Plate No. 21)

The rapid advances by the two divisions continued to be supported by the 5th Air Group, which moved its bases still farther south to keep pace with the ground troops. Air units arrived at Naguilian airfield on 26 December and at Carmen on the 27th. Naval air units continued to carry out attacks against remaining enemy fleet and air strength, and especially against transport shipping in Manila Bay.

The Fourteenth Army Commander was by this time aware that enemy forces were moving northward from Manila with the probable intention of retiring into Bataan and Corregidor. He nevertheless decided to adhere to the original operations plan, and ordered the 48th Division to advance immediately on Manila.[50] Cabanatuan was set as an intermediate objective.

On 27 December the Commanding General of the 48th Division, Lt. Gen. Yuitsu Tsuchihashi, issued a voluminous and detailed field order for the projected operation, based on the Fourteenth Army order. This order was in substance as follows:[51]

1. *(a) Three of the enemy divisions, the 11th, 71st, and 91st have been routed, and one element of the 21st Division has been met and defeated.*
   *(b) The Uejima Detachment under Army control will advance south from San Fabian and guard the Fourteenth Army right flank.*
2. *This division will drive on Manila, advancing first to Cabanatuan and clearing the area of the enemy.*
3. *At dawn tomorrow, the 28th, the 1st echelon of the left column, consisting of the 47th Infantry Regiment (reinf.), will attack the enemy southeast of San Quintin. The 1st echelon will cover the assembly of the tank brigade in that area.*
4. *The tank brigade, consisting of the 4th Tank Regiment (reinf.), the 7th Tank Regiment (reinf.), and attached troops will assemble in the area southeast of San Quintin by noon of*

---

50 (1) There were two opinions with respect to the mission of the 48th Division: (a) That it should concentrate exclusively on the occupation of Manila; and, (b) that it should advance a strong element to the right bank of the Pampanga River and begin preparations for an attack against Bataan Peninsula. A cool appraisal of the enemy situation would have revealed that serious Philippine-American resistance in the Manila area was out of the question, but indecision with regard to these conflicting opinions was allowed to determine the disposition of the division. It proved impossible to dispel the preconceptions, accompanied as they were by failure to recognize the relationship between Corregidor and Bataan and their effect on the value of Manila. Philippine Operations Record, Phase One, op. cit., pp. 115–6. (2) "A small group wished to prevent the withdrawal of MacArthur from Manila. Most of us expected the forces to flee to Mariveles and leave the Philippines. The capture of Manila was the main objective.....At that time we did not realize the value of Bataan as a defensive position." (Interrogation of Col. Motoo Nakayama, Senior Staff Officer, (Operations), Fourteenth Army.)

51 Philippine Operations Record, Phase One, op. cit., pp. 131–140.

*the 28th and will depart on the evening of the same day for Cabanatuan via Lupao, San Jose, Rizal, and Bongabon.*

5. *After covering the assembly of the tank brigade, the 1st echelon of the left column will leave the San Quintin area at 0600 29 December and advance to Cabanatuan via Lupao, San Jose, Munoz, and Baloc. The 2d echelon of the left column, consisting of the 48th Reconnaissance Regiment and the 1st (less one battalion) and 8th Heavy Artillery Regiments will move out behind the 1st echelon on the 29th. The 3d echelon of the left column, consisting of the 2d Formosa Infantry Regiment (reinf.) will move out behind the 2d echelon on the 29th.*

6. *The right column, consisting of the 1st Formosa Infantry Regiment (reinf.) will leave Rosales at 0700 29 December for Cabanatuan via Guimba and Baloc.*

On the morning of 28 December the 48th Division began its advance from the Agno River line. The 4th and 7th Tank Regiments, spearheading the advance, rolled rapidly over difficult roads through San Quintin and San Jose, reaching Bongabon at dusk on the 29th. The right and left foot columns converged on Cabanatuan through Baloc and closed up to the right bank of the Pampanga River north of Cabanatuan on the night 29–30 December. Cabanatuan was on the verge of capture.

While concentrating on the main drive toward Manila, General Homma, however, began to feel concerned over the situation on the right flank. Intelligence reports verified that the enemy forces were retiring to Bataan and Corregidor.[52] When it was reported that General MacArthur's headquarters had withdrawn to Corregidor, the air forces extended their attacks to the island in a special effort to knock out the nerve center. On 29 December the 5th Air Group carried out two heavy bombing attacks against the fortress, dropping eight tons at 1200 hours and twelve tons at 1230 hours.[53] The group was also given the mission of knocking out the bridges west of Lubao, but this was not accomplished.[54] Some support was given the ground effort in this area, however, by attacks on motorized columns moving along the roads leading into Bataan.

Ground operations in the west were also accelerated. At the time that Cabanatuan was about to fall, the main strength of the Uejima Detachment was in the Cuyapo area. At 1600 on 29 December General Homma ordered the detachment to occupy Tarlac and Angeles in an attempt to hinder any westward retirement of the enemy. To aid in this operation, the 48th Division was directed to detach an element and send it to reinforce the Uejima Detachment. The division dispatched the Kanno Detachment (3d Battalion, 2d Formosa Infantry).

Without waiting for the Kanno Detachment to come up, the Uejima Detachment advanced on Tarlac, reaching the northern outskirts of the town on 30 December. There it met bitter enemy resistance, and the town was captured only after repeated assaults in the course of which Col. Uejima was killed. Col. Takahashi, commander of the 8th Heavy Artillery Regiment, assumed command of the detachment, which thereafter took his name.

It soon became clear that the Takahashi and Kanno Detachments were not making sufficiently rapid progress to check the retirement of the

---

52   Ibid., pp. 112–3.

53   The Japanese newspaper Tokyo *Asahi* on 3 January printed a news report of unknown origin claiming that General MacArthur had been wounded in the Japanese air attack on Corregidor on 29 December.

54   (1) Philippine Operations Record, Phase One, op. cit., p. 118.  (2) "In order to prevent the withdrawal (to Bataan) we ordered air force units to bomb the bridges along the route from Manila to Bataan, and to bomb and strafe truck convoys on the road....Too few air units were assigned the task....to be effective." (Interrogation of Lt. Col. Hikaru Haba, Staff Officer (Intelligence), Fourteenth Army.)

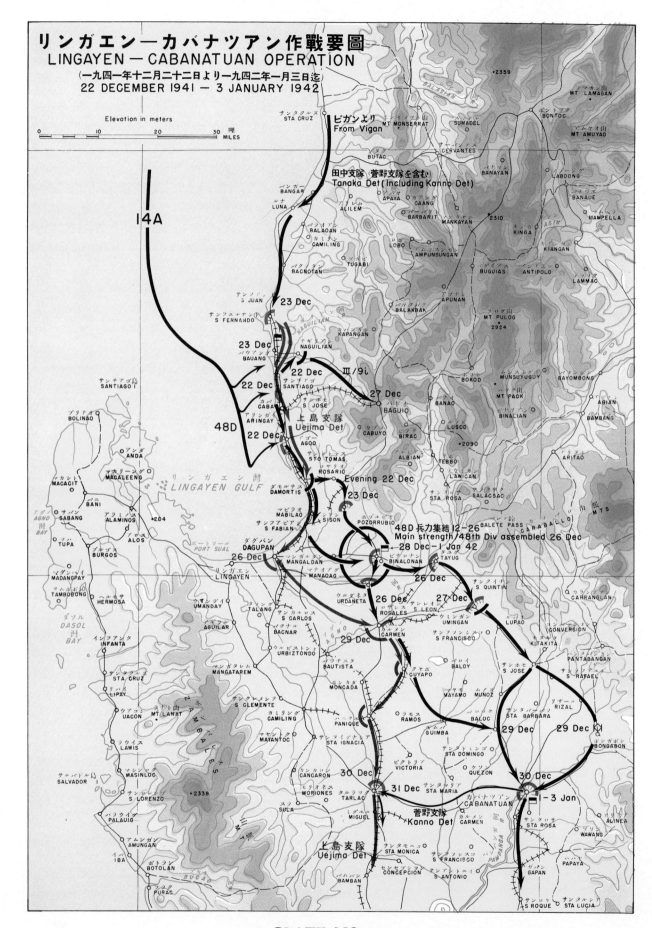

PLATE NO. 22

Lingayen-Cabanatuan Operation, 22 December 1941–3 January 1942

Original Painting by Chosei Miwa

PLATE NO. 23

Attack on Cavite Naval Base

enemy to the west. To remedy the situation, General Homma immediately ordered the 48th Division to send an infantry regiment to Guagua to seal off Bataan and Zambales provinces against further enemy withdrawals. At the same time the Takahashi Detachment was ordered to proceed to Porac as rapidly as possible. Meeting exceedingly stiff resistance the detachment advanced to Bamban on 1 January.

Concurrently with these developments, Cabanatuan had fallen on 30 December after a short, brisk engagement. The 48th Division on 1 January moved up to a line connecting Bulacan, Bocaue, and San Jose del Monte and prepared to invest Manila. In compliance with the Army order to dispatch an element to aid the Takahashi Detachment in blocking the Bataan withdrawals, the Tanaka Detachment (2d Formosa Infantry, less one battalion) was sent west from Baliuag to Calumpit, a vital bridge-point on the Pampanga River and a bottleneck on the escape route to Bataan.

Before the Tanaka Detachment could reach the bridge-site, the 7th Tank Regiment, on the initiative of its commander, drove to Calumpit and on 2 January occupied the bridges after a series of sharp encounters with enemy tank forces. The Tanaka Detachment, arriving the same day, crossed the river and advanced to San Fernando, which it entered at 1830.

On the southern front the 16th Division, encountering determined enemy resistance, cleared San Pablo and Santo Tomas and its advance guard reached Zapote on New Year's Eve. The division drew up its lines from Laguna de Bay to Cavite Harbor. Manila was besieged.

## The Fall of Manila

General Homma had hoped for a decisive battle with the Philippine-American forces in the central Luzon plain before Manila, and wished to avoid battle within the capital itself. Orders had therefore been disseminated to all troops restricting their movement across the road net encircling the city and forbidding the bombardment of the city itself.[55] However, reports from reconnaissance aircraft and observation of numerous fires within the city led the Army Commander to assume that the enemy had evacuated the city. Anxious to rescue the large Japanese population and restore public order, Gen. Homma issued orders to occupy the city.[56]

On 2 January the advance guard of the 16th and 48th Divisions entered Manila. The occupation of the city went forward efficiently, and public order was gradually restored.[57] Meanwhile, key outlying areas were being mopped up. To the south, elements of the 16th Division occupied Cavite and Batangas. To the northwest, the Tanaka Detachment joined the Kanno Detachment at San Fernando on the evening of 2 January, and all that area was

---

55 Philippine Operations Record, Phase One, op. cit., pp. 124–6.

56 (1) Vital installations in Manila and Cavite, set afire by the enemy, were burning, and looting by the native population had broken out. Ibid., p. 124. (2) Large fires were raging inside the city, and although it was reported that the Japanese residents had been released from custody, their situation was not clear. Therefore, despite the earlier Army order, it was decided to occupy the city. Statement by Col. Moriji Kawagoe, Chief of Staff, 48th Division.

57 Upon entering the city, the advance guard confirmed the fact that the Japanese colony of approximately 3,500 had already been released from custody by the American military authorities upon the evacuation of General Headquarters to Corregidor. Maj. Gen. C. A. Willoughby, Assistant Chief of Staff, G–2, who had charge of the internment of the Japanese colony, had summoned the Japanese Consul General, Mr. Katsumi Niiro, and released the internees to his custody on 27 December. (Statement by Mr. Katsumi Niiro, former Japanese Consul General in Manila.)

cleared. The Takahashi Detachment, however, was still mopping up around Mabalacat and Fort Stotsenburg, ten miles short of its goal at Porac.

With the capture of Manila only twenty-five days after the start of hostilities, the Japanese forces in the Philippines had gained possession of the foremost center of American influence in the Far East, and achieved the major objective fixed by Imperial General Headquarters. This swift victory, more apparent than real, gave the Japanese public at home the impression that the United States was not too formidable an enemy. More important, it also led the Fourteenth Army, which had expected a determined defense of Manila, to underestimate the fighting strength of the Philippine-American forces.

It was recognized that a large number of enemy troops had succeeded in withdrawing into Bataan. However, divergent opinions arose in General Homma's staff as to whether the Army's main effort should now be directed toward the establishment of military government or the continuation of field operations for the purpose of destroying General MacArthur's forces.[58] One group took the view that military government should be given first priority and that the enemy on Bataan should merely be contained and starved into ultimate surrender. General Homma, however, decided that it was best to allow the enemy no respite and to press the attack to a swift conclusion.

## Manila to Bataan

General Homma was convinced that the enemy force which had retired into the mountain fastnesses of Bataan could be easily and rapidly crushed by Fourteenth Army with forces then available in the Pampanga area. Having decided to attack, he quickly implemented his decision with orders. On the same day Manila was entered, the 48th Division was directed to move its main strength northward across the Pampanga River to Bataan and pursue the enemy down to a line running westward from Balanga. The Takahashi Detachment, then at Mabalacat, was ordered to advance rapidly to Dinalupihan through Porac to cut off further enemy withdrawals. The Tanaka Detachment, which at this time was crossing the Pampanga River at Calumpit, was to drive southwest from San Fernando to Hermosa through Lubao, Santa Cruz, and Dinalupihan. (Plate No. 21) In support of these operations the 5th Air Group was ordered to attack enemy concentrations and positions in the Bataan area.

While the 48th Division prepared to move its main force to the battle area, the Takahashi Detachment slowly forged ahead toward Bataan from the north. On 3 January the detachment attacked strong enemy defense positions at Porac and, after a brisk engagement, finally penetrated the enemy line on the night of the 4th. Meanwhile, the Tanaka Detachment (2d Formosa Infantry and one battalion, 47th Infantry) advancing from San Fernando reached Guagua against stiffening enemy resistance. The detachment then pushed on to Santa Cruz on 5 January and was there relieved by a fresh regiment (1st Formosa Infantry). On 6 January enemy resistance at Santa Cruz was broken, and the regiment pursued the enemy southwest, entering Dinalupihan at 1500 on the 6th. The next day a small element was sent forward to Hermosa, which was seized against light resistance. On the same day the 1st Formosa Infantry in Dinalupihan was joined by the Takahashi Detachment, which had taken three days to fight its way down from Porac.

---

58 Interrogation of Lt. Col. Yoshio Nakajima, Staff Officer (Intelligence), Fourteenth Army.

While the Japanese units were engaged in these preliminary operations against Bataan, Southern Army Headquarters at Saigon had reached a decision which was to profoundly affect General Homma's campaign. General Terauchi, Commander-in-Chief of the Southern Army, had become convinced that operations in the Philippines were all but completed and that the Japanese drive in the Netherlands East Indies could safely be put forward a month.

Under the Imperial General Headquarters plan for the southern operations, the 48th Division and the 5th Air Group, the backbone of the Fourteenth Army, were scheduled for redeployment to Java and Burma. A large part of the Navy's Philippines Force was also to be diverted for the attack on Dutch East Indies. The Eleventh Air Fleet, which had supplied the bulk of the air strength employed in the first phase of the Philippine operations, had already advanced most of its strength to bases in Mindanao and Jolo and was preparing for the southern drive. Only a small number of planes continued bombing operations against Bataan and Corregidor.

On the night of 2 January, within a few hours after he had ordered the 48th Division to move up for the assault on Bataan, General Homma received telegraphic orders from General Terauchi directing the execution of the basic redeployment plans.[59] This directive called for the transfer of the 48th Division to Sixteenth Army command effective 14 January, and for its embarkation from the Philippines on 1 February. The 5th Air Group was to be relieved as soon as possible in preparation for movement to Thailand after 14 January.

The loss of the 48th Division and the 5th Air Group came at an inopportune time for the Fourteenth Army. While the 1st Formosa Infantry and the Takahashi Detachment were feeling out the enemy line in the northern Bataan area, and while the 48th Division was assembling in the San Fernando area for the Bataan operation, the Army staff spent the period 4–6 January hastily writing orders to effect the necessary reshuffle of units.

To provide for the relief of the 48th Division, it was decided to bring down the 65th Brigade, which had landed at Lingayen Gulf on 1 January, as quickly as possible, and relieve the front-line units of the 48th Division in the Hermosa and Dinalupihan areas. This relief was to be effected on or about 8 January. Upon arrival the brigade was to take command of the Takahashi Detachment and advance on Balanga as soon as possible. As soon as the brigade reached Dinalupihan, an element was to be detached and sent to seize the Olongapo naval base on Subic Bay. These orders were transmitted to the 65th Brigade on 4 January.

On 5 January the 10th Independent Air Unit was reorganized at Clark Field to replace the 5th Air Group. This unit was assigned all the air strength that was not scheduled to be redeployed. On 4 January the 16 Division was designated the occupying force for the Manila area, and Lt. Gen. Susumu Morioka was named defense commander. Meanwhile the Navy, in order to provide a headquarters for the small surface contingent that was to remain in the Philippines, organized the Third Southern Expeditionary Fleet under the command of Vice Adm. Rokuzo Sugiyama, with headquarters at Manila. This fleet was to secure the seas around the Philippines and cooperate with the Fourteenth Army in all future operations.

As a result of this hasty and radical reorganization, the Army and Navy commanders in the Philippines were forced to undertake the Bataan offensive and the occupation of the

---

[59] Philippine Operations Record, Phase One, op. cit., pp. 153, 157–8.

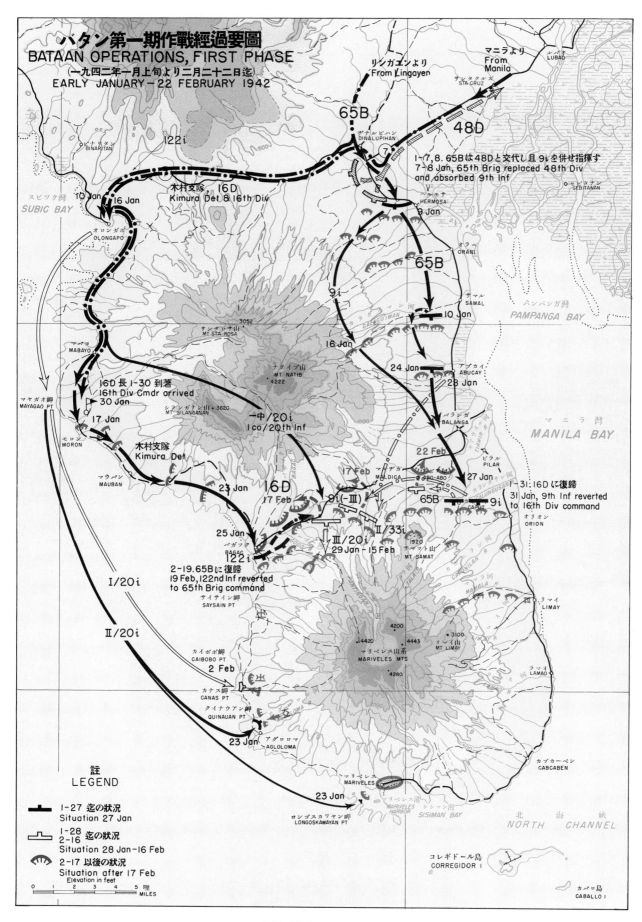

PLATE NO. 24

Bataan Operations, First Phase, Early January—22 February 1942

islands with the following principal forces:[60]

*Army Forces:*
 *Fourteenth Army*
  *16th Division*
  *65th Brigade*
   *122d Infantry Regt. (2 bns. and 1 btry)*
   *141st Infantry Regt. ( do )*
   *142d Infantry Regt. ( do )*
   *65th Signal Unit*
   *65th Engineer Unit*
   *65th Medical Unit*
  *7th Tank Regiment*
  *1st Heavy Artillery Regiment*
  *8th Heavy Artillery Regiment*
  *One mountain artillery battalion*
  *10th Independent Air Unit*
   *52d Ind. Air Company (Reconnaissance)*
   *74th Ind. Air Company (Observation and Liaison)*
   *76th Ind. Air Company (Headquarters Reconnaissance)*
   *3d Company, 50th Fighter Regiment*
   *16th Light Bomber Regiment*
   *Element, 1st Air Signal Regiment*
*Navy Forces:*
 *Third Southern Expeditionary Fleet*
  *One cruiser (Kuma, 5,100 tons)*
  *One minelayer (Yaeyama, 1,135 tons)*
  *31st Special Base Force*
  *32d Special Base Force*
  *31st, 32d Air Groups*[61]

Facing the shrunken forces of the Fourteenth Army on Bataan Peninsula and Corregidor was an enemy force which was estimated by the intelligence staff as comprising six field divisions and a variety of garrison units (mostly coast artillery) with a total strength of 40,000 to 45,000 men. Aerial reconnaissance had disclosed prepared enemy defense positions at several points on Bataan, principally in the area west of Hermosa, on the eastern slopes of Mt. Natib, and in the vicinity of Bagac. None of these positions was believed strong, and it was thought that they could be easily overrun.[62]

## Bataan, First Phase

Despite the serious reduction of the forces at his disposal, General Homma was still confident that the substantial number of enemy troops on Bataan could be defeated by a swift pursuit which would give them no breathing spell in which to reorganize and entrench themselves in strong defensive positions. Consequently, the 65th Brigade had barely relieved the forward units of the 48th Division in the Hermosa sector, when the Army Commander, on 9 January, ordered it to the attack.

Launching its drive from Hermosa the same day, the brigade main body (141st and 142nd Infantry, reinf.) advanced to the north bank of the Calaguiman River but then stalled in the face of unexpectedly severe enemy counter-fire. (Plate No. 24) Meanwhile, a separate element (one battalion, 122d Infantry, reinf.) drove unopposed across the peninsula to Subic Bay and seized the Olongapo naval base against weak resistance.[63] On the same day, 10 January, the 16th Division was ordered to dispatch a force into Cavite Province and occupy Ternate and Nasugbu in order to cut off Bataan and Corregidor from the south.

With the 65th Brigade temporarily checked on the eastern side of the peninsula, General Homma now prepared to launch a parallel drive down the west coast. To reinforce the 122d Infantry already in the Olongapo area,

---

 60 (1) Ibid., p. 206, Chart 7. (2) Table Showing Organizational Changes and Losses of Japanese Naval Forces, op. cit., p. C-6.
 61 The 31st and 32d Air Groups were organized and attached to 31st and 32d Special Base Forces from 1 February.
 62 Philippine Operations Record, Phase One, op. cit., pp. 163-4.
 63 Ibid., pp. 166, 174.

the 20th Infantry Regiment less one battalion) of the 16th Division was ordered on 13 January to move from Manila to the western sector, and the combined force was placed under command of Maj. Gen. Naoki Kimura, 16th Infantry Group commander and designated the Kimura Detachment.

Before the west coast drive got under way, developments in the eastern sector took a favorable turn. The 65th Brigade, finally breaching the Calaguiman River line, pushed south to the next enemy defense line west of Abucay. Because of the strength of these positions, General Homma adopted a plan of maneuver which called for the brigade to advance into the foothills of Mt. Natib and turn the enemy left flank with an attack from the mountainous area. The 9th Infantry Regiment, which meanwhile was sweeping around in a wider flanking movement to the west, paved the way for this maneuver on 19 January by driving a deep salient in the enemy line to a point five miles west of Balanga.

Although seriously handicapped by the Army's withdrawal on 17 January of most of its artillery support,[64] the 65th Brigade launched its flanking attack on 22 January and succeeded in forcing the enemy, on 24 January, to withdraw from the Abucay positions and retire south past Balanga under hot pursuit. The brigade, after first moving up into the area west of Balanga, extended the pursuit to the sector west of Orion, where enemy resistance again stiffened. Now handicapped more than ever by its lack of artillery, the brigade closed up to the new line and prepared for further action.

Meanwhile, on the west coast, a furious battle was in progress. Between 18 and 23 January the Kimura Detachment, advancing from Moron, met and destroyed large enemy forces between Mt. Natib and the Mauban area. It then pressed on towards Bagac. To facilitate its advance, Maj. Gen. Kimura ordered the 2d Battalion of the 20th Infantry to proceed by sea to Caibobo Point and effect a landing in the enemy rear. On 23 January the battalion was lifted at Mayagao Point, near Moron, and moved by boat down the west coast, but confused by darkness and a strong tide, the main strength landed on Quinauan Point and Agloloma by mistake, while one element continued far south and landed at Longoskawayan Point, near Mariveles. These units were immediately attacked by superior enemy forces.* On 26 January one company of the 20th Infantry Regiment was dispatched by boat from Olongapo with food and ammunition for the 2d Battalion. This company landed on Canas Point and was immediately placed under fire by American artillery, losing most of its boats and finally retiring to Mayagao Point with heavy casualties.[65] This made the position of the 2d Battalion even more desperate. Meanwhile, the Kimura Detachment pushed ahead and took Bagac on 25 January.

On the eastern sector the 65th Brigade prepared for a new offensive to dislodge the enemy forces from their positions between Mt.

---

64 Reasons for the Army's actions were: (1) the difficulty of using artillery in the heavily forested area; and (2) desire to conserve ammunition for the attack on Corregidor. Philippine Operations Record, Phase One, op. cit., pp. 179–180.

* American Editor's Note: It is of historical interest that General MacArthur's Director of Intelligence, Maj. Gen. C. A. Willoughby, became involved in this landing. On his return to Mariveles after a staff visit to General Wainwright's headquarters in Bagac, General Willoughby was in the vicinity of Agloloma Point at the time of the Japanese landing. As senior officer in the area, he took command of the sector defense forces, belonging to the 1st Provincial Constabulary Regiment, and personally led a series of sharp counterattacks to stop the Japanese advance. Aided by the dense forest terrain along this coast, he was able to deceive the Japanese as to his real strength until reinforcements entered the action on the next day.

65 Ibid., pp. 209–210.

Samat and Orion. Army artillery units began advancing into the area west of Balanga on 27 January, but because of jungle obstacles and enemy counterbattery, their efforts were largely ineffectual. The brigade nevertheless launched a coordinated attack on the 27th, failing to penetrate the American line. The battle lasted for four days and was climaxed by an attempt to take Mt. Samat on the 31st, which also failed. This sector then quieted down and attention shifted to the west coast, where the Japanese force had also fallen into serious difficulties.

The quick advance of the Kimura Detachment to Bagac had encouraged Fourteenth Army headquarters, and General Homma had decided to exploit this success by throwing fresh reserves into the area. On 28 January Lt. Gen. Morioka, 16th Division commander, joined the Kimura Detachment with two infantry battalions and took command. Attacking east of Bagac on the night of 29–30 January, the 3d Battalion, 20th Infantry, drove a salient into the enemy line but then was pinched off and pocketed by a strong enemy counterattack the following day. Attacked from all sides, the battalion suffered heavy casualties but hung on grimly while the 16th Division, attacking with the 9th Infantry and the 2d Battalion, 33d Infantry, strove to effect its relief.

General Morioka, with his operations stalled along the Bagac line and two battalions marooned behind the enemy lines, decided to effect another amphibious landing in the enemy rear. On 2 February the 1st Battalion, 20th Infantry, landed at Canas Point and also lost the greater part of its combat strength in a strong attack by a superior enemy force. The entire 20th Infantry was now threatened with destruction.

Deciding to evacuate the two battalions trapped on Quinauan and Canas Points, General Morioka on 7 February dispatched a group of landing barges from Olongapo. So intense was enemy fire at the landing points, however, that only 43 casualties could be evacuated. At this point the 10th Independent Air Unit succeeded in dropping some supplies to the beleaguered troops, but their situation remained desperate under heavy enemy attack.

It was now becoming increasingly apparent that the Fourteenth Army could progress no farther with its depleted forces.[66] The 65th Brigade and 16th Division units had fought bravely and well in driving the stubbornly resisting enemy back upon the Bagac—Orion line. The 10th Independent Air Unit and Navy air groups day after day had carried out bombing missions against enemy artillery, vehicles, strongpoints, and dumps, at the same time engaging the few remaining American aircraft in dog-fights. Nevertheless, no attack, however determined, seemed to be able to crack the line which Philippine-American troops had forged from Bagac to Orion. The possibility of success, moreover, decreased with each attack since front-line units were by this time seriously understrength.

General Homma was now placed in a difficult dilemma. His intelligence indicated that the enemy's defenses were not only strongly manned but in great depth.[67*] The Fourteenth

---

66 (1) Casualites of the Fourteenth Army between 9 January and 8 February were 6984. Philippine Operations Record, Phase One, Suppl., op. cit. (2) "By this time losses, including those not reported to Army, were so great that....only 2,500 rifles were available on the line." Interrogation of Lt. Gen. Maeda, previously cited.

67 General Homma's estimate of the enemy situation on Bataan at this time placed the Philippine-American strength at two corps, one operating in the narrow west coast sector with one division and one in the east coast sector with three divisions. The enemy troops were emplaced in a defensive position of great depth and complex organization extending from south of Bagac east along the northern slopes of Mt. Samat to south of Orion. An enemy map showing the extent of the American positions on Bataan had been found in a barracks at San Fabian, and a study of

Original Painting by Yoshinobu Sakakura

## PLATE NO. 25

Supply Train Marching Toward the Front

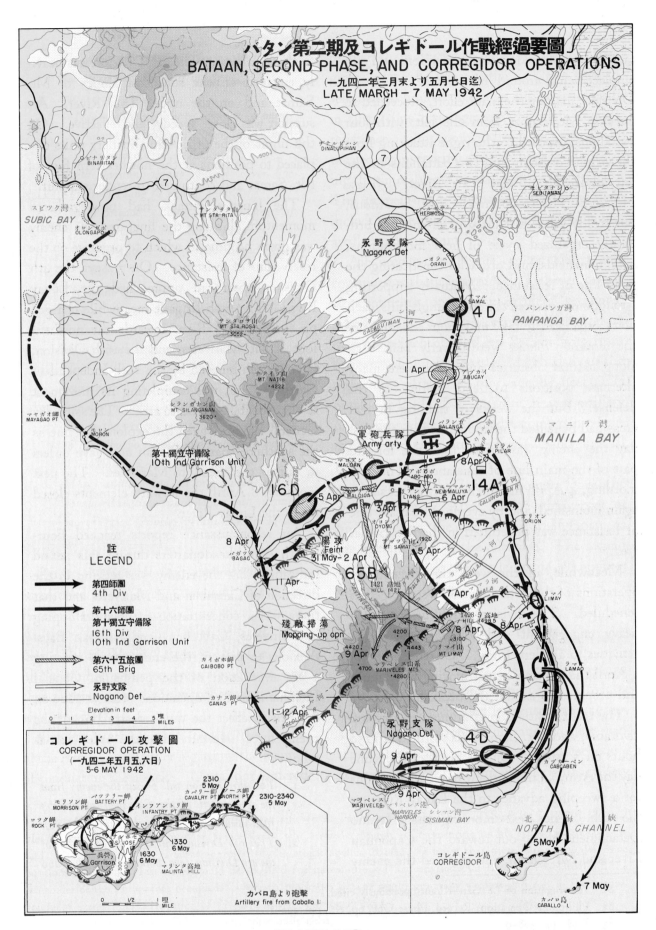

PLATE NO. 27

Bataan, Second Phase, and Corregidor Operations, Late March—7 May 1942

> *Detachment, clear the area between the Cabcaben—Mariveles road and pursue the enemy from the area south of that road towards the hills east of Mariveles.*
>
> 3. *The 65th Brigade will swing west, cross the upper reaches of the Pantingan River and prepare for further operations in the west coastal area.*
>
> 4. *Initially the main force of the artillery will move to positions in the area southwest of Limay and then gradually displace to the area north of Cabcaben. It will there support the 4th and 16th Divisions, neutralize enemy fortress guns on Corregidor, and shell enemy craft in Manila Bay.*
>
> 5. *The air units, besides continuing to render support to the 4th and 16th Divisions, will observe enemy movements along the west coast and enemy shipping in Mariveles, Sisiman, and Cabcaben Bays. They will bomb and strafe the enemy retreating along the Cabcaben—Mariveles road, enemy shipping, and Corregidor Island.*

As the Japanese forces drove forward on 9 April in pursuance of this order, enemy resistance finally collapsed. Tank forces of the 4th Division charged into Mariveles at 1300.[84] The 16th Division, echeloned to the left rear, raced along the Limay—Cabcaben—Mariveles coastal route, reaching Mariveles that night. On the same day the 65th Brigade captured the summit of Mt. Mariveles.[85]

Thus, the gallant enemy defense of Bataan, which had won the respect of even the Japanese commanders, finally ended. As the flood of sick and battle-weary prisoners increased by the hour, Major General Edward P. King Jr., American commander of the Luzon Force, sent forward a flag of truce. Hostilities on Bataan were finally brought to an end on 11 April. The final offensive had required about one week less than General Homma had expected.[86]

Japanese firepower had been the key to victory. Guns used by the Japanese forces in the Bataan operation totaled 241, of which 133 were field and mountain artillery pieces (75mm–100mm) and 108 were 120mm howitzers or larger. About 9,000 rounds of ammunition were expended by the Army artillery alone.[87] The Army air force dropped a total of 907 tons of bombs on Bataan and outlying areas, 563 tons of which were dropped during the second phase (3–11 April).[88] Casualties among the Japanese numbered about 1400.[89]

## Fall of Corregidor

The Japanese forces now turned their attention to Corregidor, the historic and formidable fortress lying at the entrance to Manila Bay. Despite the surrender of Bataan, Corregidor showed no signs of giving up. Toward the end of the Bataan campaign, Army artillery had displaced to the Cabcaben area and commenced to shell Corregidor. The air force had been bombing the island almost daily. Enemy armed boats, still active in Manila Bay, were attacked by artillery and air forces. In spite of this show of force, the defenders of Cor-

---

84 Ibid., p. 310.
85 Ibid., p. 311.
86 Interrogation of Lt. Gen. Kitano, previously cited.
87 (1) Extracted from the private papers of Col. Arao, previously cited; (2) *Dai Niji Bataan Koryakusen Sento Shoho* 第二次バタン攻略戰戰闘詳報 (Detailed Battle Report on Bataan Operations, 2d Phase) 1st Artillery Headquarters, Fourteenth Army, Jun 42, Attached Chart III.
88 Philippine Air Operations, Phase One, op. cit., pp. 6–7.
89 Philippine Operations Record, Phase One, Suppl., op. cit.

regidor appeared to be ready to make a fight of it.[90]

General Homma decided to attack the fortress of Corregidor and if necessary to invade Caballo, Carabao and El Fraile after the occupation of Corregidor. The general outline of the attack plan was formulated by 17 April, and by that date, also, approximately 80 large and small landing barges were stealthily slipped into Manila Bay. Since the operation was to be an opposed amphibious landing against a strong permanent defense installation, the preparations were carried forward with great care and secrecy.

In substance the plan was as follows:[91]

1. *Troops:*
    *Corregidor Landing Force (4th Div.)*
    *Left Flank:*
        *61st Infantry Regt. (reinf.)*
    *Right Flank:*
        *4th Inf. Gp. Hq.*
        *37th Inf. Regt.*
        *One bn. 8th Inf. Regt. (reinf.)*
2. *Operational Instructions*
    *Left. Flank: Effect landings at Infantry Point, Corregidor[92] at 2300, 5 May. Initial objective—Malinta Hill*
    *Right Flank: Effect landings between Morrison and Battery Points at 2330, 6 May. Objective—complete occupation of Corregidor.*
3. *Artillery Preparation*
    *Infantry Point, Corregidor 2230–2300, 5 May.*
    *Battery Point, Corregidor 2300–2330, 6 May.*
    *Caballo Island will be neutralized.*
4. *Army Air Units*
    *Bombing attacks will be made of the batteries and installations on Corregidor and Caballo.*
5. *16th Division*
    *Diversionary feint from the south toward Caballo and El Fraile*

While Fourteenth Army was readying its forces for the crucial assault on Corregidor, operations in the central and southern Philippines were progressing according to plan. On 19 April the Kawaguchi Detachment, transferred from Borneo, captured Cebu Island, and by about 20 April the Kawamura Detachment had overrun Panay.[93] These two detachments then moved to Mindanao and, together with the Miura Detachment, embarked on a pacification campaign throughout the island in the latter part of April.

Back on Luzon, the forces for the Corregidor offensive had completed their training in southeastern Bataan, and the necessary shipping was assembled at Lamao and Limay. On 28 April, General Homma, hoping to deceive the enemy into thinking that no attack was planned against Corregidor, staged a belated ceremonious entry into Manila. Meanwhile, the sporadic firing of the Army artillery against Corregidor was continued, together with bomb-

---

90 Fourteenth Army on 17 April estimated that the Corregidor garrison consisted of five coast artillery regiments, of which two were Filipino units. Armament ranged from 155 mm to 300 mm guns. Philippine Operations Record, Phase One, op. cit., pp. 328–30.

91 (1) Ibid., pp. 332–345. (2) Philippine Operations, Phase One, Suppl., op. cit., pp. 60–71.

92 The 4th Division commander justified the decision to make the initial landings on the narrow part of the island rather than at Morrison Point on the head of Corregidor on the ground that, since only two battalions could be lifted at one time (due to shortage of landing barges), the narrow part of the island offered the best chance of striking a concentrated blow. It was hoped to cut the island in half in this manner. Interrogation of Lt. Gen. Kitano, previously cited.

93 Philippine Operations Record, Phase One, op. cit., p. 366.

ing by the Army air force.[94]

On 29 April, Army air forces began a furious seven day preparation on Corregidor, repeatedly attacking batteries, antiaircraft positions and pillboxes. Caballo and El Fraile were also attacked during this period. On 2 May Army artillery units began three days of preliminary firing against point targets on Corregidor. By 5 May Corregidor was strangely quiet.

On the evening of 5 May the 1st and 2d Battalions of the 61st Infantry Regiment (reinf.) embarked near Limay and at Lamao. As the boat group, moving under cover of darkness, ran for the eastern tip of Corregidor, it was brought under fire from the island. Due to the darkness and a heavy inshore current in North Channel, the boat group was carried too far east, and the troops touched down on Cavalry Point and just east of North Point instead of at Infantry Point as planned. Enemy resistance was heavy, and the force took great casualties. The regiment pushed ahead, however, and at 0200 gained the high ground to the northeast of the airstrip.

At dawn a furious battle began in the narrow neck of Corregidor around Infantry Point. Air support was heavy with 88 tons of bombs dropped on 6 May in support of the 61st Infantry Regiment.[95] Between 1000 and 1100 hours a strong counterattack was mounted by the American defenders but was repulsed after fierce fighting at close quarters. All during that morning, worried about the situation, the 4th Division had been working on a plan to change the landing schedule. This change in plans was abandoned at 1330 when Lt. Gen. Jonathan L. Wainwright, USAFFE Commander since General MacArthur's departure, appeared at the front under a flag of truce and offered to surrender.

That afternoon General Wainwright was transported to Cabcaben, where he entered into surrender negotiations with General Homma. Meanwhile, the bitter struggle continued on Corregidor, and the 61st Infantry entered San Jose at 1630. During the Cabcaben interview General Wainwright could not be dissuaded from his intention of surrendering only Corregidor rather than all American forces in the Philippines. He was therefore informed that the attack would be continued.

On the night of 6 May following a sharp 15-minute artillery preparation, the right flank forces embarked at Lamao as planned, landing at 2340 slightly east of the assigned beaches on Battery Point against no resistance.[96] Sweeping inland, they quickly reached the south shore of Corregidor and, acting in conjunction with the 61st Infantry, wiped out the last pockets of resistance at 0830 on 7 May. Shortly after noon, elements of the 33d Infantry Regiment, 16th Division, occupied Carabao and El Fraile Islands after the defenders had raised surrender flags. Meanwhile the Caballo Island landing force, though seriously delayed by the necessity of beating off an attack by enemy armed boats, also proceeded to its objective, landed at 0030 on 7 May, and occupied Caballo Island. This was the last combat operation of the Philippines campaign.[97]

Due to the unexpected tenacity of the enemy defense of Bataan and Corregidor, the campaign, originally scheduled to be completed in about

---

94 The Army air units dropped 365 tons of bombs on Corregidor 12 April—5 May. Philippine Air Operations Record, Phase One, op. cit., pp. 8–9.

95 Ibid.

96 Philippine Operations Record, Phase One, op. cit., p. 360.

97 Prisoners of war taken in these operations (Bataan and Corregidor) numbered about 83,000. Philippine Operations Record, Phase One, Suppl., op. cit.

硝煙の道　（コレヒドール）　猪熊弦一郎

Original Painting by Genichiro Inokuma

PLATE NO. 28
Gun Smoke Road, Corregidor

本間ウエンライト會見圖　宮本三郎　昭和十九年

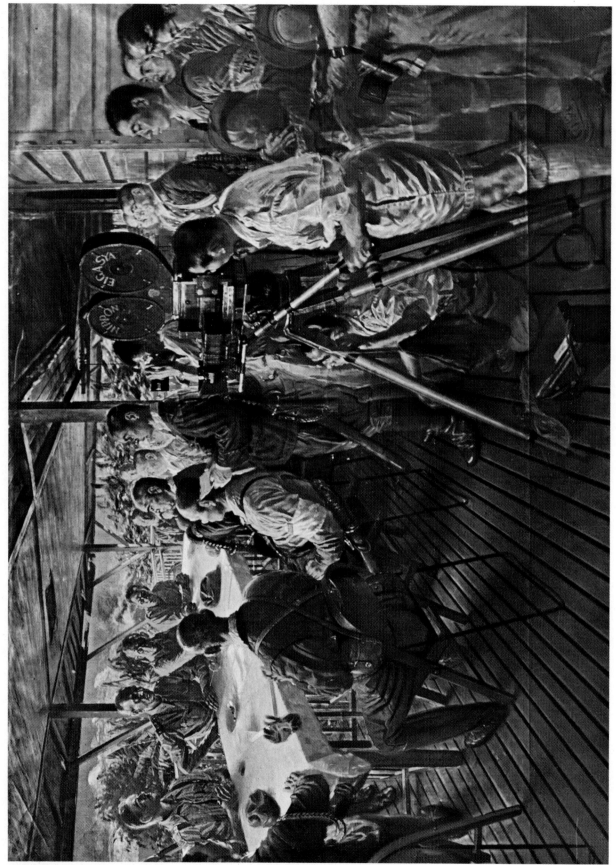

Original Painting by Saburo Miyamoto

PLATE NO. 29
Bataan Meeting of General Wainwright and Gen Homma

fifty days, had taken five months. It had also required the employment of a total, for all phases of the campaign, of approximately 192,000 army and navy personnel, a figure considerably in excess of the initial strength allotment.[98]

## Surrender

On the night of 7 May Lt. Gen. Wainwright was taken to Manila where, at 2350, he broadcast the surrender order to all American and Filipino forces throughout the islands. American staff officers were forthwith sent to the commanders of the Visayas and Mindanao areas to deliver the orders. The Fourteenth Army commander meanwhile dispatched urgent telegrams to Southern Army and Imperial General Headquarters reporting the occupation of Corregidor and the surrender of the Philippines.[99]

On 10 May Maj. Gen. Sharp, commander of Philippine-American forces in the Visayas and Mindanao, surrendered to the Kawamura Detachment. Following this surrender, General Sharp's staff officers, organized as truce teams, aided in the peaceful occupation of the

---

[98] Breakdown of total strength used in the Philippines Campaign, 1941–2, is as follows:

*Army Ground Forces*

| | |
|---|---:|
| Fourteenth Army Headquarters | 1,021 |
| Troops under direct Army command | 28,447 |
| Army Line of Communication troops | 20,956 |
| Shipping Units | 9,330 |
| 4th Division | 10,957 |
| 5th Division (Elm) | 2,667 |
| 16th Division | 14,674 |
| 18th Division (Elm) | 3,622 |
| 21st Division (Elm) | 3,939 |
| 48th Division | 15,663 |
| 56th Division (Elm) | 4,500 |
| 65th Brigade | 6,659 |
| Replacements (for 16th Div and 65th Brig) | 7,000 |
| Total | 129,435 |

*Army Air Forces*

| | |
|---|---:|
| 5th Air Group | 10,278 |
| 60th Heavy Bomber Regt. | 497 |
| 62d Heavy Bomber Regt. | 922 |
| 84th Ind. Fighter Squadron (Elm) | 82 |
| 22d Air Brigade Headquarters | 98 |
| Miscellaneous Service Elements | 875 |
| Total | 12,752 |

*Navy*

| | |
|---|---:|
| Third Fleet | 27,846 |
| Air Forces | 10,064 |
| Special Naval Landing Force | 1,236 |
| Main Body, Southern Naval Force | 7,121 |
| 3d Southern Expeditionary Fleet | 3,485 |
| Total | 49,752 |
| Grand total all forces | 191,939 |

(Statistics compiled by 1st and 2d Demobilization Bureaus, Japanese Government)

[99] An Imperial General Headquarters communique announcing the fall of Corregidor was issued on 7 May at 1910 hours. Asahi Newspaper, Tokyo, op.cit., 8 May 1942.

growth of Allied strength in the southeast area[36] led the Army to concur finally, by 28 April, in a compromise plan envisaging the occupation of strategic points in New Caledonia, the Fiji and Samoa Islands, to be carried out following execution of the deferred invasions of Port Moresby and Tulagi. As further steps to strengthen the Japanese strategic position and disrupt the flow of American supplies to Australia, the Navy had already ordered intensification of submarine warfare in the Pacific and Indian Oceans,[37] and planned the early seizure of Nauru and Ocean Islands, west of the Gilberts.[38]

Preparations by the Fourth Fleet and South Seas Detachment for the Tulagi and Port Moresby invasions were already complete, and the start of the operations waited the impending arrival at Truk of a supporting Task Force dispatched by the Combined Fleet, including the 5th Carrier Division (*Zuikaku* and *Shokaku*) and the 5th Cruiser Division.[39]

Through the subsequent conquest of New Caledonia, Fiji and Samoa, in particular, the Navy planned to establish air and submarine bases which would enable it to command both air and shipping routes from the United States to eastern Australia. Special emphasis was to be placed upon stopping the ferrying of American aircraft to Australia via the South Pacific, and the destruction of tankers transporting fuel supplies. It was estimated that such a blockade, if effective, would retard, if not prevent, Australia's development into an Allied offensive base.[40]

Before any concrete operational plans were worked out for the New Caledonia-Fiji-Samoa invasions, however, Imperial General Headquarters on 5 May issued orders for the prior execution of operations against Midway and the western Aleutians.[41] This crucial decision, which swayed the whole future course of the

---

36 Japanese intelligence estimated that Allied front-line air strength in the Australia-Papua area had increased to approximately 200 planes by April, with 30–50 aircraft of all types maintained at Port Moresby. Meanwhile, it was assumed that the American Task Force (which appeared southeast of the Solomons in February) was still in the Australian area, operating in conjunction with a battleship and two to three heavy cruisers of the British Fleet. (Statement by Comdr. Kazuo Doi, Staff Officer (Operations), Fourth Fleet.)

37 Intensification of Japanese submarine operations had already been ordered by Imperial General Headquarters Navy Directive No. 60, issued 1 March 1942. Under this order, the bulk of Japan's underseas fleet (32 submarines of the Advance Force in the Pacific, and 24 submarines of the Southern Force in the Indian Ocean) was assigned to the disruption of enemy surface traffic, particularly in the areas east and west of Australia. (1) *Daikaishi Dai Rokuju-go* 大海指第六〇號 (Imperial General Headquarters Navy Directive No. 60) 1 Mar 42. (2) Statement by Capt. Tatsuwaka Shibuya, Staff Officer (Operations), Combined Fleet.

38 Occupation of Nauru and Ocean Islands had first been ordered by the Navy Section of Imperial General Headquarters on 27 February 1942. The Combined Fleet assigned this mission to the Fourth Fleet, and the initial plans called for the execution of the operation in May, in conjunction with the seaborne attack on Port Moresby. These plans were not carried out, and the islands were not finally occupied until August 1942. (1) *Sangokai Kaisen Gaiyo* 珊瑚海海戰概要 (Summary of Coral Sea Battle) Admiral Shigeyoshi Inouye, pp. 3, 4, 6, 7. (2) Southeast Area Naval Operations, op. cit., Vol. I, pp. 2, 3, 11.

39 The *Zuikaku* and *Shokaku*, accompanied by the 5th Cruiser Division, arrived at Truk 29 April. Joint staff conferences of the Fourth Fleet and South Seas Detachment, held at Truk, had completed the operational plans by 17 April.

40 Some high-ranking Army and Navy circles thought that an air and sea war of attrition against Australia might even force that country out of the war, without the necessity of an actual Japanese invasion. (Statement by Rear Adm. Tomioka, previously cited.)

41 The invasions of Port Moresby and Tulagi, and also of Fiji, Samoa and New Caledonia, were agreed upon by the Army and Navy Operations Sections of Imperial General Headquarters in April. Immediately after this agreement was reached, the Navy Section proposed the invasion of Midway, and this proposal was subsequently included. (Statement by Col. Hattori, previously cited.)

war, was again taken at the strong insistence of the Combined Fleet and further influenced by the Doolittle raid on Tokyo of 18 April 1942.[42]

Although the decision to invade Midway and the Aleutians meant the deferment of the New Caledonia-Fiji-Samoa operations, joint staff planning for these operations continued, culminating in the issuance on 18 May of Imperial General Headquarters Army and Navy Section orders, which directed the Commander-in-Chief, Combined Fleet, and the Commanding General, Seventeenth Army, to:

> ....carry out the invasions of New Caledonia and the Fiji and Samoa Islands, destroy the main enemy bases in those areas, establish operational bases at Suva and Noumea, gain control of the seas east of Australia, and strive to cut communications between Australia and the United States.[43]

On the same day that these orders were issued, the Seventeenth Army, with a nuclear strength of nine infantry battalions, was activated in Tokyo under command of Lt. Gen. Haruyoshi Hyakutake,[44] and the Combined Fleet assigned the Second Fleet (with attached carrier forces) and Eleventh Air Fleet to the operations.[45] The New Caledonia invasion force was to assemble at Rabaul in the latter part of June, and the Fiji-Samoa forces at Truk in early July. Dependent upon Combined Fleet commitments, the operations were tentatively scheduled for the first part of July.[46]

## Abortive Sea Advance on Port Moresby

With the arrival of the 5th Carrier Division and 5th Cruiser Division at Truk on 29 April, the long-delayed seaborne invasion of Port Moresby at last got under way. Speed was essential because the Combined Fleet now planned to use these units in the subsequent Midway operation.

The final operations plans set the time of landing at dawn on 10 May, with the main South Seas Detachment forces to go ashore south-east of the Pari Mission and other elements (one battalion plus) to effect a secondary landing near the Barute Mission. (Plate No. 32) These forces were to attack the Kila Kila airfield and the Walter Peninsula immediately, while the beachheads were being secured by a battalion of the Kure 3d Special Naval

---

42 Midway and the Aleutians had been included among the possible future operations enumerated in Part IV of Combined Fleet Top Secret Operations Order No. 1, 5 Nov 41. (Cf. n. 36, Chapter V) From February 1942, the Combined Fleet began advocating definite plans for a Navy invasion of Midway, but the Navy Section of Imperial General Headquarters did not concur. Following the B-25 raid on Tokyo from the American aircraft carrier *Hornet* on 18 April 1942, Admiral Yamamoto, Commander-in-Chief of the Combined Fleet, again insisted on the Midway invasion, and the plan was finally adopted by the Army and Navy Sections of Imperial General Headquarters. "The Combined Fleet obtained the concurrence of the Navy Section, Imperial General Headquarters, by insisting that, if Midway were not occupied, the possibility of repeated American air raids could not be minimized, and the Combined Fleet would not accept responsibility for them." (Statement by Rear Adm. Tomioka, previously cited.) "The Army Section, Imperial General Headquarters, was surprised to learn of the Navy's proposal to carry out the Midway operation, but since participation of only one Army regiment was requested, it agreed to cooperate." (Statement by Col. Hattori, previously cited.)

43 *Daikairei Dai Jukyu-go* 大海令第十九號 (Imperial General Headquarters Navy Order No. 19) 18 May 42. (Text of Imperial General Headquarters Army Order No. 633, 18 May 42, was identical in substance.)

44 *Nanto Homen Sakusen Kiroku Sono Ni Dai Jushichi Gun no Sakusen* 南東方面作戰記錄其の二第十七軍の作戰 (Southeast Area Operations Record, Part II: Seventeenth Army Operations) Vol. I, pp. 5–7.

45 Ibid., pp. 12–3. It was also planned to activate the Eighth Fleet to participate in the operations and take over the defense of the islands after their occupation.

46 Statement by Rear Adm. Tomioka, previously cited.

Landing Force.⁴⁷

A source of some concern during the planning of the operations was the fact that the six Army transports assigned to carry the South Seas Detachment were old ships with a maximum speed in convoy of only 6.5 knots, which meant increased vulnerability to enemy air attack.⁴⁸ To minimize this danger, the naval escort, which consisted of only the 6th Destroyer Squadron (six destroyers, one light cruiser) with five minesweepers and one minelayer, was reinforced by the addition of a support force comprising the aircraft carrier *Shoho* and the 6th Cruiser Division (four heavy cruisers, one attached destroyer) under command of Rear Adm. Goto. This released the 5th Carrier Division (*Zuikaku* and *Shokaku*) and the 5th Cruiser Division (three heavy cruisers and seven attached destroyers), under command of Vice Adm. Takeo Takagi, to operate as a striking task force against any enemy naval units which might attempt interference.

As a further move to strengthen air support, seizure of the Deboyne Islands, east of Papua, was scheduled prior to the Port Moresby invasion, with the object of employing them as a seaplane base from which to support the later landing.⁴⁹ The Deboyne operation was assigned to a force commanded by Rear Adm. Kuninori Marumo, consisting of the 18th Cruiser Division (two light cruisers), with 12 seaplanes, two gunboats and two minesweepers.⁵⁰ In addition, two submarines were dispatched to positions in the Coral Sea, and four others were dispersed along the eastern coast of Australia to await the probable emergence of an enemy fleet.

On 25 April the 25th Air Flotilla based at Rabaul⁵¹ began attacks against northeastern Australia. Three days later, part of its strength moved up to the Shortland Islands to expand its radius of action. The Tulagi invasion, scheduled as a prelude to the operation against Port Moresby, was successfully accomplished on 3 May, the *Shoho* support force covering the invasion.

On 4 May, after Vice Adm. Inouye, Commander-in-Chief of the Fourth Fleet, had transferred his headquarters to Rabaul from Truk, the Port Moresby invasion force sailed from Rabaul. The same day, the *Zuikaku* and *Shokaku*, en route from Truk, received reports of an American carrier-plane attack on the Tulagi beachhead and convoy anchorage. They proceeded southward at top speed but were unable to spot the American carriers due to bad weather.

At 0930 on 6 May, however, a navy search plane reported an enemy task force, including a carrier and two other large units believed to be battleships, moving south 450 miles from Tulagi. Later in the day a radio report was intercepted from an Allied patrol plane to the effect that the Deboyne landing force and the Port Moresby invasion convoy had been spotted. The Japanese Task Force and convoy escort were alerted to prepare for action, but no change in the invasion schedule was ordered.

Both the Japanese and American naval groups were now committed to an engagement.

---

47 (1) *Mo Sakusen ni kansuru Riku-Kaigun Kyotei Oboegaki* 茂作戰ニ關スル陸海軍協定覺書 (Memorandum on the Army-Navy Agreement Regarding the "Mo" [Port Moresby] Operation) 25 Apr 42, p. 2. (2) File of South Seas Detachment Operations Orders, op. cit., pp. 12–13.

48 The participating naval landing troops were transported aboard six naval auxiliary vessels.

49 Summary of Coral Sea Battle, op. cit., p. 7.

50 The Deboyne Islands were occupied on 5 May, but seaplanes operated from the islands only until the following day, when the entire invasion force withdrew. Landing Operations in the Bismarck and Solomon Islands, op. cit., pp. 41–2.

51 The 25th Air Flotilla, newly organized on 1 April to include the 4th Air Group, replaced elements of the 24th Air Flotilla in the Rabaul area on 14 April. Summary of the Coral Sea Battle, op. cit., p. 4.

Early on 7 May, a Japanese scout plane reported the American Task Force only 163 miles from the Japanese carrier group. The Battle of the Coral Sea had begun. (Plate No. 32) All the attack planes of the *Zuikaku* and *Shokaku* (18 figthers, 36 bombers, 24 torpedo planes) took off at 0610 for an attack, but at 0640 another scout plane reported sighting the enemy force approximately 150 miles southeast of the Louisiade Archipelago, indicating that the first report had been erroneous (the destroyer *Sims* and tanker *Neosho* had been reported as the " enemy Task Force "). The Japanese carrier planes, which could not be recalled, attacked these ships, sinking the *Sims* and setting fire to the *Neosho*, which pilots reported abandoned by its crew.

Meanwhile at 0550 on 7 May, three B–17s flew over the Port Moresby invasion transports, and at 0700, the Port Moresby invasion transports, with the Deboyne force and part of the escort and support forces, began withdrawing to the northwest. The 6th Cruiser Division and 6th Destroyer Squadron broke off from the convoy to maneuver for a night attack on the enemy fleet in conjunction with the task force closing in from the southeast.

At 0900, 75 planes from the American Task Force struck at the *Shoho* group escorting the Port Moresby invasion convoy. Concentrated torpedo and bombing attacks sank the *Shoho* at 0930. At this time the Japanese Task Force was still about 250 miles distant from the American carrier group.

The *Zuikaku* and *Shokaku*, though unable to launch a further daylight attack on the 7th, sent up 27 torpedo and dive bombers manned by crews skilled in night-fighting to search for the enemy carriers.[52] These planes, however, were suddenly attacked by American fighters emerging from the clouds, and with darkness approaching they abandoned the search. Heading back to the carriers, the dive bombers passed directly over the enemy force but could not attack since they had already jettisoned their bombs.[53]

At dawn on 8 May, a scout plane again located the American force (now reported to include two carriers and one large unit, probably a battleship) on a bearing of 205 degrees 235 miles from the Japanese carriers. The *Zuikaku* and *Shokaku* immediately launched all 69 of their attack planes, which contacted the enemy group at 0920. Despite fierce antiaircraft fire and fighter opposition, the Japanese planes damaged the *Lexington* so severely that it subsequently was abandoned and sunk by American destroyers.[54]

Simultaneously, the Japanese Task Force underwent heavy attack by waves of enemy carrier planes from 0850 to 1020. The *Shokaku*, receiving three direct hits and eight near misses, caught fire and was unable to launch or receive planes, forcing the *Zuikaku* to accommodate all returning Japanese aircraft. By 1300, when the last plane was accommodated, it was found that only 24 fighters, nine bombers and six torpedo planes remained out of the total original complement of 36 fighters, 36 bombers and 24 torpedo planes, including both attack planes and fighters assigned to defense.

Due to these heavy losses and the fact that the Port Moresby invasion convoy was now without carrier protection, Vice Adm. Inouye

---

52 33 bombers and 11 fighters of the 25th Air Flotilla attacked the American group from Rabaul on 7 May, claiming one battleship sunk and another damaged. *Dai Toa Senso Senkun (Koku) Dai Sampen* 大東亞戰爭戰訓(航空) 第三篇 (Battle Lessons of the Great East Asia War—Air, Vol. III) Navy Battle Lessons Analysis Committee (Air Division). pp. 76–83. These claims were subsequently proven false.

53 Ibid., p. 80.

54 At the time the Japanese ascertained only that the *Lexington* had sunk. Details of the sinking were not known to them until after the war.

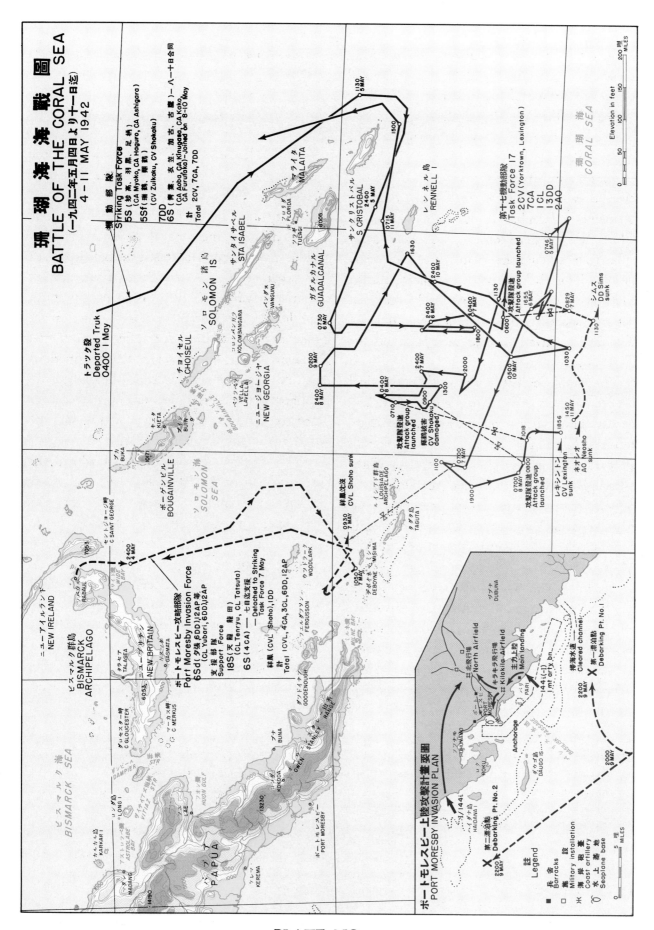

PLATE NO. 32

Battle of the Coral Sea, 4–11 May 1942

ordered the Task Force at about 1500 on 8 May to suspend the attack and head north. All units had turned about, when it was further learned that the Port Moresby landing had been postponed and the invasion convoy ordered back to Rabaul.

Vice Adm. Inouye's passive tactics, however, drew swift disapproval from Admiral Yamamoto, Combined Fleet Commander-in-Chief, who radioed orders "to make every effort to annihilate the remnants of the enemy fleet." The Task Force consequently turned south again and sought to re-establish contact, without success, until sundown on 10 May. The tactical advantage had been lost, and the Task Force withdrew.[55]

The Coral Sea battle, the world's first duel between aircraft carriers, had not resulted in a decisive naval victory for either side. However, Japanese plans for the speedy occupation of Port Moresby suffered a serious setback. Without high-speed transport and the support of powerful carrier forces, a new attempt at seaborne invasion could not be undertaken.[56] The Combined Fleet, already committed to the Midway operation in June, could not spare its carrier forces and advocated postponement of any further attempt until July.[57] Consequently, on 9 May, Imperial General Headquarters issued an army order stating:

*The South Seas Detachment shall come under the command of the Commanding General, Seventeenth Army, which will be organized shortly, and will carry out the invasion of Port Moresby during the first part of July.*[58]

## Plans for a Land Offensive

If the Battle of the Coral Sea upset the Japanese plan to tighten the encirclement of Australia, the disastrous defeat suffered at Midway[59] in the succeeding month of June 1942 dealt Japanese naval power a crippling blow that abruptly redressed the balance at sea in favor of the United States Fleet. Apart from the failure of the invasion attempt itself, the loss of the Combined Fleet's four carrier mainstays—the *Akagi*, *Kaga*, *Soryu* and *Hiryu*—against the sinking of a single American carrier, the *Yorktown*, meant the precipitate collapse of the hitherto superior Japanese carrier position and, consequently, of the combat strength of the Combined Fleet.

This disaster, the full extent of which was not revealed to the Japanese public, had swift repercussions on the southeast area front. On 11 June, four days after the Midway battle ended, Imperial General Headquarters ordered a two months' postponement of the New Caledonia-Fiji-Samoa operation, previously scheduled for early July, and one month later the operations were cancelled completely.[60] At the same time attention forcibly shifted from a direct amphibious assault on Port Moresby, now deemed impracticable, to a possible land drive from the east coast of Papua across the

---

55 General sources covering the Coral Sea battle are: (1) Landing Operations in the Bismarck and Solomon Islands, op. cit., pp. 36–42. (2) Summary of the Coral Sea Battle, op. cit. (3) Battle Lessons—Air, Vol. III, op. cit. (4) Private papers of Capt. Mineo Yamaoka, Senior Staff Officer, 5th Carrier Division.

56 This opinion was expressed in a Fourth Fleet radio dispatch to Combined Fleet headquarters, 9 May 42. (Private papers of Rear Adm. Ugaki, previously cited.)

57 Ibid.

58 Southeast Area Operations Record, Part I, op. cit., p. 26.

59 The Midway invasion force, with Admiral Yamamoto in command, left the Inland Sea on 29 May. The invasion date was set at 7 June. The Midway sea battle, like the Coral Sea battle a clash of air power without direct surface contact, began on 5 June (Japan time), continuing until the 7th.

60 *Daikairei Dai Niju-go* 大海令第二十號 (Imperial General Headquarters Navy Order No. 20) 11 Jul 42. (Text of Imperial General Headquarters Army Order No. 657, 11 Jul 42 was identical in substance.)

Owen Stanley Range.

The severe losses in carriers and aircraft suffered in the Midway battle, indeed, only served to increase the importance placed by the Army and Navy High Commands upon the capture of Port Moresby. More than ever, possession of this base was considered necessary to wrest from the Allies air control over the vital Coral Sea area, and to check the mounting threat of enemy air power not only to the Japanese outposts in eastern New Guinea but to the key stronghold of Rabaul itself.[61]

Simultaneously with the postponement of the New Caledonia-Fiji-Samoa operations, Imperial General Headquarters ordered the Combined Fleet and the Seventeenth Army (activated 18 May) to drop temporarily any plans for a second seaborne assault on Port Moresby and instead to begin formulating plans for a possible land drive. To facilitate this planning, elements of the Seventeenth Army[62] were to occupy a section of the east coast of Papua along the Mambare River as a base for reconnaissance.[63] This was designated as "Research Operation *Ri-Go*."

The Fourth Fleet, convinced that any future land or sea operations in the direction of Port Moresby required the establishment of air bases in eastern Papua, immediately began surveying the area to locate possible sites. On the basis of this survey, it was estimated that the airfield at Buna, about 60 miles south of the Mambare River mouth, could be expanded into a major base. Immediately thereafter, aerial photographic reconnaissance was made of the land route leading from Buna across the Owen Stanley Range to Port Moresby, and the Seventeenth Army, after conferring with the Fourth Fleet, ordered the South Seas Detachment on 1 July to prepare for a reconnaissance operation in the Buna-Kokoda sector to determine its suitablility as a staging area for a major land drive against the Allied base. The order stated:

*The Army will carry out the necessary reconnaissance for a land attack against Port Moresby. The South Seas Detachment commander will land a force in the vicinity of Buna; this force will advance rapidly to the pass over the Owen Stanley Range south of Kokoda and reconnoiter roads for an overland advance on Port Moresby by the main body of the Detachment. A report of this reconnaissance will be made as soon as possible.*[64]

Preparations for the movement of the reconnaissance force to Buna were still in progress when Imperial General Headquarters, going beyond the terms of its initial order, issued

---

61 A primary objective after the deployment of naval air strength to Rabaul had been to gain air supremacy in the Port Moresby area, and the seizure of Lae and Salamaua had been a step toward this objective. From March to July, Japanese naval planes (24th, later 25th Air Flotilla) kept up steady bombing attacks on Port Moresby, reaching a peak of 20 raids during May in which 403 planes were used. Allied losses as a result of these raids were constantly replaced, however, and after July improved anti-aircraft defenses at Port Moresby made the attacks more difficult. Low-altitude bombing became impossible, and the bombing level was raised to about 20,000 feet.

62 Order of battle and disposition of assigned units of the Seventeenth Army as of 1 July were as follows: Commanding General (Lt. Gen. Haruyoshi Hyakutake) and Army Headquarters at Davao (moved to Rabaul 24 July); South Seas Detachment at Rabaul; Kawaguchi Detachment (35th Infantry Brigade), Aoba Detachment (elements of 2d Division, previously in Java), 41st Infantry Regiment (Yazawa Force), and 15th Independent Engineers at Davao, in the southern Philippines. Southeast Area Operations Record, Part II, op. cit., Vol. I, pp. 6–7, 18–20, 41.

63 The Imperial General Headquarters Army Directive of 12 June stated: The Commanding General, Seventeenth Army, in cooperation with the Navy, shall immediately formulate plans for the capture of Port Moresby by employing a land route from the east coast of New Guinea and, to facilitate this planning, will occupy a section along the Mambare River with elements of the Army. Ibid., pp. 17–18.

64 Ibid., pp. 21–3.

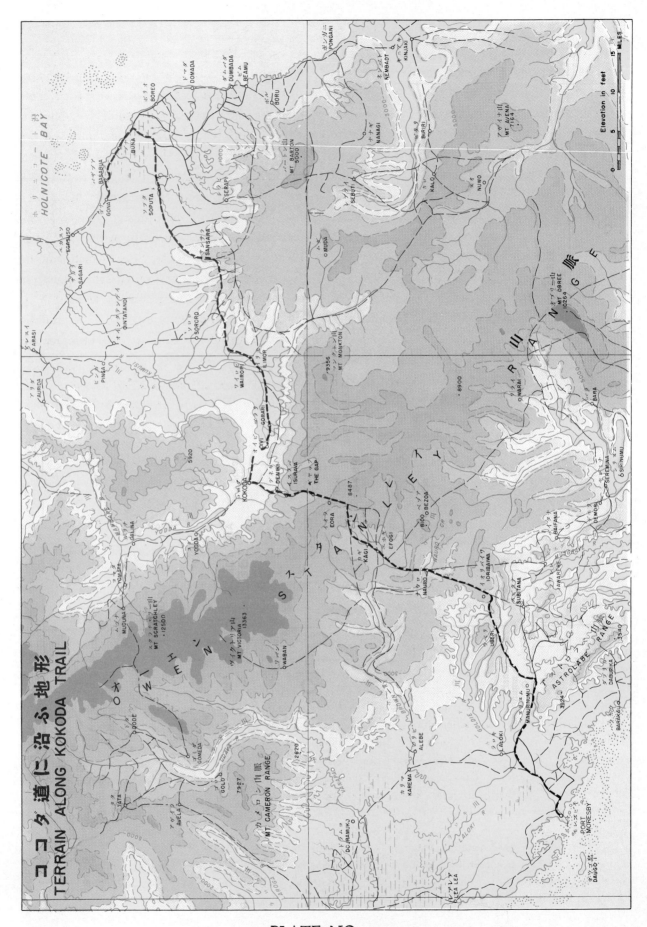

PLATE NO. 33

Terrain Along the Kokoda Trail

a new order on 11 July stating that "the Seventeenth Army, in cooperation with the Navy, shall at the opportune time capture and secure Port Moresby, and mop up eastern New Guinea."[65] This order made it clear that Imperial General Headquarters was no longer thinking in terms of a purely exploratory operation but had virtually decided upon an overland invasion of Port Moresby,[66] for which the Buna landing force was to act as a probing spearhead. Reflecting this step-up in plans, Lt. Gen. Hyakutake issued the following order at Davao on 18 July:

*The Army, in cooperation with the Navy, will promptly effect the capture of Port Moresby and strategic points in eastern New Guinea. The South Seas Detachment will speedily land at Buna, push forward on the Buna—Kokoda road, and capture Port Moresby and adjacent airfields.*[67]

Four days prior to this order, Maj. Gen. Horii, Commanding the South Seas Detachment at Rabaul, had ordered the Yokoyama Advance Force to prepare for the Buna landing and subsequent operations. This force, under command of Col. Yosuke Yokoyama, was made up of the 15th Independent Engineers, who had gained fame in the Malayan jungles, and the 1st battalion of the 144th Infantry, veterans of the Guam and Rabaul campaigns. After landing at Buna, the force's principal missions were to push to the southwestern slopes of the Owen Stanleys, secure a perimeter along this range, reconnoiter and improve roads, and build up supply depots in preparation for a drive on Port Moresby by the main body of the South Seas Detachment, to be landed later.[68]

Although the mission of the Advance Force was no more than exploratory reconnaissance, this remained necessary to later operations since the Japanese possessed virtually no information regarding the Papua interior. Military topographic surveys were non-existent, and hydrographic charts, containing data of little value to land operations, were the only operational maps available. Knowledge of terrain, climatic conditions and prevalent diseases was also lacking. The Yokoyama Advance Force therefore planned to undertake its advance relying largely upon native guides.

Aerial photographs taken by naval reconnaissance planes of the Buna-Kokoda-Moresby route and distributed to the Army units concerned were the most important contribution to pre-operation planning. These photographs, however, revealed only fragments of the jungle-hidden trail, and the information gleaned from them was pitifully inadequate.[69] For example,

---

65   Ibid., p. 24.

66   At this stage the Combined Fleet informed the newly activated Eighth Fleet that Port Moresby, even if taken by a land advance, would be difficult to hold and of dubious strategic value unless heavy equipment, including antiaircraft and naval defense guns, not transportable over the Owen Stanley Range, were moved in by sea. To accomplish this, the Combined Fleet proposed new amphibious operations around the southeastern tip of New Guinea as a step toward setting up a coastal supply route over which high-speed transport vessels might operate. (Statement by Vice Adm. Gunichi Mikawa, Commander-in-Chief, Eighth Fleet.)

67   Southeast Area Operations Record, Part II, op. cit., Vol. I, pp. 29–32.

68   Ibid., pp. 41–2.

69   Later staff analysis of the New Guinea operations freely acknowledged the lack of careful preliminary reconnaissance as a contributory cause of Japanese failure. One of these studies stated: " Before the start of any operation, reconnaissance and investigation must be made in detail. Study of terrain and communication routes from military geographies and aerial photographs is of vital importance. It is preferable to rely upon Army rather than Navy planes for reconnaissance for land operations. The Navy, because of its special characteristics, lacks the proper experience to estimate and reconnoiter routes and terrain." *Lessons from New Guinea Operations*, Japanese document translation published in ATIS Enemy Publication No. 285, 18 Jan 45, p. 3.

the orders to the Yokoyama Advance Force directed that the road north of the Stanley Range be improved for motor, or at the very least, for wagon traffic, and the road to the south of the range for pack-horse and if possible wagon traffic. Actually, the Buna-Moresby "road" was nothing but a native trail which alternately ran through jungle swamps and over precipitous mountains. Throughout the entire campaign the use of vehicular transport was out of the question.[70]

### Landing at Buna

The final operational plans agreed upon between the South Seas Detachment and Fourth Fleet commands at Rabaul called for the execution of the Buna landing with a strength of about 3,600 Army and Navy personnel. The Yokoyama Advance Force, comprising 1,002 men of the 15th Independent Engineers, 855 men of the 1st battalion, 144th Infantry, a mountain artillery battery (200 men) and service units was to embark on two Army transports, while a third transport, the *Kinryu Maru*, was to carry a company of the Sasebo 5th Special Naval Landing Force (about 300 men) and the 15th Naval Construction Unit (about 800 men). The naval landing force was assigned the mission of securing the beachhead and the Buna Village area, and the construction unit was to begin immediate enlargement of the airfield.

The Rabaul-based 25th Air Flotilla, with 60 fighters and 48 bombers under its command, was ordered to provide air cover for the operation, while the 18th Cruiser Division (two light cruisers) under Rear Adm. Koji Matsuyama, with three destroyers and other smaller units, was assigned as naval escort. The landing was scheduled for 21 July.

On 20 July the convoy weighed anchor from Rabaul, undergoing an attack by American B-17 bombers during the run across to New Guinea. The *Kinryu Maru* sustained slight damage from five near misses but was able to continue in convoy to the Buna anchorage, where the invasion force arrived on schedule at 1600 on the 21st. The naval landing force went ashore three miles northwest of Buna at 1730, while the Yokoyama Advance Force began disembarking at Basabua, a short distance farther to the northwest, at 1900. No resistance was encountered, and within 14 hours of the landing, Buna Village and the airfield were securely in Japanese hands.[71]

Simultaneously with the Buna landing, the 82d Naval Garrison Unit holding Lae and Salamaua[72] launched attacks on the Australian strongpoints at Gabmatsung Mission, about 18 miles west of Lae, and Mubo, about 15 miles southwest of Salamaua. The operations had been decided upon to put a stop to increasing guerrilla activities in these areas and were also timed to serve as a feint covering the Buna landing. At Gabmatsung the Australians promptly withdrew south of the Markham River, and the Japanese discontinued the action

---

70 "At the beginning of the present action (New Guinea Operations), both the Army and Navy were defective in the interpretation of aerial photographs and estimated that roads would permit the passage of practically all motor vehicles....We found many errors. Many places in the jungle (steep slopes, swampy ground, and narrow sections of road) were not seen in the photographs. Therefore, in mountain areas, particular care is necessary in photographic interpretation." Ibid., p. 3.

71 (1) Southeast Area Operations Record, Part II, op. cit. Vol. I, p. 41. (2) New Guinea Naval Operations, op. cit., p. 16.

72 The 82d Naval Garrison Unit had approximately 1,300 troops and was under command of Comdr. Kashin Miyata. The attacks of Gabmatsung and Mubo were begun on 21 July by forces of one company each. Major New Guinea Naval Operations, op. cit., pp. 7–8.

on 22 July, returning to Lae. At Mubo, however, the Japanese attack force encountered resistance by about 200 Australian troops, who inflicted some losses before retreating to the south. The Japanese returned to Salamaua on 23 July, ending the operation.[73]

Meanwhile, at the Buna beachhead, the Japanese landing forces underwent heavy air attack on 22 July by approximately 100 Allied planes, including B-17's, B-26's and P-39's. These attacks continued daily thereafter, inflicting damage to one transport and a destroyer of the invasion convoy. Despite these attacks, additional troops were successfully put ashore from a destroyer on 26 July, and from two transports, a light cruiser and a destroyer on 29 July, but the unloading of war materials ended in failure.[74]

Due to the steady intensification of Allied air attacks, however, the transport of reinforcements to Buna became rapidly more perilous. On 30 July the transport *Kotoku Maru* had to be abandoned after receiving hits in a strike by eight Flying Fortresses. On 31 July, another transport en route to Buna under naval escort was forced to turn back to Rabaul due to air attack.[75]

## Advance to Kokoda

Initial reports to Seventeenth Army headquarters by the Yokoyama Advance Force were optimistic. Immediately after the landing, a spearhead patrol of company strength set out along the trail to Kokoda, meeting only sporadic resistance from a small Australian force about 100 strong, which retreated before them. After dispersing these remnants, the patrol advanced as far as Oivi Hill, about nine miles east of Kokoda, where it was shortly joined by the main body of the Yokoyama Advance Force. Launching an attack on the night of 28 July, the Force routed the Australian 39th Battalion and moved into Kokoda at dawn on 29 July, occupying the nearby airfield at noon the same day.[76]

Although the capture of Kokoda was effected earlier than anticipated, the advance had not been without hardship. The Yokoyama Advance Force, in its rapid drive along the arduous jungle trail, had shed all excess equipment and rations, and when it reached Kokoda, the problem of bringing up food and ammunition assumed prime importance. To the rear, the engineers doggedly worked to widen and improve the trail and could not be spared to move supplies. This vital task therefore had to be undertaken by elements of the naval construction unit, which was engaged in improving the Buna airfield.

## Final Plans Against Moresby

On the basis of the early reports sent in by the Yokoyama Advance Force, Lt. Gen. Hyakutake, whose headquarters had now moved to Rabaul, hastily advised Imperial General Headquarters that an overland attack on Port Moresby was feasible and recommended adop-

---

73 Native reports at this time claimed that the Australians, in anticipation of further attacks, had moved farther inland, setting fire to installations at their Wau, Bulolo and Bulwa air bases. Ibid., pp. 7–8.

74 In connection with the Buna landing, the Fourth Fleet dispatched two submarines to points off Townsville and Port Moresby to hinder the transport of troop reinforcements from Australia.

75 Major New Guinea Naval Operations, op. cit., p. 25.

76 A radio dispatch from Col. Yokoyama to Maj. Gen. Horii, South Seas Detachment commander, reported that the Yokoyama Advance Force "reached the vicinity east of Kokoda on the morning of 28 July. Defeated an enemy force of about 1,200 men and attacked Kokoda the same night. Kokoda occupied at dawn 29 July." Southeast Area Operations Record, Part II, op. cit. Vol. I, pp. 41–2.

オーエンスタンレー山脈に於ける軍隊の苦難　伊原宇三郎　昭和二十三年三月十七日

Original Painting by Usaburo Ihara

PLATE NO. 34
Hardships of the Troops in the Owen Stanleys

tion of definite plans to move the main forces of the Seventeenth Army across the Owen Stanleys to effect the capture of this important base.[77] The Army and Navy Sections of Imperial General Headquarters thereupon drew up the main lines of an operational plan, issued in the form of orders to the Combined Fleet and the Seventeenth Army on 28 July. The essential portion of these orders read:

4. Outline of Operations

   a. ....the main force of the Army will capture the Buna-Kokoda road area extending to Port Moresby and the airfields adjacent to Port Moresby as rapidly as possible.

   b. ....if necessary, to facilitate the operation, elements of the Army will land in the vicinity of Port Moresby from the sea at an appropriate time.

   c. The Navy will destroy enemy air power in the Port Moresby area and sweep enemy vessels from the northern Coral Sea to protect the amphibious forces. It will cooperate closely with all land operations.

   d. Concurrently with the attack upon Port Moresby and subsequent to its capture, other strategic points in eastern New Guinea will be occupied. The capture of islands and strategic points along the north shore will be effected by Special Naval Landing Forces, and the Army will be responsible for other areas.[78]

Upon receipt of these orders, the Eighth Fleet[79] and Seventeenth Army immediately began working out the details of the final plan reaching agreement on 31 July. The essential points of the plan were as follows:

1. The main body of the South Seas Detachment, reinforced by the 41st Infantry Regiment, to land at Buna on 7 August, move up along the Buna-Kokoda road and join the Yokoyama Advance Force for further operations toward Port Moresby.

2. One battalion of the 35th Infantry Brigade, reinforced by a special naval landing force, to move toward Port Moresby by sea, using as transport seven patrol boats and a few destroyers, and effect a landing to the east of Port Moresby in coordination with the final phase of the South Seas Detachment land drive[80]

3. Naval forces to occupy Samarai, at the southeastern tip of New Guinea, as speedily as possible and establish a seaplane base.

4. The 82d Naval Garrison Unit in the Lae-Salamaua area to launch a feint attack toward Wau as cover for the Port Moresby operation.

5. The Navy to assure the protection of troop convoys to Buna, support the amphibious operations around the southeast tip of New Guinea, and provide necessary air cover by naval air units.[81]

These plans were ready to be put into execution when last-minute delay in the completion of the Buna airfield, followed by the sudden landing of American marines on Guadalcanal on 7 August, forced a postponement of the scheduled date for the debarkation of the South Seas Detachment main body at Buna until 16 August. It was estimated that, by that date, preparations for the recapture of Guadalcanal would be complete, and sufficient air strength would be available to cover the Buna landing operations.

---

77 Southeast Area Operations Record, Part III, op. cit., Vol. I. p. 15.

78 *Daikaishi Dai Hyakujugo-go* 大海指第百十五號 (Imperial General Headquarters Navy Directive No. 115) 28 Jul. 42. (Text of Imperial General Headquartes Army Directive No. 1218, 28 Jul 42 was identical in substance.)

79 The Eighth Fleet, according to original plan, was to be activated for participation in the New Caledonia-Fiji-Samoa operations. When these operations were cancelled following the Midway Battle, the activation was delayed until 14 July, when it was carried out with the object of replacing the Fourth Fleet as the Navy's operating force in the Southeast area. (The Fourth Fleet was then assigned only to defense of the mandated islands and the Gilbert and Wake Islands areas.) The Eighth Fleet formally took over on 27 July, when its headquarters reached Rabaul.

80 This amphibious force was to time its departure to follow the break-through of the South Seas Detachment to the southern side of the Owen Stanley Range. Southeast Area Operations Record, Part II, op. cit. Vol. I, p, 39. It was estimated that this would occur by the end of August.

81 Ibid., pp. 34-9.

Up to 7 August, when the American invasion of the Solomons began, the total number of Japanese troops and naval personnel put ashore at Buna for the Port Moresby operation approximated 7,430.[82] Of these, 430 were naval landing troops, and 2,000 naval construction personnel. The remaining 5,000 represented the original Yokoyama Advance Force plus reinforcements and replacements sent in subsequent to the 21 July landing.

### Fighting on Guadalcanal

A radio dispatched to Eighth Fleet headquarters in Rabaul at 0530 on 7 August reported both Guadalcanal and Tulagi[83] under heavy enemy naval and air bombardment. From the strength and make-up of the enemy naval force—two aircraft carriers, one battleship, three cruisers, 15 destroyers and 30 to 40 transports—it was evident that landings were contemplated.

The Japanese forces on Guadalcanal at that time numbered only 250 naval garrison troops and two construction units of about 1,600, stationed near Lunga Point.[84] Before communications ceased, the Eighth Fleet received a report that they were retreating into the interior after engaging the enemy landing forces. Meanwhile, on Tulagi, the Japanese naval garrison of approximately company strength was believed annihilated.

Despite the success of the American landings, Imperial General Headquarters in Tokyo took the optimistic view that the operation was nothing more than a reconnaissance in force, and that, even if it were the beginning of a real offensive effort, Japanese recapture of Guadalcanal would not be excessively difficult.[85] Reports were lacking from the Japanese forces on the spot, and the situation was vague.[86]

On the other hand, Admiral Yamamoto, Commander-in-Chief of the Combined Fleet, regarded the American counterthrust more seriously and promptly appointed Vice Adm. Nishizo Tsukahara, Commander of the Eleventh Air Fleet, as Commander of the Southeast Area Force, a new intermediate fleet command.[87] First priority was given to the recapture of Guadalcanal, and all available ships and planes were assembled for an immediate and decisive counterattack.

While the 25th Air Flotilla threw all its operational strength[88] into a series of damaging air assaults, the Eighth Fleet's most powerful combat ships, under personal command of Vice Adm. Gunichi Mikawa, Eighth Fleet Commander-in-Chief, sailed from Rabaul at 1430 on 7 August to attack the enemy vessels off Guadalcanal. In this attack (Plate

---

82  On 6 August three transports sailed from Rabaul under naval escort, carrying reinforcements for Buna. With the Guadalcanal attack, however, their fighter cover was diverted to the Solomons, and the convoy returned to Rabaul. Japanese losses thus far in transporting troops to eastern New Guinea were one transport lost and one transport and three escort vessels damaged. New Guinea Naval Operations, op. cit., pp. 16–18.

83  Following the occupation of Tulagi in May 1942, it was found that a portion of Guadalcanal Island was equally suitable for the construction of an air base, and the Fourth Fleet dispatched two naval construction units on 1 July to undertake this project. By 3 August, one airstrip and a dummy field had been roughly completed. (Statement by Rear Adm. Yano, previously cited.)

84  Extracted from the private papers of Rear Adm. Masao Kanazawa, Commander, 8th Base Force.

85  Southeast Area Naval Operations, op. cit. Part I, p. 5.

86  Statement by Lt. Gen. Seizo Arisue, Chief, 2d Bureau (Intelligence), Imperial General Headquarters, Army Section.

87  Southeast Area Naval Operations, op. cit. Part I, p. 6.

88  Operational strength at this time was: 39 fighters, 32 land-based attack planes, 16 bombers.

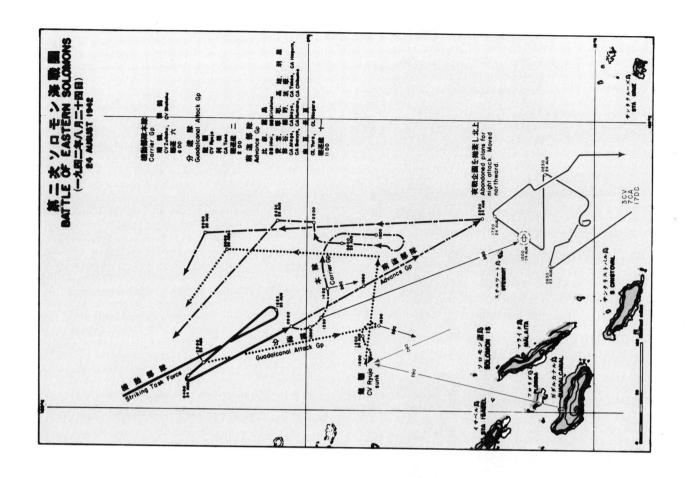

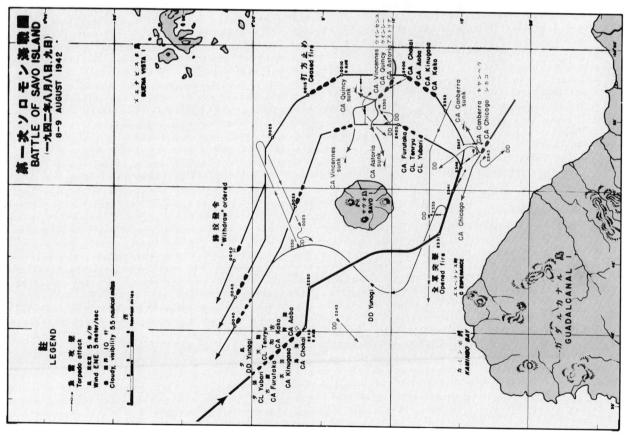

PLATE NO. 35

First and Second Battles of Solomon Sea, 8–9 & 24 August 1942

No. 35), carried out on the night of 8 August,[89] heavy losses were inflicted on the Allied convoy force, and although the marine beachhead remained intact, active reinforcement efforts did not immediately develop.

The absence of further American attempts to send in troops was interpreted by Imperial General Headquarters as confirming the first estimate that a major offensive was not developing,[90] and that recapture of Guadalcanal could be speedily achieved. Accordingly, on 13 August, a new operational directive was issued for the southeast area, stipulating that elements of the Seventeenth Army should be dispatched immediately to recapture Guadalcanal and Tulagi, and that "the invasion of Port Moresby shall be speedily carried out in accordance with previous plans."[91]

In order to seize the tactical opportunity before the enemy foothold on Guadalcanal could be consolidated, the Army General Staff advised employment of small forces which could be swiftly moved to the scene of action by destroyers, rather than an attempt to transport the Seventeenth Army's larger uncommitted units—the 35th Inf. Brigade at Palau, and Aoba Detachment at Davao.[92] The Ichiki Detachment[93] of approximately regimental strength, which had been placed under Seventeenth Army command on 10 August for use on Guadalcanal, was already at Truk. Hence it was decided to use this detachment plus a naval landing force in the initial recapture attempt.

Embarking from Truk on six destroyers, the main strength of the Ichiki Detachment landed on 18 August at Taivu Point, while a force of naval troops landed simultaneously at Lunga Point. The Ichiki Detachment launched a vigorous attack on the airfield area but were driven back in retreat west of the Tenaru River, Col. Ichiki himself having been killed. Not only had the attempt failed, but on 21 August it was confirmed that American planes had begun to operate from Henderson Field, while increasing numbers of troop reinforcements were landing from transports. Abruptly the situation darkened.

On 19 August, the day following the landing of the Ichiki Detachment main strength, the Seventeenth Army had issued new orders for the immediate advance to Guadalcanal of the remainder of the detachment plus the 35th Inf. Brigade,[94] whose Commanding General (Maj.

---

89  This naval engagement was known to the Japanese as the "First Battle of the Solomons." The American designation is the "Battle of Savo Island."

90  About this time an intelligence report from Moscow to the effect that, owing to heavy losses, the Americans were contemplating withdrawal from Guadalcanal, reached the Navy General Staff. This report was relayed to the Southeast Area Force in a radio sent 16 August by Rear Adm. Shigeru Fukudome, Chief, First Bureau (Operations), Imperial General Headquarters, Navy Section.

91  *Daikaishi Dai Hyakuniju-go* 大海指第百二十號 (Imperial General Headquarters Navy Directive No. 120) 13 Aug. 42. (Text of Imperial General Headquarters Army Directive No. 1235, 28 Jul 42 was identical in substance).

92  A radio dispatch from Lt. Gen. Moritake Tanabe, Deputy Chief of the Army General Staff, to the Commanding General, Seventeenth Army, on 12 August, stated: "The scope of operations for the recapture of strategic points in the Solomon Islands will be decided by the Army Commander on the basis of his estimate of the enemy situation. The Army General Staff believes that it is feasible to use the 35th Infantry Brigade and Aoba Detachment if the situation demands. However, since tactical opportunity is a primary consideration under existing conditions, it is considered preferable, if possible, to recapture these areas promptly, using only the Ichiki Detachment and Special Naval Landing Forces." Southeast Area Operations Record, Part II, op. cit. Vol. I, p. 52.

93  This detachment was originally assigned to the Midway invasion, after the failure of which it was held at Guam until its assignment to Seventeenth Army on 10 August. It then moved to Truk.

94  Under earlier plans for the Port Moresby campaign, a battalion of the 35th Inf. Brigade was to make an amphibious landing east of Moresby. Cf. section on Final Plans Against Moresby.

Gen. Seiken Kawaguchi) was assigned to command all Army and Navy forces in the Guadalcanal-Tulagi area.[95] As the transport groups carrying these forces moved south from Truk, powerful naval screening forces, including three carriers (*Zuikaku, Shokaku, Ryujo*), swept around the eastern side of the Solomons to divert and crush enemy naval and air forces. In the ensuing Second Battle of the Solomons,[96] fought on 24 August, damage was sustained on both sides, the *Ryujo* going down under heavy attack by carrier and land-based aircraft. The transport convoy, which also underwent attack by enemy land-based bombers, was forced to turn back, and the reinforcement attempt failed.

In view of these developments, Imperial General Headquarters began to show increased concern over the Guadalcanal situation and, on 31 August, issued orders giving first priority to the recapture of the Solomons. In accordance therewith, efforts were pushed by the Seventeenth Army and Southeast Area Force Commands to get in reinforcements in preparation for a general offensive. Little by little, between 30 August and 7 September, night runs by destroyers and landing barges succeeded in putting ashore the remainder of the Ichiki Detachment and the entire 35th Brigade. In all, a total of about 5,200 troops were transported by this means subsequent to the enemy landing of 7 August.[97]

Despite the limited strength of these forces, plans were laid for the start of a general offensive on 11 September. The 35th Brigade was to launch a surprise attack on the American perimeter guarding Henderson Field, while a powerful naval force made up of the Eighth Fleet supported by elements of the Second and Third Fleets was to move directly up to the Lunga anchorage to cut off both possible reinforcements and enemy retreat.[98]

Owing to delay in bringing up artillery support and maneuvering all forces into position through the jungle, the general offensive did not begin until 2000 on 13 September. As the 35th Brigade troops pressed forward, they met increasingly severe resistance by the entrenched marines, and heavy losses finally forced them into retreat. Following this debacle, Maj. Gen. Kawaguchi decided to reassemble all the Japanese forces on the west bank of the Matanikau River. (Plate No. 37)

Despite two failures, the High Command still remained determined to recapture Guadalcanal at any cost. Preparations were consequently begun for the dispatch of still further reinforcements, upon the arrival of which another general offensive was to be attempted.[99]

### Build-up of Forces in New Guinea

In spite of the American invasion of the Solomons, the over-all strategic plan decided upon by Imperial General Headquarters on 28 July for the capture of Port Moresby remained unchanged during August. The Seventeenth Army, although obligated to furnish

---

95 Southeast Area Operations Record, Part II, op. cit. Vol. I, p. 64.

96 Referred to in American accounts as the Battle of the Eastern Solomons.

97 *Nanto Homen Kaigun Sakusen* 南東方面海軍作戰 (Southeast Area Naval Operations) 2d Demobilization Bureau, Oct 46. Vol. I, p. 16.

98 (1) Southeast Area Operations Record, Part II, op. cit. Vol. I, pp. 62–3, 96–8. (2) Southeast Area Naval Operations, op. cit. Vol. I, pp. 17–19, 21–2.

99 The new plan called for the commitment of the Seventeenth Army's Aoba Detachment, previously scheduled for use in the New Guinea operations, together with the main body of the 2d Division, transferred from Java. Every effort was to be made to land heavy artillery for the support of ground operations, but if sufficient heavy weapons could not be transported, naval units were to lay down a bombardment of enemy positions to pave the way for the general attack. (1) Southeast Area Operations Record, Part II, op. cit. Vol. I, pp. 96–8. (2) Southeast Area Naval Operations, op. cit. Vol. I, pp. 21–2.

ガダルカナル島に於ける陸海軍の協同　中村研一　昭和一九・一・一

Original Painting by Kenichi Nakamura

PLATE NO. 36
Army-Navy Cooperation on Guadalcanal

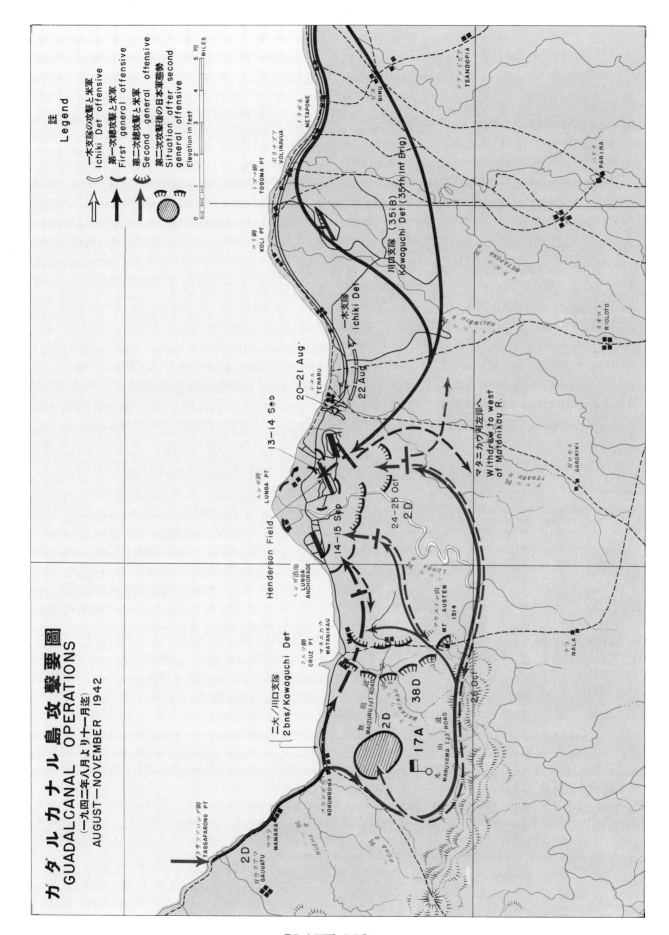

PLATE NO. 37

Operations on Guadalcanal, August—November 1942

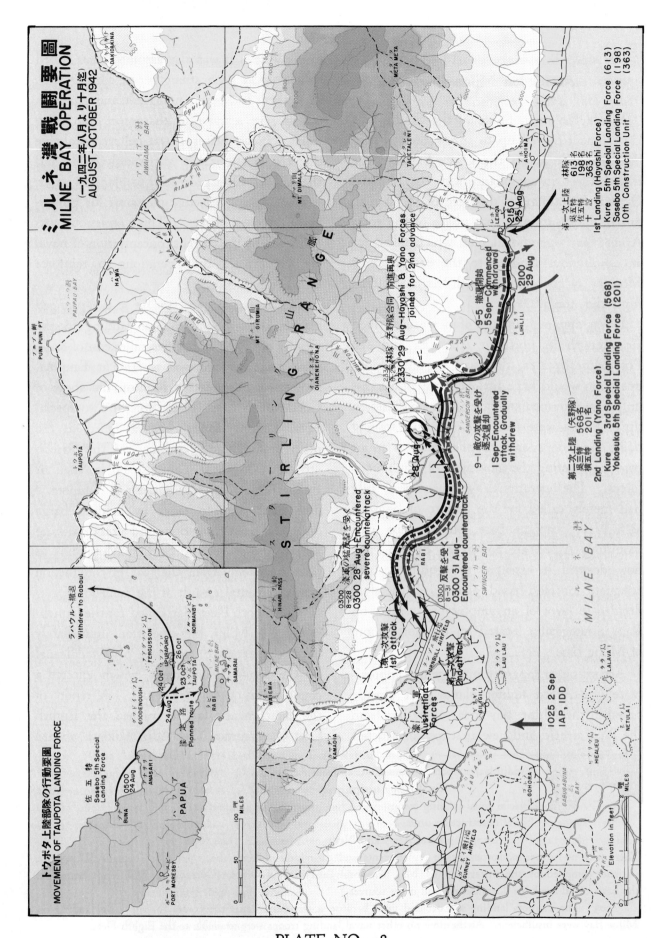

PLATE NO. 38

Landings on Milne Bay, August—October 1942

his flagship during the subsequent operations to move in the 35th Brigade. However, still confident that the recapture of Guadalcanal would be effected at an early date, Vice Adm. Mikawa dispatched orders to the 18th Cruiser Division, held at Rabaul in readiness for the Milne Bay operation, to begin execution of the attack plan without awaiting the outcome on Guadalcanal.

The forces assigned to the main landing sailed from Rabaul on 24 August, undergoing a light attack by about ten Allied planes as they neared Milne Bay on the afternoon of the 25th. The convoy, which entered the Bay late on the 25th, included two transports carrying 811 naval landing troops[104] and 363 personnel of the 10th Naval Construction Unit, all under command of Comdr. Shojiro Hayashi. Two light cruisers, five destroyers and two submarine-chasers composed the naval escort.

The Lehoa landing was carried out with reasonable ease at 2150 on 25 August. The next morning, however, Allied planes suddenly attacked the beachhead, destroying a large part of the food and ammunition supplies which had been unloaded. As the landing force began moving forward toward the Rabi airfield, its main objective, enemy air strikes increased in intensity, making daylight movement impossible. Advancing by night over unknown terrain, the troops floundered through jungle swamps to reach the eastern edge of the airfield on the night of 27 August. Here they met such unexpectedly savage resistance by the Allied troops defending the airstrip that, at dawn on the 28th, it was decided to retire into the jungle and await reinforcements.

At 2100 on the 29th, 769 additional naval landing troops under[105] Comdr. Minoru Yano landed slightly to the west of Lehoa and joined the initial force in a second advance on Rabi, which began at 2330 the same night. A diary account portrayed the optimistic mood in which the Japanese moved up for the attack:

*29 August: Waited in the jungle at Milne Bay. The concerted attack has been ordered....All of us are in good spirits....Nothing but serving the Emperor....We make our sortie, all hopeful of success.*[106]

On the following day, however, constant air attacks again pinned down the Japanese advance in the jungle, but with nightfall the attacking force succeeded in advancing to the eastern perimeter of the airfield, where it again met withering fire from the Australian defenders. Beginning at daybreak on 31 August, Allied tactical aircraft joined in the battle, and Japanese Navy fighters, hurriedly dispatched to the scene, were unable to gain air control.[107]

Eighth Fleet headquarters, recognizing the need of sending further reinforcements, alerted a newly-arrived naval landing force at Rabaul, but on 2 September, just as preparations began for their embarkation, Allied reinforcements were put ashore at Giligili, threatening the Japanese force from the rear. A radio dispatched to Eighth Fleet headquar-

---

104 These comprised 613 men of the Kure 5th Special Naval Landing Force and 198 men of the Sasebo 5th Special Naval Landing Force.

105 These comprised 568 men of the Kure 3d Special Naval Landing Force and 201 men of the Yokosuka 5th Special Naval Landing Force.

106 Extract from diary of a member of the Milne Bay landing force. (ATIS Current Translations No. 4, 25 Nov 42, p. 35.)

107 A company commander of the Kure 3d Special Naval Landing Force recorded in his diary: "We have to take constant cover in the jungle. We cannot send up any smoke at all, for if detected by the enemy, we can't escape bombing and machine-gun strafing.... We are soaking wet from head to foot and so uncomfortably cold that we are at our wits' end." (Ibid., p. 11.)

ters reported: "Situation most critical. We shall defend key position to the last man."[108]

It was now evident that the piecemeal commitment of small naval forces would not retrieve the situation, and that the circumstances called for large-scale army reinforcements. Accordingly, agreement was reached between the Eighth Fleet and Seventeenth Army to dispatch the Aoba Detachment, approximately 1,000 strong, to relieve the hard-pressed naval force at Milne Bay. However, the situation at Guadalcanal made it impossible for this plan to be put in to effect despite the above local Army-Navy agreement. Consequently the naval forces were ordered to avoid major action and resort to delaying action.

Relentless Australian counterattacks, however, soon produced a situation in which it was clear that the battle was irretrievably lost. On 4 September, the Eighth Fleet issued orders for the withdrawal of all forces from Milne Bay, and the Seventeenth Army subsequently relieved the Aoba Detachment of its reinforcement mission, assigning it instead to Guadalcanal.[109]

Evacuation of the naval landing forces by one light cruiser and three patrol boats began on 5 September and was carried out with reasonably satisfactory results. Of the 1,943 troops landed on 25 and 29 August, 1,318 were eventually withdrawn.[110] All of the survivors, however, were incapable of further combat due to sickness, wounds or battle fatigue, a testimonial to the bitter hardships they had been through.[111]

A contributory cause of the failure at Milne Bay was the fact that the planned rear attack from Taupota never materialized. The naval landing force assigned to this operation had sailed from Buna on 24 August aboard seven landing barges, but while temporarily anchored off the shore of Goodenough Island, the group underwent a concentrated Allied air attack which destroyed all seven barges, together with their radio equipment.[112] Suffering from near-starvation and malaria, the survivors remained stranded on the island for two months, twice making desperate efforts to con-

---

108   Southeast Area Naval Operations, op. cit. Vol. I, p. 15.
109   Southeast Area Operations Record, Part II, op. cit. Vol. I, p. 88.
110   Major New Guinea Naval Operations, op. cit., p. 15.
111   Some members of rear-guard units and isolated groups could not be evacuated and were left in the area to find their way through the jungle back to Buna. The horrible suffering experienced by these men was told in a diary picked up by the Allied forces, which read: "30 Aug. Beginning of the retreat....into the mountains with a grenade splinter through my right hand. ....rotting of the feet makes it difficult to walk....sleeping in the mountains with the rain falling almost incessantly. It is harder to bear than death. 15 Sep. Our troops have not arrived for 14 days. I have been waiting patiently, but I am beginning to lose consciousness....potatoes....potatoes....my wife....my mother. 22 Sep. Engaged a large enemy force....lost all our weapons and have only the clothes we wear. Nothing to eat. ....25 Sep. I have fever and am nearly unconscious but holding on. 26 Sep. Our forces haven't arrived yet....no use waiting....I'm mad. ....28 Sep. I detest rain...." (This is the last entry.) Extract from diary of 3d Class Petty Officer Morita, Yokosuka 5th Special Landing Force. (ATIS Current Translations No. 2, 11 Nov 42, p. 17.)

112   An unidentified member of the Sasebo 5th Special Naval Landing Force recorded the details of the attack in his diary, as follows: "At 1130 we anchored at the mouth of a certain river. A single enemy plane flew over. We were making up enough food to last for three meals and were scheduled to sail at 1530, when at about 1230 ten enemy fighters came over and attacked both the landing craft and our troops with machine-gun fire and bombs. The seven landing craft caught fire simultaneously. We fought back, but what could we do against fighter planes? Eight dead, six badly wounded, 30-40 slightly wounded. With all the landing craft destroyed by fire, our future movement is a problem. We should retire immediately, but without radio we have no means of communication, so the only thing we can do is wait for assistance." (ATIS Current Translations No. 14, 18 Jan. 43, p 4.)

tact the Japanese forces at Buna by sending messengers across to the New Guinea coast by canoe.

When approximately 300 Allied troops began landing on the eastern and southern shores of Goodenough Island on 23 October, the plight of the Japanese unit became still more serious, although it retained sufficient strength to repulse the enemy landing force in the southern shore sector.[113] Finally, using two landing barges which had been sent in response to its appeals to Buna, the unit withdrew to Upurapuro, on adjacent Fergusson Island, where it was picked up on 26 October by the cruiser *Tenryu* and evacuated to Rabaul.[114]

With the collapse of the Milne Bay invasion attempt and the steady deterioration of the situation on Guadalcanal, Japanese hopes on the southeast area front now centered on the advance of the South Seas Detachment, which had forged its way over the Owen Stanley Range almost within striking distance of Port Moresby.

## Owen Stanleys Offensive

After capturing Kokoda on 29 July, the Yokoyama Advance Force pressed forward into the Owen Stanley range on 7 August, meeting almost continuous resistance by a force of about 200 Australian troops. (Plate No. 39) From about 20 August there was a marked intensification of Allied air reconnaissance, followed by severe bombing and strafing attacks against the advancing Japanese column. On 26 August the Yokoyama Advance Force ran into stiff opposition by an enemy force of estimated battalion strength on the heights near Isurava, and as soon as the first elements of the South Seas Detachment, newly-arrived from the Buna beaches, were able to move up to the front line, the battle was joined.

Due to stubborn enemy resistance and difficulties encountered in moving up supplies over the steeply mountainous and jungle-covered terrain, the initial attacks failed to dislodge the Australians from their strong position, and the advance was stalled until the arrival on 27 August of the 2d Battalion of the 41st Infantry, led by the regimental commander, Col. Kiyomi Yazawa. On the 29th the 144th Infantry threw its full strength into a combined frontal assault and enveloping attack around the enemy's right flank, succeeding after eight hours of bitter fighting in overrunning the Australian outer perimeter and part of the main enemy positions, but only at the cost of heavy casualties. On 30 August the 2d Battalion of the 41st Infantry successfully enveloped the enemy's left flank by a difficult advance over the mountains, and by the 31st the Australians were encircled and defeated with heavy losses. On 1 September the South Seas Detachment forces entered Isurava.[115]

Despite its own heavy casualties, the South Seas Detachment was heartened by its success and pushed on beyond Isurava, its progress becoming ever more difficult as it penetrated deeper into the Stanley Range. The hardships

---

113 Major New Guinea Naval Operations, op. cit., p. 16.

114 When the unit was ready to board the rescue ship, its commander, Comdr. Tsukioka, addressed the men as follows: "We are all thin with lack of food, but when we board the ship, do not show a haggard countenance. There is a saying that the Samurai displays a toothpick even when he has not eaten. This is an example worth emulating at the present time." (ATIS Current Translations No. 14, op. cit., p. 6.)

115 Details of the Isurava action are as related to the writer by Maj. Mitsuo Koiwai, commander, 2d Battalion, 41st Infantry Regiment, one of the few surviving officers of the South Seas Detachment. Maj. Koiwai stated that two companies of the 144th Infantry's 2d Battalion, which carried out the flank assault on 29 August, lost the majority of their officers, both commissioned and non-commissioned.

PLATE NO. 39

Owen Stanley Penetration, 21 July—26 September 1942

of the advance were graphically recorded in this passage from a soldier's diary:

*The road gets gradually steeper. Bushes cover the countryside. Cicadas and birds are singing. We are in a jungle area. The sun is fierce here. One party of troops crawled up and scaled the mountains and continued its advance. Troops are covered with dirt and sweat so much that it is difficult to tell one man from another. We make our way through a jungle where there are no roads. The jungle is beyond description. Thirsty for water, stomach empty. The pack on the back is heavy. My arm is numb like a stick. My neck and back hurt when I wipe them with a cloth. No matter how much I wipe, the sweat still pours out and falls down like crystals. Even when all the water in your body has evaporated, the sun of the southern country has no mercy on you. The soldiers grit their teeth and continue advancing, quiet as mummies. No one says anything unnecessary. They do not even think but just keep on advancing toward the—front....*

*"Water, water!" all the soldiers are muttering to themselves. Those who believe in miracles are whispering, "I want water, I want water." We reach for the canteens at our hips from force of habit, but they do not contain a drop of water. Yet the men still believe in miracles. The fierce sun makes them sleepy. The weeds and trees are snatching a peaceful sleep under the burning sun....The sound of the enemy planes and our marching seem to lull us to sleep. The men sleep while they walk and sometimes bump into trees. Enemy planes fly over the jungle and repeatedly attack.*[116]

After suffering from the intense heat of the lower altitudes, the troops, as they climbed toward the summit of the Stanley Range, now began to suffer from the frigid nights and icy rains, against which their tropical battle-dress gave little protection. Still more serious was the appearance of shortages of rations and ammunition, partially resulting from a tendency among the foot-weary troops to lighten their packs during the gruelling advance. So alarming was the situation that Maj. Gen. Horii, in a special order issued 1 September, enjoined strict measures of economy:

*Although the loss of time caused by the difficulties of the range and by enemy action had been foreseen, we are concerned at the small quantity remaining of the ammunition and provisions originally carried. Although economy precautions were previously ordered, these directions have regrettably not yet driven home, perhaps owing to continued action and prolonged marches. All unit commanders and officers, of whatever rank, must exercise the most painstaking control and supervision, so that every bullet fells on enemy and every grain of rice furthers the task of the Detachment. They must also see that full use is made of captured ammunition and provisions.*[117]

Indeed, the problem of supply was the most critical one facing the Japanese forces pushing toward Port Moresby. It was now recognized that roads capable of accommodating vehicular traffic were non-existent. At best, in the jungle, troops could follow the native trails, tangled with undergrowth, but in the mountains these trails narrowed down to nothing more than forgotten tracks clinging precariously to the sides of cliffs, or vanishing perpendicularly into steep canyons.

In the course of its advance, the South Seas Detachment carved 20,000 steps in the mountainsides to facilitate its march, yet at the end of the operation it was still impossible to use even pack horses south of Isurava. Rains held up all transport for days at a time. Moreover, in good weather, Allied air attacks soon reached a degree of intensity which made it

---

116 Extract from diary of an unidentified soldier, belonging to a unit of the South Seas Detachment. (*ATIS Current Translations No. 64, 13 Jul 43, pp. 7–8.*)

117 Directions Regarding Economy in the Use of Ammunition and Provisions, South Seas Detachment Hq., 1 Sep 42. (*File of Nankai Shitai Operations Orders, 16 Aug–15 Oct 42, ATIS Enemy Publications No. 33, 12 Aug 43, p. 10.*)

impossible to move men or supplies during daylight hours.[118]

In addition to the obvious fact that the South Seas Detachment was outstripping its supply lines, Seventeenth Army headquarters in late August recognized the unwisdom of attempting a headlong advance on Port Moresby without regard for the progress of the parallel operation at Milne Bay, which was to pave the way for a coordinated amphibious assault. The situation at Milne Bay was critical, and plans were being formulated to send in the Aoba Detachment to turn the tide of battle. Dependent upon the success of this operation, it was planned to move part of the main strength of the 2d Division, scheduled to be transferred from Java, to the area for the projected amphibious assault on Moresby.[119]

Pending the execution of these plans, the Seventeenth Army decided to slow the advance of the South Seas Detachment so as to conserve and build up its combat strength for the final push from the Owen Stanleys. Accordingly, on 28 August, the following order was received by Maj. Gen. Horii from Seventeenth Army headquarters at Rabaul:

*Should the South Seas Detachment succeed in destroying the enemy in the Owen Stanley Range and reach the strategic area on the south side of the range, elements of the Detachment will secure this line, while the main body of the Detachment will concentrate north of the range to prepare for subsequent operations. The advance beyond this line will be by separate order.*[120]

Meanwhile, in Tokyo, the thinking of Imperial General Headquarters had also undergone a marked change since its 13 August directive calling for swift execution of the Port Moresby campaign "in accordance with previous plans". The failure of the initial attempts to recapture Guadalcanal, which indicated that the American foothold was considerably stronger than at first estimated, led to a shift of emphasis to the Solomons, where Imperial General Headquarters now foresaw the probable necessity of committing a large portion of Seventeenth Army strength previously intended for New Guinea.

The Army Section of Imperial General Headquarters therefore reached the decision that the strength of the Seventeenth Army should not be further expended in New Guinea until the recapture of the Solomons was assured. In accordance with this decision, a radio to the Seventeenth Army on 29 August explicitly directed that the South Seas Detachment halt its advance at the southern edge of the Owen Stanleys,[121] and on 31 August a formal Imperial General Headquarters directive stipulated that major operations in New Guinea

---

118 The importance of the supply problem in causing the failure of the land drive on Port Moresby was stressed in subsequent Army staff studies analyzing the New Guinea operations. One of these studies stated: "This operation has been greatly influenced by insufficient transport. Inability to move supplies seriously diminished front-line combat strength. Transport activity decreased because it was impossible to move by day, and halts were unavoidable after rains. The difficulty of transporting supplies by pack-horse increased, and the number of horses decreased. When men were used, only small quantities could be carried because of the large number of sick and the difficulty of negotiating the mountain inclines. Also, the carriers themselves consumed half of the provisions they carried while on the way." (*Lessons From New Guinea Operations, Jul 42–Apr 43*, ATIS Enemy Publications No. 285, 18 Jan. 45, pp. 13–14.)

119 The 4th Infantry Regiment of the 2d Division was already contained in the Aoba Detachment. The main body of the Division (16th and 29th Infantry Regiments) was assigned to the Seventeenth Army on 29 August and completed its movement from Java to Rabaul on 29 September, by which date its commitment on Guadalcanal had been decided. Under the tentative plans formulated in August, one portion of its strength was to be used in the amphibious operation against Moresby, while the remainder was to reinforce the South Seas Detachment land drive.

120 Southeast Area Operations Record, Part II, op. cit. Vol. I, pp. 85–6.

121 Ibid., p. 86.

would be held in abeyance pending the clean-up of the Solomons. The latter order stated:

*After the recapture of the Solomons is nearly complete, naval forces will be diverted to the New Guinea area. Naval air units will destroy enemy air strength in this area, enabling the land attack forces to proceed southward from the Kokoda sector. The Army, in cooperation with the Navy, will support the land attack with an amphibious operation near Port Moresby to capture the airfields in its vicinity.*[122]

This order came as a severe blow to the South Seas Detachment. In particular, it meant that the Detachment could continue to expect no naval air support of its operations. Although Navy fighters had operated briefly from the Buna airstrip to cover the landings of the main Detachment forces, they had soon been diverted to Guadalcanal, with the result that Allied aircraft dominated the skies over New Guinea, hammering constantly at the Japanese ground troops as they advanced and disrupting their supply lines to the rear.

Nevertheless, the troops on the Isurava front, encouraged by their successes of late August, pushed on toward the summit of the Stanley Range with the main body of the 41st Infantry spearheading the advance. On 1 September the 41st vigorously attacked the Australian position at Camp Gap, held by a force of approximately battalion strength, and succeeded in taking it by 0400 the following day. At dawn on 3 September, Eora was penetrated after three separate charges. The 41st Infantry drove on, meeting scattered opposition from the Australians who retreated to still another defense position estimated to be at the summit of the range. Throughout 4 September the 41st Infantry prepared for the assault, which was launched after nightfall. By 0200 5 September, the summit was in Japanese hands.

In the fighting from Camp Gap to the summit, the 41st Infantry had lost 44 killed and 62 wounded, but it had also inflicted considerable losses on the Australians, whose combat strength was estimated to be greatly reduced. Large amounts of abandoned stores were found during the advance from Camp Gap, and it was noted that the Australians were no longer using the main road in their retreat.

At this juncture a sudden intensification of Allied air strikes against the Buna area, coupled with intelligence reports of the presence of American air borne troops in Australia, led to fears that the enemy might be contemplating amphibious or airborne landings in that area in order to cut off the South Seas Detachment from the rear. To meet this threat, the Seventeenth Army on 8 September ordered Maj. Gen. Horii to reassemble the 41st Infantry Regiment at Kokoda, and a further order on 14 September directed that one battalion of the 41st Infantry be stationed near Buna to assure the defense of that area.

Meanwhile, the 144th Infantry Regiment, which had relieved the 41st at the front on 5 September, pushed off from the summit in the wake of the retreating Australians. Morale was high with the goal of the Moresby plain not far away, and the men advanced thinking of the ancient battle of Hiyodorigoe, famed in Japanese history and legend.[123]

On a hill south of Efogi, the 144th Infantry encountered its first stubborn opposition from the Australians, who were supported by heavy mortar fire. Sustaining considerable losses, the regiment was held up for three days until 8 September, when enemy resistance was finally overcome. After the battle, a company com-

---

122 *Daikaishi Dai Hyakunijushichi-go* 大海指第百二十七號 (Imperial General Headquarters Navy Directive No. 127) 31 Aug 42. (Text of Imperial General Headquarters Army Directive No. 1246, 31 Aug 42, was identical in substance)

123 The battle of Hiyodorigoe was fought in 1184 A.D., when Yoshitsune, leading the Genji forces, attacked the Heike forces from the mountains to their rear.

*It is now over one month since this Detachment left Rabaul and took over from the Yokoyama Advance Force, which had put up a brave fight prior to our arrival. We first reduced the strong position at Isurava, and continued on, crushing the enemy's resistance on the heights north of Isurava, at the Gap, Eora, Efogi, etc. Repeatedly we were in hot pursuit of the enemy. We smashed his final resistance in the fierce fighting at Ioribaiwa, and today we firmly hold the heights of that area, the most important point for the advance on Port Moresby.*

*For more than three weeks during that period, every unit forced its way through deep forests and ravines, and climbed scores of peaks in pursuit of the enemy. Traversing knee-deep mud, clambering up steep precipices, bearing uncomplainingly the heavy weight of artillery ammunition, our men overcame shortages of supplies and succeeded in surmounting the Stanley Range. No pen or words can depict adequately the magnitude of the hardships suffered. From the bottom of our hearts we appreciate these sacrifices and deeply sympathize with the great numbers killed and wounded.*

*We realize that the enemy on Tulagi and Guadalcanal has not yet been annihilated. We have not yet won back the Samarai and Rabi air bases. But the Detachment will stay here and firmly hold its position in order to perfect its organization and replenish its fighting strength. We will strike a hammer-blow at the stronghold of Port Moresby. However, ahead of us the enemy still crawls about. It is difficult to judge the direction of his movement, and many of you have not fully recovered your strength. I feel keenly that it is increasingly important during the present period, while we are waiting for the opportunity to strike, to strengthen our positions, reorganize our forces, replenish our stores, and recover our physical fitness.*

*Now, all must bear in mind the vital situation and the role of the Detachment in the South Pacific, and your increasingly heavy responsibilities. Strengthen your morale, replenish your vigor, and prepare for battle. When next we go into action, the unit will throw in its fighting power unreservedly.*[128]

## Retreat from the Owen Stanleys

Pursuant to Maj. Gen. Horii's message of instructions, the South Seas Detachment prepared to consolidate its hard-won positions on the southern slopes of the Stanley Range and simultaneously regroup its forces for the later assault on Moresby. However, crucial developments on other sectors of the southeast area front outdated this plan even before its execution began.

On Guadalcanal, the first general offensive of the 35th Inf. Brigade on 13 September had failed, and the Seventeenth Army was now preparing to commit its remaining reserves—Aoba Detachment and main strength of the 2d Division, previously intended for the final campaign against Moresby—in a second general offensive in October. At the same time, in New Guinea, Japanese supply difficulties became more acute, and there were mounting indications that General MacArthur planned early landings in the Buna area, which would, if successful, seal the fate of the South Seas Detachment and doom the entire Port Moresby invasion plan.[129]

Faced by this new situation, Seventeenth Army headquarters saw no alternative but to divert a substantial portion of South Seas Detachment strength back to the Buna area to counter enemy landing attempts. Consequently, on 23 September, the following order was dispatched to Maj. Gen. Horii:

*1. On the basis of intelligence reports it appears*

---

128 ATIS Current Translations No. 2, 11 Nov 42, p. 23.

129 Japanese intelligence indicated a marked reinforcement of Allied strength in New Guinea after the beginning of September. Approximately 10,000 Australian troops (2d Division) and 2,000 American infantry and marines were estimated to be in the Port Moresby area, while the transport of ground troops to the Milne Bay—Samarai area also showed a sharp increase. This, coupled with stepped-up air reconnaissance over the Buna area, pointed to the strong possibility of a landing attempt in that area. Southeast Area Operations Record, Part II, op. cit. Vol. I, pp. 87, 184.

*that the enemy is contemplating new landings in eastern New Guinea and the Solomons.*

*2. The Army will continue its preparations for the recapture of Guadalcanal, and at the same time will readjust its front in the South Seas Detachment area and strengthen defenses in the vicinity of Buna.*

*3. The Commanding General, South Seas Detachment, will reassemble his main strength in the Isurava-Kokoda sector, secure bases for future offensive operations, and reinforce defenses in the Buna vicinity. For this purpose, a force composed principally of the 41st Infantry Regiment will be dispatched without delay to the Buna sector, where it will smash enemy invasion plans and, in particular, secure the vicinity of the airfield. Further, one element of the Detachment will endeavor to hold a position on the southern slope of the Stanley Range.*[130]

Two main points of operational policy were clear from this order. The first was that the Seventeenth Army intended to suspend positive operations in the Owen Stanleys sector until Guadalcanal had been recaptured. The second was that the Army still desired to hold the Isurava-Kokoda sector, north of the Owen Stanleys, as a staging area from which to mount an ultimate attack on Port Moresby; but its major concern was now to secure the vital Buna area against threatened enemy attack, if necessary at the cost of relinquishing the vantage points gained by the South Seas Detachment almost within striking distance of Port Moresby.

Maj. Gen. Horii and his staff, conscious of the sacrifices paid to win possession of these vantage points, doubted the wisdom of relinquishing them. However, the Army order to move the 41st Infantry Regiment, which now composed the main combat strength of the Detachment,[131] immediately back to the Buna area, rendered it necessary to pull the Detachment front line back from Ioribaiwa to a point closer to Isurava and Kokoda. After a nightlong staff conference, Maj. Gen. Horii on 24 September ordered the weakened remnants of the 144th Infantry to begin the withdrawal to Eora.[132]

On the 25th, headquarters personnel, supply and hospital units began moving to the rear, followed on the 26th by the combat troops, who withdrew under constant mortar fire from enemy positions. Active pursuit by the Australians did not begin immediately, however, and no enemy ground attack was received until the 144th Regiment had pulled back to the area south of Eora. (Plate No. 42) There, the 2d Battalion, with a mountain artillery battery and engineer company attached (henceforth designated as the Stanley Detachment) took up a strong position near Eora and covered the withdrawal of the remaining elements of the regiment. The latter completed their movement to Kokoda by 4 October.

Meanwhile, the 41st Infantry Regiment, less the 2d and 3d Battalions, had left Kokoda on 25 September arriving in Buna on 28 September, at which time the 3d Battalion rejoined the Regiment and by 4 October the Regiment had taken up defensive positions in the Buna-Gona-Giruwa area. The 2d Battalion remained at Kokoda with the main strength of the 144th Infantry to serve as a reserve force. These new dispositions were substantially

---

130  Ibid., p. 184.

131  The 41st Infantry Regiment had been held in reserve at Kokoda since its relief by the 144th on 5 September. As of 20 September, it had over 1,700 effectives, whereas the 144th Regiment, after fighting to Ioribaiwa, was down to less than one-half its original strength of 2,932.

132  The conference on the Seventeenth Army order lasted from 1700 on 23 September until 0400 the next day. Some members of Maj. Gen. Horii's staff strongly opposed withdrawal on the ground that the supply situation was no better in the Kokoda area, and even favored pressing on toward Port Moresby in the hope of capturing enemy food supplies. (Statement by Maj. Koiwai, previously cited.)

オーエンスタンレー山脈よりポートモレスビーを望む　小早川篤四郎　昭和十三年三月十七日

Original Painting by Atsushiro Kobayakawa

PLATE NO. 41
Looking at Port Moresby from Owen Stanleys

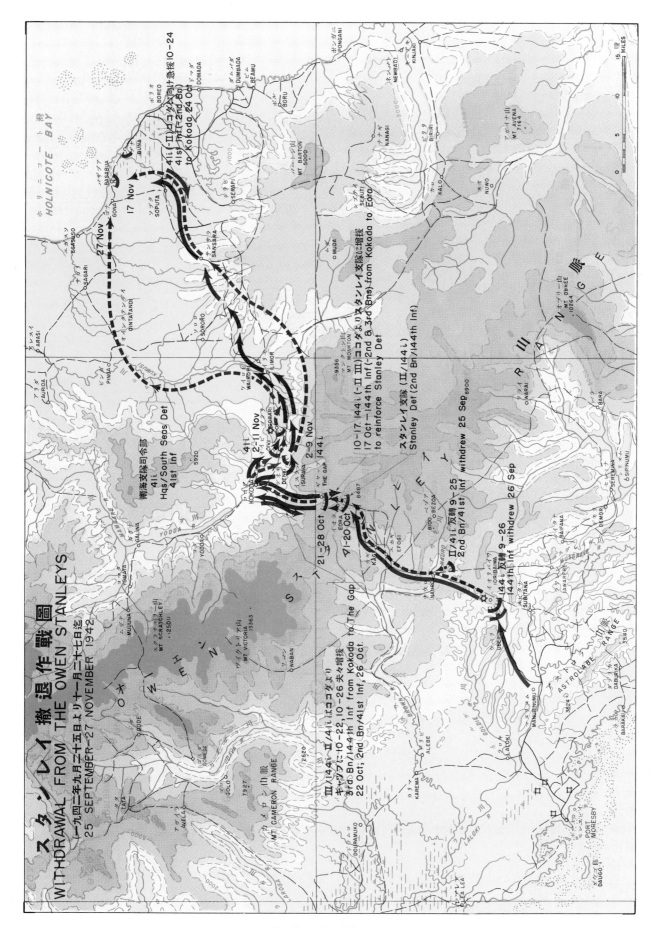

PLATE NO. 42

Withdrawal from the Owen Stanleys, 25 September—27 November 1942

confirmed by a Seventeenth Army order on 30 September, which directed Maj. Gen. Horii:

*1. To secure an offensive base near Isurava and a defensive base in the Buna area with elements of the South Seas Detachment.*

*2. To employ his main strength in improving the Giruwa-Kokoda road for vehicular transport, and the Kokoda-Isurava trail for pack-horse transport, by the end of October.*

*3. To complete improvement of the Kokoda airstrip into an operational air base.*[133]

However, beginning early in October, Australian attacks on the position held by the Stanley Detachment at Eora mounted in severity. Mercilessly pounded by enemy mortar fire and constant attacks from the air, the unit suffered extremely heavy casualties,[134] and its position was rapidly becoming untenable. On 14 October, to save the Detachment from annihilation, Maj. Gen. Horii ordered the 144th Regiment less 2d and 3d Battalions back to the front, from which they had been withdrawn only ten days earlier. Even after these troops had reached Eora, however, the situation was so critical that the 3d Battalion of the 144th Regiment and 2d Battalion of the 41st Regiment, still in reserve at Kokoda, was ordered to prepare to cover the further withdrawal of all front-line forces by 25 October.

Already forced to withdraw to a new position north of Camp Gap on 21 October, the Stanley Detachment again attempted to make a stand, only to sustain further Australian attacks which steadily mounted in ferocity. The defense of the Gap soon became so precarious that Maj. Gen. Horii, on 24 October, ordered the main body (1st and 3d Battalions) of the 41st Infantry guarding the Buna-Gona area, to move up to the front again for the purpose of relieving the battered Stanley Detachment. Two days later, on 26 October, a further order by Maj. Gen. Horii conceded the probable necessity of a retreat as far as Oivi, relinquishing Kokoda to the enemy. The order stated:

*The enemy in the Stanley Detachment area has an approximate strength of two battalions. After the 144th Infantry Regiment fell back to Eora, fresh enemy troops appear to have come up. Their trench mortars are active. The Stanley Detachment, although endeavoring to improve its position, has had to fight over a long period, transporting supplies under great difficulty, and with inadequate rations its fighting strength is at its lowest.*

*The Detachment will exert the utmost effort to hold firmly its present position, but if this is impossible, it will withdraw after dusk on 28 October at the earliest. It will withdraw in the direction of Oivi, delaying the enemy advance as long as possible, especially in the sector between Isurava and Deniki.*[135]

Two days following this order, a Seventeenth Army dispatch on 28 October advised an even further withdrawal to the east bank of the Kumusi River, but Maj. Gen. Horii and his staff rejected this as unwise on the ground that the low terrain east of the Kumusi was unfavorable both for defense and as the starting-point of future offensive operations.

In accordance with Maj. Gen. Horii's order of 26 October, the Stanley Detachment and the 1st and 3d Battalions of the 144th Infantry hastily pulled out of their position near the Gap on 28 October, the 2d Battalion of the 41st Infantry covering their withdrawal. The plan was now to establish a strong north-south

---

133 Southeast Area Operations Record, Part II, op. cit. Vol. I, p. 185.

134 On 14 October, a non-commissioned officer of one of the companies of the Stanley Detachment recorded in his diary that he had become acting company commander after its four officers had all been killed or wounded. The company had only four out of 17 non-commissioned officers left, and was down to a total strength of 42 men out of its normal strength of 178. (ATIS Current Translations No. 15, 22 Jan 43, p. 22.)

135 South Seas Detachment Operations Order, 26 Oct 42. *File of Nankai Shitai Operations Orders,* ATIS Enemy Publications No. 39, 1 Sep. 43, p. 6.

defense line at Oivi, with the 41st Infantry on the right flank around Oivi itself and the 144th Infantry on the left flank slightly to the south, guarding a secondary trail from Kokoda to the Buna area. South Seas Detachment headquarters was to be at Gorari, approximately three miles east of Oivi.[136]

All dispositions were complete by 2 November, when the 41st Infantry's 2d Battalion fell back to Oivi, joining the 1st and 3d Battalions which had moved up from the Buna-Gona area. The Stanley Detachment and most of the remaining strength of the 144th Infantry meanwhile took up their left-flank position to the south of Oivi. The stage was set for battle.

The Australian forces now split into two elements, the first advancing on Oivi in a frontal assault while the second swept to the south in a flanking movement, launching a surprise dawn attack on 5 November against the 144th Infantry position. Driven back by the unexpected weight of the enemy assault, the 144th began retreating eastward on 9 November, crossing the Kumusi and heading toward Buna. The Australian force then wheeled swiftly northward to attack the 41st Regiment from the rear and cut off its retreat at Gorari.

Caught between the closing Australian pincers and cut off from contact with the 144th Regiment, the 41st, together with South Seas Detachment headquarters, found itself under heavy fire and facing imminent danger of encirclement. On 10 November Maj. Gen. Horii decided to withdraw toward the Kumusi under cover of a daring night attack against the eastern prong of the enemy pincers at Gorari to open a retreat passage. Preparations for a full-scale attack could not be completed in time, however, and a preliminary attack by two companies on the night of the 10th failed to breach the enemy positions. Maj. Gen. Horii then ordered the 41st Regiment to cross Oivi Creek, skirt around the northern enemy flank, and recross the creek farther to the east to get back on the trail to Buna. The withdrawal began at 0900 on 11 November.

Although the Japanese troops successfully crossed to the north bank of Oivi Creek on the 11th, heavy rains on the afternoon of that day flooded the creek to such an extent that attempts to recross to the south bank after skirting around the enemy at Gorari proved unsuccessful. Maj. Gen. Horii and his troops therefore continued to retreat along the north bank of Oivi Creek toward its juncture with the Kumusi River. They still had not succeeded in effecting a crossing when the approach of pursuing Australian troops on 13 November forced them to turn northward and flee along the trackless west bank of the Kumusi toward Pinga.

As the Japanese troops approached Pinga, the sound of gunfire was heard from across the river in the direction of Gona, and it was feared that the anticipated Allied landings had already taken place. Maj. Gen. Horii, gravely concerned over the situation, decided to attempt to reach the Buna area by canoe down the Kumusi. Setting out on 19 November with a staff officer and a runner, he succeeded in reaching the mouth of the river, and from there the canoe headed down the seacoast toward Buna. When directly off Gona, a sudden squall arose and capsized the canoe. Attempting to swim ashore, both Maj. Gen. Horii and his staff officer were drowned.

The 41st Infantry Regiment at Pinga, under the regimental commander, Col. Kiyomi Yazawa, had meanwhile built rafts, crossed the Kumusi River and started overland toward Gona. Losing many additional men and abandoning a large part of their weapons and equipment in the difficult overland trek, the depleted rem-

---

[136] At the time, Gorari was mistakenly referred to by the Japanese as Ilimo, which lies a few miles farther to the east.

nants did not reach the Gona area until 27 November.[137]

The 144th Regiment, after its hasty retreat from the Oivi sector on 9 November, withdrew northeastward toward Giruwa. Most of its remaining troops reached the Giruwa area by 17 November, only two days before strong American and Australian forces suddenly attacked from the sector south of the Buna airstrip.

Without time in которых to reorganize its command and regroup its scattered, demoralized and weakened forces, the South Seas Detachment was now called upon to resist a powerful Allied pincers assault which threatened to wrest from the Japanese forces their last remaining foothold in Papua.

---

[137] Statement by Maj. Koiwai, previously cited.

# CHAPTER VIII
# DEFENSE OF PAPUA

## Eighth Area Army Activated

With the sudden reversal of Japanese fortunes in New Guinea and the parallel failure of the second general offensive mounted by the Seventeenth Army Guadalcanal, Imperial General Headquarters for the first time began to assess the full gravity and implications of the situation which was developing on the southeast area front.[1]

It was evident that the Seventeenth Army, its major forces already expended in the futile attempts to retake the Solomons,[2] could not cope with the added menace presented by General MacArthur's thrust against the Japanese right flank in Papua. To repulse these twin Allied drives and pave the way for ultimate resumption of the offensive toward Port Moresby, a drastic reorganization of command and an immediate reinforcement of fighting strength were imperative.

Therefore, on 16 November, Imperial General Headquarters ordered the activation of the Eighteenth Army to take over the conduct of operations in New Guinea, restricting the operational sphere of the Seventeenth Army exclusively to the Solomons. Both armies were simultaneously placed under a new theater command designated as the Eighth Area Army, and Lt. Gen. Hitoshi Imamura, Commanding General of the Sixteenth Army in Java, was ordered to Rabaul to assume command. These command dispositions were to become effective on 26 November.

Upon activation, the Eighth Area Army consisted of the Seventeenth Army, the newly activated Eighteenth Army, the 6th Division, 21st Independent Mixed Brigade, and 12th Air Brigade. Elements of the 5th Division were also assigned on 20 November. On 27 November, the Army's 6th Air Division[3] commanded

---

[1] This chapter was originally prepared in Japanese by Comdr. Masataka Chihaya, Imperial Japanese Navy. Duty assignments of this officer were as follows: Antiaircraft Gunnery Officer, battleship *Musashi*, 15 Sep 41—10 Oct 42; Staff Officer (Operations), 11th Battleship Division, 10 Oct–30 Nov 42; Staff Officer, Third Section (Military Preparations), Imperial General Headquarters, Navy Section, 20 Jan—1 Jul 43; Naval War College, 1 Jul 43—5 Mar 44; Staff Officer (Operations), Fourth Southern Expeditionary Fleet, 15 Mar 44—18 Jan 45; Staff Officer (Operations), Combined Fleet, 1 Feb—1 May 45; Staff Officer (Operations), General Navy Command, concurrently attached to headquarters, First and Second General Armies, 1 May—6 Sep 45. All source materials cited in this chapter are located in G-2 Historical Section Files, GHQ FEC.

[2] The second general offensive on Guadalcanal began with a night attack on the airfield on 24 October 1942. The American positions were penetrated, but the 2d Division sustained heavy casualties and was unable to hold its gains. A second attack was ordered for the night of 25 October, but could not be carried out due to strong American counterattacks, although one brigade on the Japanese left flank carried out an abortive suicide assault. At 0600 on 26 October, orders were issued to suspend the attack and withdraw. *Nanto Homen Sakusen Kiroku Sono Ni: Dai Jushichi Gun no Sakusen* 南東方面作戦記錄其の二：第十七軍の作戦 (Southeast Area Operations Record, Part II: Seventeenth Army Operations) 1st Demobilization Bureau, Sep 46, Vol. I, pp. 166–73.

[3] Strength of the 6th Air Division at its activation was: 54 light bombers, 84 fighters, and 9 reconnaissance planes. Subsequent assignment of additional units brought the division up to its maximum operational strength at the end of May 1943, when it had 77 light bombers, 114 fighters, and 26 reconnaissance aircraft, a total of 217 planes. Losses reduced this strength to 165 aircraft at the end of June, and 99 aircraft as of 26 July 1943. *Nanto Homen Koku Sakusen Kiroku* 南東方面航空作戦記録 (Southeast Area Air Operations Record) 1st Demobilization Bureau, Sep 46, pp. 6, 24.

by Lt. Gen. Giichi Itahana was activated and placed under the Commander-in-Chief, Eighth Area Army, to be used in support of both Seventeenth and Eighteenth Army operations.

To command the new Eighteenth Army in charge of New Guinea operations, Imperial General Headquarters appointed Lt. Gen. Hatazo Adachi, then chief of staff of the North China Area Army. At the date of its activation, the Eighteenth Army's combat forces comprised only the remnants of the South Seas Detachment, the 41st Infantry Regiment, the 15th Independent Engineers Regiment and some small units, all badly battered in the Owen Stanleys campaign. On 20 November, however, elements of the 65th Brigade[4] in the Philippines were transferred to Eighteenth Army by Imperial General Headquarters order. This was followed by an Eighth Area Army order on 26 November, placing the major portion of the 21st Independent Mixed Brigade,[5] as well as one infantry battalion and one mountain artillery battery of the 38th Division,[6] under Eighteenth Army operational control.

Parallel with this regrouping and replenishment of forces, Imperial General Headquarters on 18 November issued an operational directive to the Commanders-in-Chief of the Eighth Area Army and the Combined Fleet, clarifying future objectives on the southeast area front. The directive continued to give priority to the recapture of the Solomons, but at the same time it called for a strengthening of Japanese bases in New Guinea with the ultimate objective of resuming the offensive toward Port Moresby and sweeping the Allies from Papua. Essential points of the directive were as follows:

*1. The Army and Navy will cooperate in hastily reinforcing and equipping air bases in the vicinity of the Solomon Islands for employment in subsequent operations, and will devote special attention to strengthening air defenses in this sector. Army forces on Guadalcanal will immediately secure key positions in preparation for offensive operations, while recovering their strength. The Navy, during this period, will use every means at its disposal to check enemy reinforcements to the Solomons and will cooperate with the Army in curbing enemy air activity. The Army and Navy will intensify air operations as they extend their air bases and, when enemy air strength has been neutralized, will seize the opportunity to transport reinforcements to Guadalcanal for the Army's offensive.*

*2. After these preparations have been completed, the Army, in cooperation with the Navy, will recapture the airfield on Guadalcanal and annihilate the enemy forces on that island. At the earliest opportunity, Tulagi and other key positions in the Solomons will also be occupied.*

*3. During the operations in the Solomons, the Army and Navy will secure strong operational bases at Lae, Salamaua and Buna, will strengthen air operations by extending and fitting out air bases, and will prepare for future operations. The Army, in cooperation with the Navy, will occupy Madang, Wewak and other strategic areas. Preparations for future operations in the New Guinea area will embrace every feasible plan for the capture of Port Moresby, Rabi, and the Louisiade Archipelago.[7]*

Reaching Rabaul on 22 November after a hasty air journey from Tokyo,[8] Lt. Gen. Ima-

---

4 These elements included the Brigade headquarters, the 141st Infantry Regiment, and half of the brigade's service units.

5 The 21st Independent Mixed Brigade, previously on guard duty in French Indo-China, was transferred to Rabaul in November. (For composition, cf. n. 28)

6 The 38th Division had been assigned to Seventeenth Army on 17 September, for use on Guadalcanal. Initial elements were transferred to Guadalcanal in early November. Cf. p. 177.

7 *Daikaishi Dai Hyakugojukyu-go* 大海指第百五十九號 (Imperial General Headquarters Navy Directive No. 159) 18 Nov 42. Army directive was identical in substance.

8 Lt. Gen. Imamura flew via Truk, where he conferred with Admiral Yamamoto, Commander-in-Chief of the Combined Fleet, regarding future Army-Navy cooperation in the southeast area. On 24 December, an Imperial Gen-

one mile south of Buna Village in the so-called Triangle area.[26] The naval garrison troops, thus far not engaged in battle, were fresh and commanded by an officer specially trained in land warfare.[27] The 3d Battalion, 229th Infantry, which had moved up to the Buna front immediately after its arrival from Rabaul on 17 November, was a crack unit, eager for battle. Occupying positions which took every possible advantage of the difficult terrain, these forces fought stubbornly, further aided by the fact that the enemy had not yet been able to bring up heavy equipment. Until early December the situation on this front remained stalemated.

To the west, the Australian frontal attack against the 144th Infantry positions at Southern Giruwa, launched about 20 November, was also successfully checked. However, on 24 November, enemy elements succeeded in infiltrating around these positions to drive a wedge between South Giruwa and the central position to the north. An attempt to eliminate this wedge by a force of battalion strength, dispatched from the central position, failed, and by early December all communication with the hard-pressed Japanese force in Southern Giruwa had been cut off.

On the right flank, the heterogenous Japanese force defending the Gona—Basabua sector, hastily reinforced on 19 November by an infantry unit sent from Giruwa, succeeded in repelling Australian advance elements which penetrated the area on 20 November. On 26 November additional reinforcements arrived, but the weak defending forces were soon bottled up, and an attempt at a rescue made by 41st Infantry troops from the central Giruwa position proved unavailing. Hemmed in on all sides and suffering heavily from intense artillery bombardment, the Gona force nevertheless continued to resist. By early December it appeared doomed to annihilation unless fresh attempts to send in reinforcements from Rabaul, then already under way, succeeded in bringing immediate relief.

### Reinforcement Attempts

Upon the entry into effect on 26 November of the new command dispositions ordered by Imperial General Headquarters for the southeast area, Eighth Area Army Commander Lt. Gen. Imamura promptly ordered the Eighteenth Army to dispatch strong reinforcements to the Buna area in order to swing the tide of battle in favor of the Japanese forces. The 21st Independent Mixed Brigade,[28] only recently arrived in Rabaul from Indo-China, was assigned this mission by Eighteenth Army order, and preparations were hastily completed for its shipment to New Guinea aboard destroyers allotted by the Navy's Southeast Area Force.

Making the first reinforcement attempt, four destroyers left Rabaul on 28 November carrying Maj. Gen. Tsuyuo Yamagata, brigade commander, with the headquarters and a portion of the brigade strength. Despite air cover provided by six Navy fighters, enemy B-17's

---

26 Ibid., pp. 37-8.

27 Capt. Yoshitatsu Yasuda, commander of the Buna naval garrison, was known in the Japanese Navy as an expert in land warfare, having received training with the Army. (Statement by Capt. Toshikazu Ohmae, Staff Officer (Operations), Southeast Area Fleet.)

28 The 21st Independent Mixed Brigade was composed of the 170th Infantry Regiment, one artillery battalion, one antiaircraft battery, one tank company and one engineer company. Upon arrival in the Buna area, the brigade commander, Maj. Gen. Yamagata, was to assume command of all Army forces in the Buna—Giruwa area, grouped together under the designation, Buna Detachment. Southeast Area Operations Record, Part III, op. cit. Vol. I, pp. 37-9.

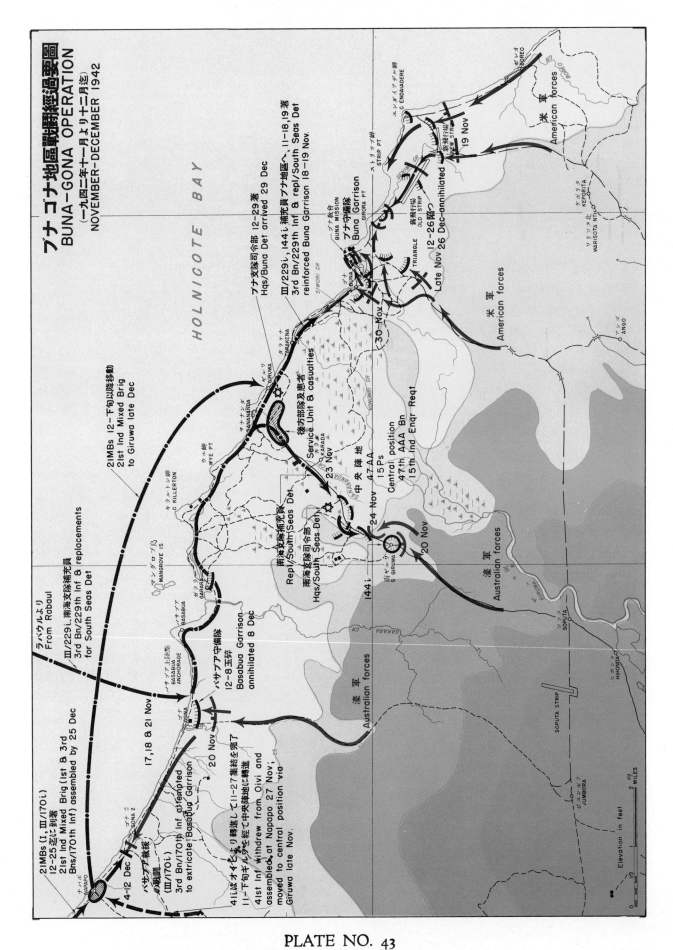

PLATE NO. 43

Buna—Gona Operation, November—December 1942

attacked the convoy north of Dampier Strait on the morning of the 29th, and two of the destroyers sustained damage, forcing the convoy to turn back to Rabaul.[29]

On 30 November a second attempt was launched. The convoy of four destroyers, again carrying the brigade headquarters, the 3d Battalion of the 170th Infantry Regiment, and signal units, totalling in all 720 officers and men, this time took a course skirting south of New Britain, with a reinforced air escort of 13 planes. Despite sporadic enemy air attacks, the convoy safely reached the anchorage near Gona on the evening of 1 December.

At this point, however, Allied planes launched an attack of such intensity that it was impossible for the troops to board landing craft, and the destroyers were obliged to move on to the mouth of the Kumusi River, 18 miles northwest of Gona. Here, with enemy planes still dropping flares, 425 of the troops succeeded in transferring to landing craft, but in the movement to the shore they became dispersed and landed at widely-separated points between the Kumusi River mouth and Gona. The remaining 295 troops could not be landed and returned with the destroyers to Rabaul.[30]

It was 6 December before Maj. Gen. Yamagata was able to reassemble his scattered forces and move them into a concentration area about two miles west of Gona. From this point he ordered his troops forward in an attempt to break the Australian envelopment of Gona, but the enemy lines held, and the fighting entered a stalemate. With succor so near and yet unable to reach them, the Japanese forces in Gona were finally overwhelmed on 8 December, only a handful of survivors escaping by sea or through the jungle to the Giruwa area.[31]

Meanwhile, a third reinforcement convoy of five destroyers had set out from Rabaul on 7 December, carrying additional elements of the 21st Independent Mixed Brigade. This time, however, vigilant enemy planes spotted the ships after they had barely emerged from the St. George Channel, a few hours out of Rabaul. Under severe attack, the convoy was forced to put back into port immediately.

With the Japanese defenses in the Buna sector also beginning to crack under intensified enemy pressure, it was now more imperative than ever to move the remaining strength of the 21st Brigade to the battle area without delay. Hence, on 12 December, a fourth and final reinforcement attempt by destroyer was begun. Five ships with an escort of nine fighters left Rabaul on that date, taking a roundabout course to the north of the Admiralty Islands in an attempt at deception. Despite this maneuver, the convoy underwent heavy bombing as it neared its destination.[32]

With the Gona–Basabua anchorage already under enemy control, the mouth of the Kumusi had been fixed as the landing point. However, already behind schedule due to the intense air attacks, the destroyers put in at the mouth of the Mambare River, 40 miles short of the goal, and disembarked the troops before dawn on 14 December. The 1st Battalion of the 170th Infantry, one company of the 3d Battalion, the regimental gun company and 25th Field Machine Gun Company, aggregating 870 troops, were successfully put ashore.[33] Between 18 and 25 December, these troops moved southward along the coast by small craft and

---

29 Major New Guinea Operations, op. cit., pp. 56–7.
30 (1) Ibid., p. 55. (2) *Yamagata Heidan* Troop Transport Plan. ATIS Current Translations No. 16, 25 Jan 43. p. 27.
31 Southeast Area Operations Record, Part III, op. cit. Vol. I, p. 41.
32 Major New Guinea Operations, op. cit., p. 56.
33 Southeast Area Operations Record, Part III, op. cit. Vol. I, pp. 42–3.

joined the 21st Brigade units already in the Napapo area west of Gona.

Owing to steadily increasing Allied air domination of the sea approaches to the Papuan coast, no further attempts to dispatch reinforcements to the Buna area by destroyer were undertaken. Only two of the four attempts made between 28 November and 14 December had been halfway successful, and the effort had cost damage to four destroyers of the dwindling naval forces in the southeast area.[34]

## Fall of Buna

The stalemate which had prevailed on the Buna sector front since the initial enemy attacks in late November finally ended on 5 December, when powerful offensives were launched by the American forces against both the Senimi Creek—Cape Endaiadere position and the Buna Village area. (Plate No. 43) In the former sector, the enemy again failed to breach the strong outer perimeter,[35] but, on the right flank, enemy troops which had gradually infiltrated past the Japanese strongpoints in the Triangle area toward Buna Village succeeded in driving a wedge to the sea between the village and Buna Mission, at the same time capturing some of the positions on the southern perimeter of the village. By 14 December, the small defending force of Army troops and naval construction personnel in Buna Village had been overcome.[36]

On 18 December the enemy again struck with renewed vigor at the Triangle area on the right flank, and the Senimi Creek—Cape Endaiadere line on the left. The troops in the Triangle area, resisting fierce bombardment by enemy mortars and artillery, again held their positions. However, on the left flank, a powerful assault, spearheaded for the first time by tanks, broke through the Japanese defenses in the coastal sector and drove a salient northward past Cape Endaiadere.[37]

The Japanese naval unit defending the Senimi Creek bridge-crossing southeast of the airstrip was now forced to pull back to the airstrip defenses, where it prepared to make a suicide stand.[38] Enemy tanks were soon brought across the creek to support the ground troops' assault, and the Japanese defenses slowly gave way in heavy fighting. By 26 December the last naval antiaircraft battery emplaced near the central portion of the strip was wiped out after firing its last remaining rounds of ammunition against oncoming enemy tanks.[39] Three days later, on the right flank, the Japanese positions in the Triangle area were finally overcome, and enemy elements, in another drive to the sea, cut off the Buna garrison headquarters, northwest of the airstrip, from Buna Mission.

In view of the increasingly critical situation, the Eighteenth Army Commander at Rabaul had already dispatched urgent orders to Maj. Gen. Yamagata on 26 December, directing

---

34  Major New Guinea Operations, op. cit., pp. 55–7.

35  On 9 December fierce enemy shelling of the Japanese positions in the Senimi Creek area knocked out one mountain gun and smashed a number of bunkers. The positions were restored during the night, however, and successfully held. Ibid., p. 38.

36  (1) Ibid., p. 38.  (2) Papuan Campaign, compiled by the Military Intelligence Division, U. S. War Department, p. 43.

37  Daily Operations Log of *Tsukioka* Unit (Sasebo 5th Special Naval Landing Force) Headquarters, 5 Oct—24 Dec 42. ATIS Current Translations No. 27, 19 Apr 43, p. 15.

38  Diary of 1st Class Seaman Masaji Konagaya, Yokosuka 5th Special Naval Landing Force, 9 June—23 Dec 42. Entry for 19 December states: "No. 3 Sentry Post withdrawn to the airdrome. The *Tsukioka* Unit intends to resist to the last." ATIS Current Translations No. 60, 3 Jul 43, p. 18.

39  Major New Guinea Operations, op. cit., p. 39.

him to move the 21st Brigade troops, still held up west of Gona, to Giruwa by sea and from there launch an attack toward Buna to relieve the Japanese forces cut off in that sector.[40] It seemed improbable, however, that Buna itself could be saved. Hence, on 28 December, the Army and Navy commands at Rabaul ordered the withdrawal of all forces from the Buna sector to join in the defense of Sanananda—Giruwa.

Between 27 and 29 December, Maj. Gen. Yamagata with a relief unit of one battalion (reinf.)[41] successfully moved from Napapo to Giruwa by small landing craft. After setting up his headquarters at Giruwa, Maj. Gen. Yamagata placed the relief detachment under command of the 41st Infantry regimental commander, Col. Yazawa, and on 31 December ordered it to move up for an attack on the enemy left flank above Buna.[42]

Even before the relief unit had started, however, Maj. Gen. Yamagata's headquarters received a report from the Buna front to the effect that, on 1 January, enemy troops, spearheaded by six tanks, had penetrated into the isolated headquarters area northwest of the airstrip. There, the Army and Navy commanders of the Buna garrison forces, Col. Yamamoto and Capt. Yasuda, were reported leading the last handful of survivors of the headquarters personnel in a suicide stand.[43] (Plate No. 44)

The Yazawa relief unit, still hoping to rescue the Japanese troops holding out in the Buna Mission area, started its movement from Giruwa on 2 January. Upon reaching Siwori Creek on the night of the 4th, the unit was held up by a bloody encounter with about 300 enemy troops, but it pushed on across the creek to a point about one mile west of Buna, where by 8 January it had received a total of a few hundred Army and Navy personnel, the sole survivors of the force which had so ably defended the Buna sector. The relief unit then fell back under constant enemy harassment to the Konombi Creek line, where it occupied positions for the defense of Giruwa.[44]

While the rescue operation was in progress, the last Japanese positions in the Buna Mission area had fallen to the enemy. The battle for Buna was at an end.

## Sanananda-Giruwa

With the final collapse of Japanese resistance in the Buna sector, the full weight of the Allied assault immediately shifted to the front west of the Giruwa River, where the Japanese still clung tenaciously to a five-mile strip of coast extending from Konombi Creek, above Buna, to Garara, west of Cape Killerton. The nerve-center of the Japanese defenses was situated in the vicinity of Sanananda Point and Giruwa, on the coast, protected on the inland side by the already isolated outpost at Southern Giruwa and the so-called central position between Southern Giruwa and the coast.

---

40 Southeast Area Operations Record, Part III, op. cit. Vol. I, p. 44.

41 *Nishi* Operations Order A No. 39, issued 27 December at Napapo. ATIS Current Translations No. 29, 28 Apr 43, p. 15.

42 *Nishi* Operation No. 44, issued 31 December at North Giruwa. ATIS Current Translations No. 29, op. cit., pp. 15–6.

43 The adjutant of the naval garrison force, on orders from Capt. Yasuda, made his way out of the encircled headquarters position on the night of 1 January and eventually reached Giruwa to report on the final situation. Col. Yamamoto and Capt. Yasuda planned to lead the surviving Army and Navy personnel in a suicide attack on 2 January. Major New Guinea Operations, op. cit., p. 37.

44 Diary of Maj. Nojiri, commander, 1st Battalion, 170th Infantry Regiment, 2 Dec 40–15 Jan 43. ATIS Current Translations No. 29, op. cit., pp. 18–9.

ニューギニヤに於ける安田部隊の最期　藤田嗣治　昭和一八・一・一

Original Painting by Tsuguji Fujita

PLATE NO. 44

Fate of Yasuda Force on New Guinea Front

The condition of the Japanese forces holding this area was now desperate in the extreme. The flow of supplies from Rabaul had been stopped with the exception of small amounts of provisions and ammunition brought to the mouth of the Mambare River by submarine and thence moved to Giruwa by small landing craft under cover of night. By the end of the first week in January, all food supplies had been exhausted, and the troops were eating grass and other jungle vegetation. Deaths from tropical diseases exceeded battle casualties. Enemy artillery fire and air bombardment had razed the protective jungle covering around the Japanese bunkers and trenches, and rains flooded these positions as fast as they could be drained.[45]

Enemy air supremacy over the battle area was virtually complete. During the first phase of the Buna fighting, Japanese naval air units based on New Britain had carried out a series of effective attacks against the enemy advance air base at Dobodura and against Allied supply shipping, but by mid-December control of the air had definitely passed to the Allies. The arrival in Rabaul at about this time of the 6th Air Division, the first Army air unit to be dispatched to the southeast area, came too late to exert much effect.[46] Enemy observation craft flew unhindered over the Japanese positions, increasing the effectiveness of artillery fire to a point of deadly accuracy.

By late December, Imperial General Headquarters in Tokyo had reluctantly recognized the inevitable pattern of defeat that confronted the Japanese forces both in eastern Papua and on Guadalcanal. Therefore on 23 December, Imperial General Headquarters modified its 18 November directive and placed the decision of withdrawal from the Buna area to the discretion of the local commander, dependent upon the local situation. However, Imperial General Headquarters placed such great significance on the evacuation of Guadalcanal that it was not until the Imperial conference of 31 December that its final decision was reached. On 4 January, therefore, an order was dispatched to Lt. Gen. Imamura, Commander of the Eighth Area Army, and to Admiral Yamamoto, Commander-in-Chief of the Combined Fleet,[47] directing the first major withdrawal of Japanese troops since the sartt of the Pacific War. The order stated:

*1. In the Solomons area, the fight to recapture Guadalcanal will be discontinued, and the Army will evacuate its forces immediately. Henceforth, the Army will secure the northern Solomons, including New Georgia and Santa Isabel, and the Bismarck Archipelago.*

---

45 Maj. Kempo Tajima, South Seas Detachment staff officer, in a statement on the condition of the Japanese forces in the Giruwa area prior to evacuation, wrote: Japanese officers and men presented a gruesome sight. Their skin had turned pale, their eyes were sunken, their clothing was in shreds, and only a few wore shoes. The sword alone was a heavy burden for those who carried them. ....The hospital was filled with dead and wounded, and hundreds of corpses were left on the ground uncollected. ....It was difficult even to obtain a few sheets of paper on which to write orders, and communications were so disrupted that it was frequently impossible to transmit messages by field telephone. (Statement by Maj. Kempo Tajima, Staff Officer, South Seas Detachment.)

46 (1) On 26–27 December a fighter unit of the 6th Air Division made its first sorties over the Buna area in support of the naval air forces. Southeast Area Operations Record, Part III, op. cit. Vol. I, pp. 50–1. (2) No Army air strength had previously been sent to the southeast area since operations in this area were primarily the Navy's responsibility. After the enemy reinvasion of Guadalcanal, however, the Navy requested the dispatch of Army air units for employment in the Solomons. The Army Section of Imperial General Headquarters at first declined on the ground that this would seriously weaken air operations on the Burma and China fronts, but as the Guadalcanal situation worsened, the High Command finally agreed to dispatch the 6th Air Division to Rabaul. (Statement by Col. Takushiro Hattori, Chief, Operations Section, Imperial General Headquarters, Army Section.

47 *Daikaishi Dai Hyakuhachijuichi-go* 大海指第百八十一號 (Imperial General Headquarters Navy Directive No. 181) 23 Dec 42.

*2. In the New Guinea area, the Army will immediately strengthen its bases of operation at Lae, Salamaua, Madang and Wewak. The strategic area north of the Owen Stanley Range will be occupied, and thereafter preparations will be made for operations against Port Moresby. The forces in the Buna area will withdraw to the vicinity of Salamaua, as required by the situation, and will secure strategic positions there.*[48]

Studying the situation at Rabaul, Lt. Gen. Imamura decided to delay the issuance of implementing orders for two reasons. First, he thought that the situation was not so grave as to warrant an immediate evacuation. Second, it was essential to delay relinquishment of the Giruwa area until reinforcements, then preparing to leave Rabaul, had reached Lae–Salamaua and strengthened that area against possible Allied attack.

Meanwhile, however, the final disintegration of the Japanese forces on the Giruwa front was already beginning. On 12 January, three days after its last food supplies had been exhausted, the isolated Japanese force in Southern Giruwa, unable to communicate with rear headquarters, launched an independent break for freedom through the enemy lines. Heading southwest into the jungle, the troops found their way to the Kumusi River, and thence retreated northward. A small number of survivors reached the Japanese positions at the mouth of the Kumusi in the latter part of January.

Finding the resistance before them ended, the Australians quickly moved up the Soputa–Sanananda track and joined the enemy force already blocking the track to the north in an assault on the Japanese central positions. (Plate No. 45) At the same time, elements swung around to the west of these positions in a flanking movement, one force advancing to the coast to capture the Japanese right flank outpost at Garara on 13 January, and another cutting in from the west to split the central positions from South Seas Detachment headquarters near Sanananda Point. The force which took Garara immediately drove eastward along the coast, capturing Wye Point on 15 January.

While the Australians closed in from the south and west, the American forces pushing up the coast from Buna launched an attack on the Japanese forces, left flank along Konombi Creek on 12 January. Here, the remaining strength of the 1st Battalion, 170th Infantry, put up a determined fight which held up the enemy advance until about 20 January.[49]

On 13 January, following the landing of 51st Division reinforcements at Lae, Lt. Gen. Imamura ordered Eighteenth Army to begin the evacuation of the Japanese forces from Giruwa. In compliance with this order, Lt. Gen. Adachi, Eighteenth Army Commander, dispatched an immediate order to Maj. Gen. Yamagata, commanding all forces in the Giruwa area, directing that the evacuation be carried out as follows:

*1. The Buna Detachment Commander will abandon his present positions and divert his troops as follows:*

*a. Diversion of the main force will commence about 25 January and end about 29 January.*

*b. The main force of the 21st Independent Mixed Brigade will assemble in the Mambare area, and one element of the brigade in the Zaka—Morobe area.*

*c. Main strength of other units will be dispatched to Lae, and the remainder to Salamaua.*[50]

Although 25 January was fixed as the starting date of the general evacuation, the rapid closing of the Allied pincers on Giruwa and the

---

48 Southeast Area Operations Record, Part III, op. cit. Vol. I, p. 52.
49 Nojiri Battalion Order, issued 20 Jan. ATIS Current Translations No. 32, 1 May 43, p. 10.
50 Southeast Area Operations Record, Part III, op. cit. Vol. I, pp. 59–61.

increasing disorganization of the Japanese forces led Maj. Gen. Yamagata to advance the date to 20 January.[51] Communications were so disrupted that it was only with great difficulty that the withdrawal order was transmitted to the units in the front lines. Japanese forces holding inland positions along the Sanananda–Soputa track were instructed to withdraw independently by land to the mouth of the Kumusi River, while troops in the coastal sector around Giruwa, including Buna Detachment headquarters, were to evacuate by sea.

With destroyer movement impossible due to Allied air domination of the Solomons Sea and the Papuan coast, the sea evacuation had to be carried out by small landing craft. A number of these was dispatched from Lae but had only reached the mouth of the Kumusi River by 20 January, when the evacuation was scheduled to take place. On the night of 19–20 January a total of only 250 personnel, including Maj. Gen. Yamagata, the headquarters staff and casualties, was successfully evacuated aboard landing craft already available in the Giruwa area.[52]

The Japanese remnants along the Sanananda–Soputa track meanwhile succeeded in breaking through to the west, heading for the assembly point at the mouth of the Kumusi River. Although favored by slow enemy pursuit, the battleworn, half-starved survivors experienced extreme hardship moving through the jungle, and many stragglers were left along the route of retreat. (Plate No. 46)

On 18 January two companies of the 102d Infantry Regiment, 51st Division, had been dispatched by landing craft from Lae to cover the withdrawal of the troops evacuated from Giruwa. One of these companies landed at the mouth of the Mambare, while the other reached the mouth of the Kumusi on 24 January, there helping to repulse an attack by enemy troops pushing up from the Gona area.

By 7 February a total of approximately 3,400 survivors of the bloody Buna–Gona campaign had assembled at the mouth of the Mambare River.[53] At the time of the evacuation order, Lt. Gen. Adachi's plan had been to hold the Mambare River line as an advance offensive base, using the 21st Brigade forces withdrawn from the Giruwa area. It was now obvious, however, that the decimated remnants of the Giruwa forces were unequal to any further combat mission, and the intervening failure of the Japanese offensive against Wau made it necessary to retract the first line still farther to the Mubo—Nassau Bay area. Eighteenth Army therefore ordered the troops assembled at the mouth of the Mambare to continue their withdrawal by sea to Lae and Salamaua. This movement was completed early in March.[54]

The loss of the Buna–Gona area rang down the curtain on the six months long Papuan campaign, which in September 1942 had seen the South Seas Detachment with Port Moresby almost in its grasp. Between the initial landing at Buna in July 1942 and the end of the Buna–Gona battle in January 1943, a total of approximately 18,000 to 20,000 troops had

---

51 Maj. Gen. Yamagata estimated that it would be impossible to prevent the Japanese forces from falling into a rout if the evacuation were delayed beyond 20 January. (1) Statement by Lt. Col. Tanaka, previously cited. Col. Tanaka at this time was Eighteenth Army staff officer attached to Buna Detachment Headquarters.) (2) *Nishi* Operations Order Nos. 65 and 66. ATIS Current Translations No 32, op. cit., pp. 24–5.

52 Between 12 and 18 January, 1,000 hospital patients had already been successfully evacuated. In the final evacuation, Maj. Gen. Kensaku Oda, newly-appointed Commanding General of the South Seas Detachment, was killed. (1) Southeast Area Operations Record, Part III, op. cit. Vol. I, pp. 59. 60 and 63 (2) Statement by Lt. Col. Tanaka, previously cited.

53 Ibid., p. 67.

54 Between April and June, most of the troops evacuated from Giruwa were shipped back to Rabaul for recuperation and reorganization. The 21st Independent Mixed Brigade was deactivated at Rabaul, and the South Seas Detachment and 41st Infantry remnants were transferred to other theaters.

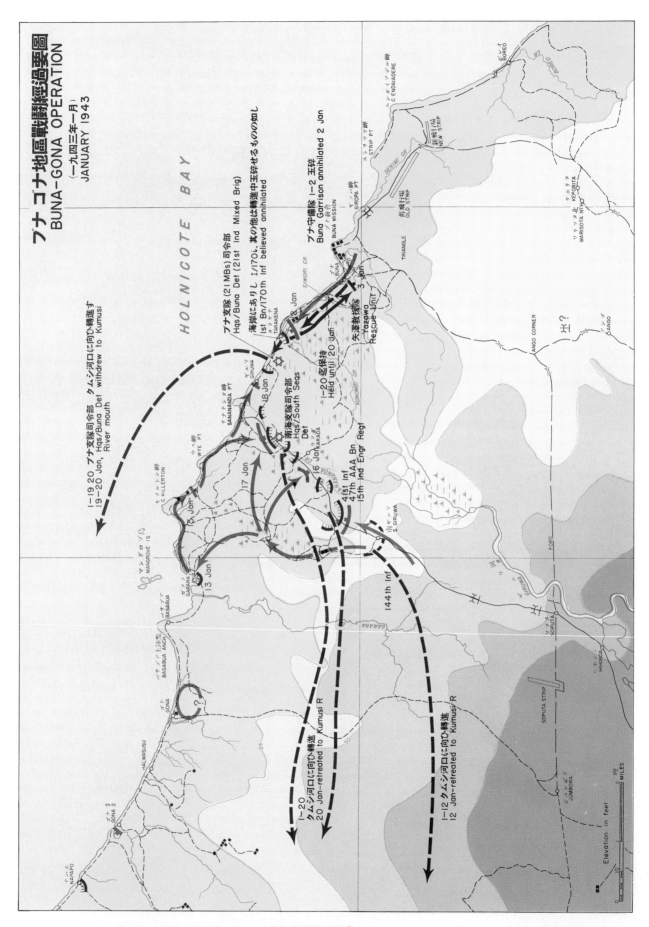

PLATE NO. 45

Buna—Gona Operation, January 1943

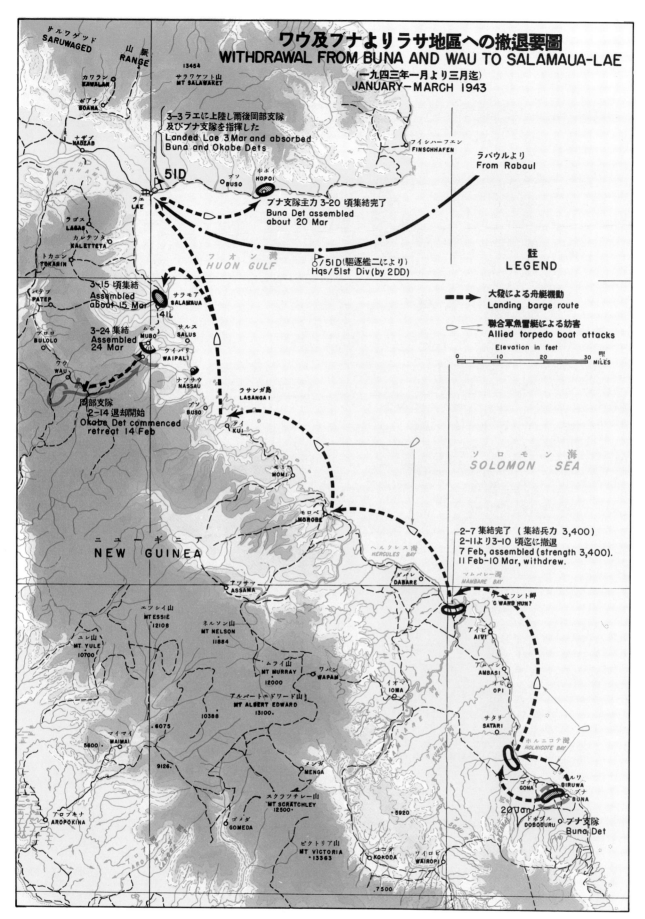

PLATE NO. 46

Withdrawal from Buna and Wau to Salamaua-Lae, January—March 1943

been thrown into the Owen Stanleys drive and the subsequent effort to stop the Allied counteroffensive. About 15,000 had been lost in the whole campaign.[55] Of this number, the bitter fighting in the Buna–Gona area alone had cost between 7,000 and 8,000 lives, of which over 4,000 were killed in battle and the remainder succumbed to disease.[56] Despite this costly effort, Papua had been lost, and with it the strategic area north of the Owen Stanley Range, the key to Port Moresby.

### Strengthening of Bases in New Guinea

While the Japanese forces in Papua were still carrying on their stubborn but hopeless fight to retain possession of the vital Buna–Gona area, hasty action was being taken by the Eighth Area Army and the Combined Fleet to reinforce the general Japanese strategic position in New Guinea through the seizure of new bases on the northeast New Guinea coast and on both sides of the Vitiaz Strait.

By its directive of 18 November 1942, Imperial General Headquarters had recognized the necessity of building up the New Guinea flank against General MacArthur's advance by establishing bases in the vacuum areas to the rear of the vulnerable Japanese advance outposts at Lae–Salamaua and Buna. The Eighth Area Army had therefore been ordered, as one of its initial missions, to effect the early occupation of Madang, Wewak and other strategic points.[57]

By early December, when plans and preparations for execution of this mission were under way, the increasing probability that the Buna-Gona area could not be held made it doubly essential to effect an immediate strengthening of Japanese defenses to the north. The Eighth Area Army and Southeast Area Fleet commands therefore decided to supplement the occupation of Madang and Wewak with the simultaneous seizure of Finschhafen, on the Huon Peninsula, and Tuluvu, on western New Britain, both of which were considered necessary to safeguard Japanese control of the Dampier and Vitiaz Straits and thus strengthen the defense of Lae–Salamaua.

Although the initial characteristic of these plans was defensive, they were also designed to lay the groundwork for the ultimate resumption of the offensive by the Japanese forces in New Guinea, after the American assault on the Solomons had been successfully parried. Emphasis was placed upon the development of operational air bases at Tuluvu, Wewak and Madang, and Wewak was to be transformed into a big rear supply base for the support of future operations.

On 12 December, Lt. Gen. Imamura, Eighth Area Army Commander, assigned the mission of occupying Madang, Wewak and Tuluvu to the Eighteenth Army, placing under its command for this purpose three newly-arrived infantry battalions of the 5th Division[58] and the 31st Road Construction Unit. Under final Eighteenth Army plans, two infantry battalions were allotted to the occupation of Madang, one to Wewak, and the 31st Road Construction Unit (less two companies) to Tuluvu.[59] The

---

55 Cf. Vol. I, SWPA Series: *MacArthur's Campaigns in the Southwest Pacific*, p. 96, GHQ SWPA Communique No. 271, 8 Jan 43.

56 Statement by Lt. Col. Tanaka, previously cited.

57 Cf. text of Imperial General Headquarters directive, already quoted on p. 159.

58 Transfer of three battalions from the 5th Division, then operating under Sixteenth Army command in the Dutch East Indies, to Eighth Area Army was effected by Imperial General Headquarters order. These battalions were taken from the 11th, 21st and 42d Infantry Regiments. (Statement by Col. Hattori, previously cited.)

59 (1) Major New Guinea Operations, op. cit., pp. 45-7. (2) Southeast Area Operations Record, Part III, op. cit. Vol. I, pp. 172-4.

occupation of Finschhafen, by local Army-Navy agreement, was assigned to a small force of special naval landing troops.[60]

Naval convoys carrying the Madang and Wewak occupation forces sailed from Rabaul on 16 December, while a surface support force including one aircraft carrier headed south from Truk to cover the operation. The Wewak force reached its destination without mishap on 18 December, but the Madang force underwent both air and submarine attack off the New Guinea coast, the escort flagship *Tenryu* sinking as a result of torpedo hits and one converted cruiser carrying troops receiving bomb damage. Despite these attacks, the convoy continued to Madang and unloaded its troops early on 19 December.[61]

While the Madang and Wewak operations were in progress, the Tuluvu occupation force completed its movement from Rabaul aboard a single destroyer on 17 December.[62] The Finschhafen force left Kavieng, New Ireland, on two destroyers the following day, executing a successful landing on 19 December.[63] Work began immediately at the occupied points to prepare airstrips for operational use and set up base installations.

Immediately upon completion of these new occupation moves, the Eighth Area Army turned its energy to the urgent problem of strengthening the defenses of the Lae-Salamaua area, now seriously jeopardized as a result of the deteriorating situation on the Buna–Gona front to the south. This area was tenuously held by a naval garrison force of 1,300 men, which had never been able to do more than secure the immediate vicinities of Lae and Salamaua against enemy guerrilla forces. At Wau, 30 miles southwest of Salamaua, the Allied forces possessed a strategically located base of operations, with an airfield capable of accommodating at least light planes.

On 21 December, Lt. Gen. Imamura ordered the Eighteenth Army to strengthen its strategic position for future operations " by securing important areas to the west of Lae and Salamaua."[64] A further order on 28 December directed the immediate dispatch of troops to the Lae–Salamaua area, and on 29 December Lt. Gen. Adachi, Eighteenth Army Commander, ordered the Okabe Detachment, composed of one reinforced infantry regiment of the 51st Division,[65] to proceed from Rabaul to Lae. The missions assigned to the detachment were specified as follows:

*1. The detachment, in cooperation with the Navy, will land in the Lae area, and a portion of its strength will secure that area.*

*2. The main strength will immediately advance to Wau, and elements to Salamaua, in order to secure those areas and establish lines of communication.*

*3. The detachment will thereafter be responsible for the land defense of the Lae–Salamaua area, and*

---

60  The Finschhafen force was one company (270 men) of the Sasebo 5th Special Naval Landing Force. Major New Guinea Operations, op. cit., p. 48.

61  Ibid., p. 46.

62  Greater East Asia War Summary, op. cit., p. 73.

63  Major New Guinea Operations, op. cit., pp. 48–9.

64  Southeast Area Operations Record, Part III, op. cit. Vol. I, pp. 88–9.

65  Maj. Gen. Tooru Okabe, 51st Infantry Group Commander, was placed in command of the detachment. Detailed composition of the force was as follows:  51st Inf. Gp. Hq.; 102d Infantry Regt.; 2d Battalion (less one company) 14th Field Artillery Regiment; one engineer company; one transport company; and one field antiaircraft machine-gun company. The 51st Division had reached the Rabaul area early in December from South China, and was placed under Eighteenth Army command. (1) Southeast Area Operations Record, Part III, op. cit. Vol. I, pp. 95–6; (2) *Maru* Operations Order A No. 270, 1 Jan 43. ATIS Bulletin No. 260, 27 Jul 43, p. 1.

*it will also carry out preparations for offensive operations against the Buna area.*[66]

Five transports carrying the Okabe Detachment sailed from Rabaul on 5 January with a surface escort of five destroyers. On 6 January, enemy B-17's spotted the convoy as it proceeded through the Bismarck Sea, and the transport *Nichiryu Maru*, carrying most of the 3d Battalion, 102d Infantry Regiment, caught fire and sank after receiving a direct bomb hit.[67] Though badly battered, the remainder of the convoy proceeded on to Lae, where it arrived on 7 January and began discharging troops and supplies under continuous Allied air attack.

Despite efforts to break up the enemy air assault by fighters which had moved forward from Rabaul on 7 January to operate from Lae, bombing of the anchorage became so severe on the 8th that unloading had to be discontinued and those ships which were still navigable sent back to Rabaul.[68] With the exception of the troops aboard the *Nichiryu Maru*, all personnel of the Okabe Detachment had been put ashore, but only half of the supplies had been safely unloaded.[69]

## The Wau Offensive

Notwithstanding this initial setback, Maj. Gen. Okabe decided to proceed according to plan and immediately ordered the detachment to prepare to move against Wau. The general plan of attack called for the main strength of the detachment to move by landing craft to Salamaua, and from there advance on Wau via Mubo, Waipali, and the mountain track through Biaru. (Plate No. 47) This route was chosen in preference to the easier track leading from Mubo along the Buisaval River Valley, which offered little cover against enemy air attack.[70]

Amphibious movement of the detachment from Lae to Salamaua was completed between 10 and 16 January. Two days before its completion, on 14 January, the advance echelon of the attack force, composed of the 1st Battalion, 102d Infantry (reinf.), had already moved out of Salamaua on the first leg of its advance toward Wau. Maj. Gen. Okabe, with detachment headquarters and the main body of the attack force, followed on 16 January. Total effective strength of the force as it started out from Salamaua was approximately 3,000 officers and men.[71]

First enemy ground reaction developed as the advance echelon moved south from Mubo on 16-17 January. In the vicinity of Waipali, a small enemy force of about 40 men, equipped with mortars, offered light resistance, retiring

---

66 Southeast Area Operations Record, Part III, op. cit. Vol. I, pp. 90-1.

67 Escorting destroyers rescued 739 of the 1,100 troops aboard. Most of these returned to Rabaul, but some disembarked at Lae and were kept there as a supply depot unit. Ibid., p. 97.

68 A second transport, the *Myoko Maru*, was so badly damaged by Allied bombing that she had to be beached to prevent sinking.

69 All medical supplies were lost with the sinking of the *Nichiryu Maru*. Report on Medical Situation during the Wau Operation. ATIS Current Translations No. 73, 10 Aug 43, p. 6-C.

70 Three alternative routes had been considered by Eighteenth Army headquarters during the preliminary planning, but final decision was left to Maj. Gen. Okabe and his staff to be made after arrival at Lae. These routes were: Lae—Markham Point—Wampit—Bulolo—Wau; Salamaua—Misim—Wau; and Salamaua—Mubo—Wau. (Statement by Lt. Col. Tanaka, previously cited.)

71 Only seven of the 102d Infantry Regiment's 12 infantry companies participated, since two companies of the 3d Battalion were lost with the *Nichiryu Maru*, two companies (2d and 8th) were dispatched south to cover the withdrawal of the Japanese forces from Giruwa, and one company was assigned to garrison duty at Mubo. (Statement by Lt. Col. Tanaka, previously cited.)

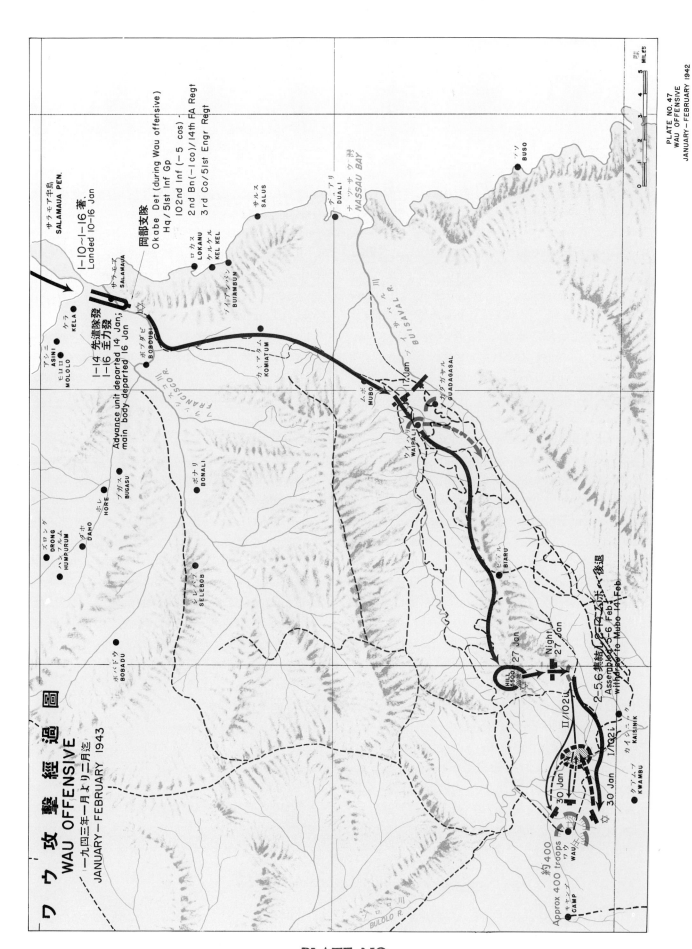

PLATE NO. 47

Wau Offensive, January—February 1943

southward after a brief encounter.[72] The advance echelon then pushed on to the southwest, the steadily increasing difficulty of the mountainous terrain slowing its rate of progress at times to less than three miles a day. It was 27 January before all units of the attack force had finally assembled at Hill 5500, about six miles northeast of Wau, whence the attack was to be mounted.

From the vantage point of Hill 5500, Wau and its adjacent airfield were clearly visible and appeared to be within a few hours' march of the assembly area. Maj. Gen. Okabe, estimating enemy strength at no more than 400 and anxious to gain the advantage of surprise, immediately ordered the 102d Infantry regimental commander to launch the attack on the night of the 27th.[73] The final attack plan called for the regiment's right wing (2d Battalion, reinf.) to strike at the airfield defenses from the east and northeast, while the left wing (1st Battalion, reinf.) was to launch the main attack from the southeast. Both attacks were scheduled to begin at 0100, 28 January, and all objectives were to be occupied by dawn.[74]

Right and left wings began moving into position for the attack at dusk on 27 January. A small enemy patrol encountered two miles south of Hill 5500 was rapidly dispersed, but progress through the unknown jungle terrain in darkness was so slow that, even by dawn, neither force had reached its scheduled attack position. Movement was stopped until late afternoon of the 28th to guard against attack by enemy aircraft. During the evening, as the advance resumed, a further encounter with an enemy elements delayed progress, and the morning of the 29th found the attack columns still bogged down in the jungle.

A sharp increase in enemy fighter activity kept the Japanese troops pinned down again until the night of the 29th, when both columns pressed forward once more. Again the advance was so slow that, by dawn of 30 January, the left wing force was still about two and a half miles from the airfield. Meanwhile, the enemy was profiting from the delay to fly in reinforcements.

Deciding that any further delay might spell failure, Maj. Gen. Okabe personally took command of the left wing force and ordered it forward on the night of 30 January to attack the southwest perimeter of the airstrip. The attack failed, however, as the assault units, moving up in the darkness, suddenly ran into fierce automatic weapons fire from enemy positions and were thrown into confusion.

Meanwhile, on the right flank, the 2d Battalion had launched an attack on the morning of 30 January and succeeded in capturing a segment of the enemy positions at the northeast corner of the airfield. Due to severe losses, however, the battalion was unable to hold its ground and fell back east of the airfield to reorganize. The strength of both 1st and 2d Battalions was now badly depleted. Average company strength was down to 50 in the 1st Battalion, and 40 in the 2d. Artillery units were at one-third and engineer units at one-half of normal strength.[75]

The reinforced enemy troops in Wau, with heavy air support, now launched a counteroffensive, which resulted in sharp fighting just southeast of the airfield. By 4 February, the 102d Infantry was threatened with encirclement, and on the 6th Maj. Gen. Okabe ordered all units to retire to a concentration area two and a half miles east of the airfield to reorganize.

---

72 Southeast Area Operations Record, Part III, op. cit. Vol. I, p. 113.
73 Ibid., pp. 15–6.
74 102d Infantry Regiment Operations Order, 27 Jan 43. ATIS Current Translations No. 27, op. cit., p. 2.
75 Southeast Area Operations Record, Part III, op. cit. Vol. I, p. 119.

On the same day, ten Japanese fighter planes sent from New Britain attacked the Wau airfield in an effort to curb enemy air activity, but the effort could not be maintained and consequently failed to improve the situation appreciably.[76]

On 12 February, Maj. Gen. Okabe ordered a further withdrawal to a provisions storage dump about a mile and a half to the rear. The troops, on short rations since an early stage of the advance from Salamaua, had exhausted their food supplies during the protracted campaign and were now existing on wild potatoes (taros) and a small amount of captured enemy provisions.

With its hopes of taking Wau completely shattered, the Okabe Detachment on 13 February received orders from the Eighteenth Army command at Rabaul to abandon the attempt and withdraw its forces to Mubo and the Nassau Bay area. The withdrawal began on the 14th and was completed in ten days without enemy pursuit. Out of 3,000 troops which had set out from Salamaua for the Wau offensive, only 2,200 survivors returned to Mubo. More than 70 per cent of these, moreover, were suffering from malaria, malnutrition, dysentery and other diseases, and were unfit for combat duty.[77]

The failure of the attempt to take Wau had serious consequences for the Japanese situation in New Guinea. Not only had the major strength of the Okabe Detachment been expended in futile fighting, but the Eighteenth Army's plans to strengthen the flank defenses of the Lae–Salamaua area were seriously unhinged.

## Evacuation of Guadalcanal

While the Eighteenth Army in New Guinea was being forced to pull back its front line to the Lae-Salamaua area following the loss of Papua and the failure of the Wau offensive, a withdrawal on a much larger scale and of considerably greater difficulty was being carried out from Guadalcanal, in the Solomons, under the Imperial General Headquarters directive of 4 January.[78]

Following the collapse of the second general offensive on Guadalcanal in late October 1942, the Seventeenth Army and Southeast Area Naval Force had continued efforts to move in reinforcements for a new offensive planned for January.[79] Initial elements of the 38th Division were successfully transported by destroyers from the Shortland Islands in early November, but the main reinforcement effort in mid-November met disaster when Allied planes sank or set afire all but four of eleven transports en route from Bougainville.[80] In a series of accompanying surface actions between 11 and 15 November, moreover, the Japanese naval forces lost two battleships, one cruiser and three destroyers,

---

76 Eighth Area Army headquarters ordered the air attack on Wau only with extreme reluctance, since all available Army and Navy aircraft on New Britain were needed to support the current withdrawal operations from Guadalcanal. The Wau sortie was ordered, however, in view of reports from Maj. Gen. Okabe indicating the serious situation of his forces. (Statement by Col. Sugita, previously cited.)

77 Southeast Area Operations Record, Part III, op. cit. Vol. I, pp. 127–8, 131.

78 Cf. text of 4 January directive, quoted on p. 169.

79 The decision in favor of a new offensive in January was strongly influenced by reports regarding the damage inflicted on the American naval forces in the Santa Cruz sea battle of 26 October. These reports claimed a heavy blow to enemy carrier strength, which it was thought would facilitate the movement of large-scale reinforcement to Guadalcanal. (Statement by Capt. Ohmae, previously cited.)

80 The four transports which reached Guadalcanal were damaged by bomb hits and had to be beached to permit unloading of troops. *Nanto Homen Kaigun Sakusen Sono Ichi* 南東方面海軍作戦其の一 (Southeast Area Naval Operations, Part I) 2d Demobilization Bureau, Jun 49, p. 40.

Original Painting by Toraji Ishikawa  Photograph by U. S. Army Signal Corps

## PLATE NO. 48

Sea Battle in South Pacific

佐野部隊長還らざる大野挺身隊と訣別す　田村孝之介　昭和一九年

Original Painting by Konosuke Tamura  
Photograph by U. S. Army Signal Corps

PLATE NO. 49

Suicide Unit Bidding Farewell to Commanding General Sano

with three cruisers and three destroyers heavily damaged.[81]

With aerial supremacy over the southern Solomons already in Allied hands and the combat effectiveness of the naval forces reduced by ship losses, Imperial General Headquarters reluctantly decided that the fight to retake Guadalcanal must be abandoned and all Japanese forces withdrawn.[82] The directive of 4 January accordingly ordered the Eighth Area Army and Combined Fleet to make immediate preparations for the withdrawal.

Evacuation of approximately 18,000 troops still exchanging fire with the enemy from the immediate vicinity of an enemy airfield was a formidable task which required careful planning and preparation. Land-based naval air units on New Britain were weakened by extended combat. Carrier aircraft strength, seriously depleted in the Santa Cruz sea battle, had not yet been replenished, and the withdrawal operation involved risking virtually all the remaining destroyer forces of the Combined Fleet.

Despite these handicaps, the Eighth Area Army and Southeast Area Fleet jointly worked out plans which called for the employment of all available aircraft in a sustained offensive designed to neutralize enemy air and sea strength long enough to permit seaborne evacuation operations. Following a preliminary series of night raids, mass daylight attacks were to begin from about 28 January. The ground forces were to begin gradual withdrawal to embarkation points from 25 or 26 January, and the evacuation itself was to be effected by destroyers in three separate runs on the nights of 31 January, 3 and 6 February.[83] The plans called for participation of 212 Navy and 100 Army aircraft, predominantly fighters, while 22 destroyers and several submarines were made available by the Navy.[84]

Night raids by small numbers of Navy aircraft on Henderson Field began on 21 January and continued almost without interruption until the end of the month. The first mass daylight attack was staged on 25 January by 91 Navy planes, followed on the 27th by Army fighters of the 6th Air Division. On 29 January naval aircraft reported inflicting heavy damage on an enemy naval force, including cruisers and battleships, between San Cristobal and Rennell Islands.[85]

Due to the appearance of the enemy naval force, the evacuation schedule was retarded one day, the first evacuation taking place on the night of 1–2 February. Eighteen destroyers drew in at Kaminbo, on the northwestern tip of Guadalcanal, and successfully took aboard 4,940 troops who were put ashore the following day in the Shortland Islands. One destroyer sank upon hitting an enemy mine near Kaminbo, while another was damaged by air attack and had to withdraw.

On 2 February 56 Navy planes carried out another heavy strike on Henderson Field to keep enemy air power neutralized. The second evacuation followed on the night of the 4th, when 17 destroyers took aboard and carried to the Shortland Islands 3,902 troops. In this operation one destroyer was hit by an enemy bomb and forced out of action.

In the final evacuation on 7 February, 1,730 troops were removed from the island, bringing

---

81  Ibid., 38–40.
82  This decision was formally made in the Imperial conference on 31 December 1942. (Statement by Col. Joichiro Sanada, Chief, Operations Section, Imperial General Headquarters, Army Section.)
83  Southeast Area Operations Record, Part II, op. cit. Vol. II, p. 12.
84  Southeast Area Naval Operations, Part I, op. cit., pp. 51–2.
85  Ibid., pp. 55–6.

the total number of troops evacuated to 10,572.[86] Including damage sustained by one destroyer in this operation, total naval losses for the whole evacuation amounted to only one destroyer sunk and three damaged.[87]

With the termination of the fight for Guadalcanal, the Solomons area entered a period of temporary quiescence, during which both sides prepared for the next phase of battle. The Japanese front line was withdrawn to New Georgia and Santa Isabel Island. These were only lightly garrisoned by about three infantry battalions and a few antiaircraft units, and airfields were still in process of construction. To remedy this situation, the Southeast Area Fleet, in the latter part of February, directed the Eighth Fleet to move two units of the 8th Combined Special Naval Landing Force to Munda as a preliminary reinforcement measure.[88] In April these were augmented by elements of two infantry regiments (13th Inf. Regt., 6th Division, and 229th Inf. Regt., 38th Division), and on 3 May all Army forces in the New Georgia area were combined in a newly-activated Southeast Detachment under the operational command of the Eighth Fleet.[89]

Ground defense of the northern Solomons was left in the hands of the Seventeenth Army. Army headquarters was established on Bougainville, and the 6th Division, already moved to Bougainville in January, was newly placed under Seventeenth Army command. The battered units evacuated from Guadalcanal were gradually moved back to Rabaul, where the 38th Division was reorganized for defense of New Britain. The 2d Division and 35th Infantry Brigade were transferred to other theaters.[90]

## Menace of the B-17's

Various factors were responsible for the parallel setbacks suffered by the Japanese forces in Papua and the Solomons, but the most important of these was the gradual loss of air supremacy over the areas of battle to the Allies.

At the time of the American invasion of the Solomons in mid-summer of 1942, the outcome of the battle for aerial supremacy still hovered in the balance. Japanese naval aircraft based at Rabaul, chiefly Zero fighters and land-based medium bombers, were still able to operate with a certain degree of effectiveness over Papua and the Solomons, where the Allies did not yet possess superiority in numbers of aircraft.

However, Allied plane strength in the southeast area soon began to increase at a rate with which the Japanese could not keep pace. Numerical superiority passed to the hands of the enemy, and in addition, his ability swiftly to construct and expand forward bases increased the effectiveness of his air forces. Similar Japanese efforts to develop forward air bases, though they made some progress, were retarded by shortages of manpower and equipment, with the result that sorties were still being flown chiefly from Rabaul in the fall of 1942. The distances involved seriously curtailed the effectiveness of the air effort over Papua and the

---

86 Exact number of troops evacuated from Guadalcanal is difficult to determine due to the contradictions found in available wartime documents. Figures used in this chapter are as accurate as can be determined from the existing documents. (1) Situation Report (Summary) of the Seventeenth Army, p. 4. (2) Report to the Emperor (Draft) by the Seventeenth Army Commander, p, 17. (3) The Number of Troops Retreating to Erventa. Extracted from the Private Papers of Col. Haruo Konuma, Staff Officer (Operations), Seventeenth Army.

87 Southeast Area Naval Operations, op. cit. Vol. I, pp. 56-7.

88 These units were the Kure 6th Special Naval Landing Force and the Yokosuka No. 7 Special Naval Landing Force. *Nanto Homen Kaigun Sakusen Sono Ni* 南東方面海軍作戰其の二 (Southeast Area Naval Operations, Part II) 2d Demobilization Bureau, Jun 49. pp. 6 and 14.

89 Maj. Gen. Minoru Sasaki was appointed to command the Southeast Detachment and arrived on Kolombangara Island 31 May to take command. Southeast Area Operations Record, Part II, op. cit. Vol. II, pp. 105-6.

90 Ibid., pp. 98-9.

Solomons.[91]

In the Solomons, it was primarily the enemy's expanding carrier-borne air forces which captured control over the Guadalcanal battlefield and thwarted Japanese reinforcement attempts. In the battle for Papua, a major factor was the long-range B-17 bomber. From the autumn of 1942, these powerful craft intensified their attacks on Japanese troop and supply shipping in the Solomon and Bismarck seas, and by December were carrying out regular night raids on Rabaul itself.

In an attempt to elude B-17 attack, Japanese vessels on transport and supply missions began moving as much as possible at night or in bad weather, but enemy radar equipment made even such movement risky. Japanese destroyers, despite their speed and maneuverability, often could not elude the extremely accurate bombing of the B-17's, and escort fighters offered little protection.[92] The Zero fighter, armed with two 20-millimeter automatic cannon, was then a relatively powerful craft, but repeated engagements indicated that two or three Zeros still were no sure match for a single B-17.[93] Attempts to develop new fighter types capable of combatting the B-17's were only partially successful.[94]

The gradual loss of the air campaign over the Solomons and eastern New Guinea underlined the urgent necessity of infusing fresh air strength into the southeast area. This in turn demanded accelerated mass production of aircraft and training of air crews in the homeland. During the bitter battle for Guadalcanal, the Navy had poured in a large portion of its available land-based air strength, but this had been so rapidly expended that the Japanese air potential in the southeast area actually showed little or no increase.[95] The Army's 6th Air Division, though activated in November to reinforce the naval air forces in the southeast area, did not begin operating from Rabaul until late in December, when both the Guadalcanal and Buna campaigns were already virtually lost.

To alleviate one of the major handicaps which had reduced the effectiveness of the Japanese Air forces in these campaigns, the Army and Navy commands at Rabaul began early in 1943 to concentrate special effort on the construction of new air bases and the reinforcement of air defenses in northeast New Guinea and western New Britain. At Wewak, Madang and Tuluvu, lack of airfield construction personnel and equipment neces-

---

91  The strips at Lae and Buna, though improved for operational use, were inadequate and subject to frequent enemy air attack. In the Solomons, construction of bases in the Buin area on Bougainville and on New Georgia had not been completed until after the Allied invasion of Guadalcanal. (Statement by Capt. Ohmae, previously cited.)

92  During the one-month period from 15 November to 15 December 1942, B-17's sank one destroyer carrying troop reinforcements to Buna and damaged six others. By 15 December all destroyer movement to the Buna area had to be abandoned. Greater East Asia War Summary, op. cit., pp. 45–72.

93  At Rabaul it was extremely rare for a B-17 to be shot down either by antiaircraft fire or defending fighters. Vice Adm. Masao Kanazawa, 8th Naval Base Force Commander, recorded that he first saw a B-17 shot down over Rabaul on 9 August 1942, and that all personnel were " wild with joy." Extracted by the writer from personal papers of Vice Adm. Kanazawa.

94  In the spring of 1943, the new *Gekko* (月光) night fighter, armed with a fixed machine gun mounted at an angle of about 30 degrees to the fuselage axis, was pitted against the B-17, but it was only partially effective in checking night raids on Rabaul. (Statement by Capt. Ohmae, previously cited.)

95  Between the American invasion of Guadalcanal in August and the end of 1942, nearly 800 naval planes were expended in the Solomons campaign. Approximately one-third of this total represented carrier-borne aircraft. (Statistical data compiled by 2d Demobilization Bureau, Liquidation Division.)

働く軍隊（南方戦區）　寺内萬次郎　昭和二十三年三月十七日

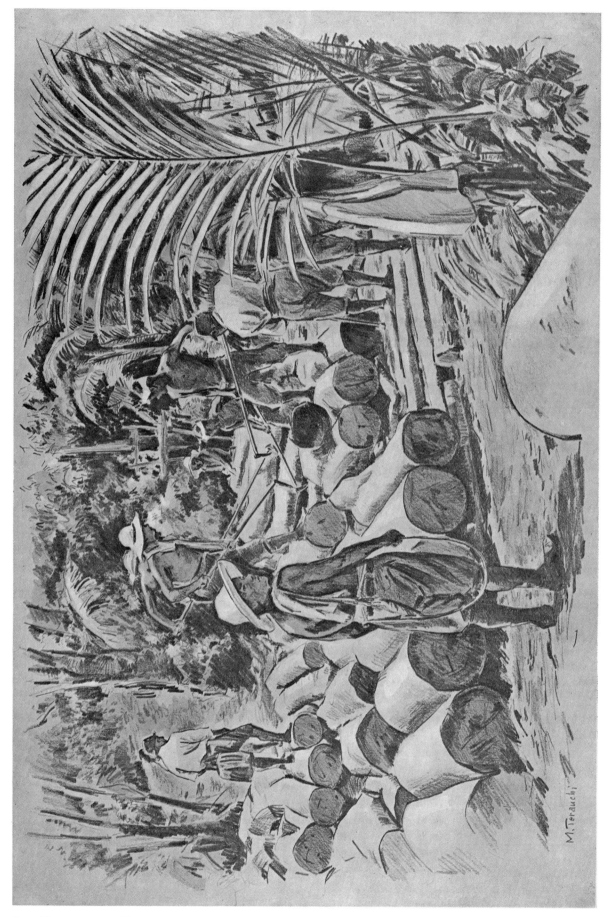

Original Painting by Manjiro Terauchi

PLATE NO. 50

Troops at Work, Southern Area

brilliant victory for Allied air power, was over.

Out of slightly over 6,900 troops badly needed for the defense of Lae–Salamaua, 3,664 had been lost. Only about 800 troops had actually reached Lae, while 2,427 survivors were brought back to Rabaul. Supplies and heavy equipment aboard the transports had gone down with the burning ships, and all survivors with the exception of those which reached Lae by destroyer had lost even their small arms.[108]

With the Bismarck Sea disaster, the Army and Navy commands in the southeast area were forced to relinquish all hope of sending troops or supplies directly to Lae by regular transport vessels or by destroyers. Henceforth, ships could proceed only as far as Finschhafen, whence troops or supplies destined for Lae had to move overland or by small landing craft. The first transport run to Finschhafen was carried out on 20 March by four destroyers carrying approximately one reorganized battalion of the 115th Infantry Regiment, 51st Division. Two further attempts were made on 2 and 10 April to transport 66th Infantry units, but on both occasions enemy air attacks forced the destroyer convoys to turn back before reaching Finschhafen.

With destroyer movement even as far as Finschhafen rapidly becoming perilous under the menace of Allied air power, resort was made to transport by small landing craft, which moved only at night along a chain of bases from Tuluvu, on northwest New Britain, to Lae. The loading capacity of these small craft was generally between five and ten tons, but by using approximately 200 of them, it was possible to transport more than 3,000 troops and a considerable amount of supplies from Rabaul to Lae over a period of about four months.[109] Later, a similar transport system was established along the New Guinea coast linking Hansa and Madang with Finschhafen in order to facilitate the movement of 20th and 41st Division troops to Lae. Submarines were also used extensively after March to move medical supplies, rations, and vital equipment to the Lae area from New Britain.[110]

Although these makeshift measures were partially effective, the ever increasing difficulty of transporting troops and supplies by sea in the New Guinea area strengthened the demand for developing overland transport routes linking Lae with rear bases at Madang and Hansa. Already at the end of January, Eighteenth Army had ordered the 20th Division to undertake construction of a road from Madang to Lae via the Mintjim-Faria Divide in the Finisterre Range, and the Ramu and Markham River Valleys.[111] Work actually was not begun until April, however, and the difficulties encountered were so much greater than anticipat-

---

108 Foregoing account of the Bismarck Sea battle is based on the following sources: (1) Greater East Asia War Summary, op. cit., pp. 142–50. (2) Statements by Comdr. Nikichi Handa, Staff Officer (Signal), Lae Transport Escort Force and Capt. Ohmae and Col. Sugita, previously cited. (Col. Sugita, accompanying the Eighteenth Army headquarters, was aboard the destroyer *Tokitsukaze*, sunk on 3 March.)

109 Two alternative routes were used between Tuluvu and Lae: (1) Tuluvu—Busching—Finschhafen—Lae; and (2) Tuluvu—Umboi—Sio—Finschhafen—Mange—Lae. The landing craft were able to complete each leg of their voyage during ten hours of darkness each night, remaining inactive during the day to escape enemy air attack. Southeast Area Operations Record, Part III, op. cit. Vol. I, pp. 259–61.

110 From March to September 1943, submarines made 81 supply runs from Rabaul to Lae. Major New Guinea Operations, op. cit., p. 30.

111 A survey of the projected Madang–Lae road, carried out between late December and early February by units stationed at Madang, found that construction of a road suitable for motor traffic would require four to five months using a labor force of 3,000 men. Even before the survey was completed, however, Eighteenth Army had decided that the project must be undertaken at any cost. Southeast Area Operations Record, Part III, op. cit. Vol. I, pp. 188, 191 and 199.

ed that by July the road had been completed only as far as Mablugu, 40 miles from Madang. Allied aircraft also hampered the project by bombing the supply base at Erima and bridges to the south.

## Shift of Emphasis to Papua

The virtually complete destruction in the Battle of the Bismarck Sea of the 51st Division forces counted upon to hold the Lae–Salamaua area against anticipated Allied attack shocked Imperial General Headquarters into realization of the extremely tenuous situation on the Japanese right flank in the southeast area. This served as a decisive reason for a vital revision of joint Army-Navy operational policy, whereby first priority was definitely shifted from the Solomons area to New Guinea.

The terms of the revised policy were stipulated in a new Army-Navy Central Agreement on Southeast Area Operations, issued by Imperial General Headquarters as an operational directive to the Eighth Area Army and Combined Fleet on 25 March 1943. This directive stated:[112]

> *1. Operational Objective: To establish a strong strategic position by occupying and securing key points in the southeast area.*
> *2. Operational Plan:*
> *a. Army and Navy forces, acting in complete coordination, will concentrate their main effort on operations in New Guinea and will secure operational bases in that area. At the same time, defenses will be strengthened in the Solomon Islands and the Bismarck Archipelago, key points will be secured, and future enemy attacks will be crushed at the opportune time.*
> *b. New Guinea Operations:*
> *(1) Strategic points in the Lae–Salamaua area will be held against enemy airborne, ground or sea attack. The Army and Navy will take all necessary measures to maintain supplies to this area and increase the combat strength of the forces there.*
> *(2) Air operations will be intensified, and enemy air strength destroyed as far as possible. Every effort will be made to check increased enemy transport, especially along the east coast of New Guinea, and at the same time to provide thorough protection of our own supply routes.*
> *(3) Army and Navy forces will cooperate in immediately strengthening air defense installations, air bases and supply transport bases in New Guinea and New Britain. Efforts will also be made, principally by Army forces, to complete the construction of necessary roads and speed the accumulation of military supplies. New operational bases will be developed in New Guinea and western New Britain.*
> *(4) Troop strength in the Lae-Salamaua area will be reinforced, and various military installations improved. Preparations will subsequently be made for the resumption of operations against Port Moresby.*
> *3. Air Operations:*
> *a. In order to facilitate general operations, the Army and Navy will speedily reinforce their air strength and expand air operations on a large scale.*
> *b. Special effort will be made to increase the effectiveness of these operations by close cooperation between Army and Navy Air forces.*

Under further stipulations governing air operations, missions of the Army Air forces were restricted principally to the New Guinea area, while the Navy Air forces, in addition to supporting New Guinea air operations, remained primarily responsible for defense of the Bismarck Archipelago and solely responsible for air operations in the Solomons. Army air strength in the southeast area was to be stepped up to 240 aircraft of all types by September 1943, and Navy air strength to 357 planes, exclusive of carrier-borne aircraft, by the end of June.[113]

---

112 *Daikaishi Dai Nihyakujusan-go* 大海指第二百十三號 (Imperial General Headquarters Navy Directive No. 213) 25 Mar 43. (Imperial General Headquarters Army Section directive was couched in identical terms.)
113 Ibid.

To implement the Imperial General Headquarters directive, Lt. Gen. Imamura, Eighth Area Army Commander, summoned a conference at Rabaul on 12 April, attended by the commanders of the Seventeenth and Eighteenth Armies, the 6th Air Division, and units under direct Area Army command. At this conference, the following Area Army order was issued, specifying the missions of the Seventeenth and Eighteenth Armies and 6th Air Division:[114]

*1. In cooperation with the Navy, the Area Army will endeavor to achieve the following objective: In the New Guinea area, to consolidate its strategic position and carry out preparations for subsequent offensive operations. In the Solomons and Bismarck Archipelago, to consolidate and strengthen present positions.*

*2. The Seventeenth Army, in cooperation with the Navy, will conduct operations in the Solomons area in accordance with the following:*

  *a. The Army will assume responsibility for the defense of the northern Solomons. It will consolidate and, as far as possible, strengthen existing positions.*

  *b. In matters pertaining to operational preparations, the Army will direct Army units operating under Navy command in the zone of naval responsibility in the central Solomons....*

*3. The Eighteenth Army, in cooperation with the Navy, will conduct operations in the New Guinea area in accordance with the following:*

  *a. The Army will first secure the strategic sectors of Lae and Salamaua, and by assuring the flow of supplies to these sectors, establish a firm basis for strengthening the Army's strategic position. To facilitate these objectives, the Army will speedily formulate plans for the establishment of overland and coastal supply routes linking Madang and western New Britain with the Lae area.*

  *b. To strengthen transport and supply operations, line of communications and naval transport bases will be established and improved at important points along the eastern New Guinea coast west of Madang. Air bases will also be established as required.*

  *c. Along with the consolidation of the Army's strategic position as outlined above, all positions will be strengthened and preparations made for future operations.*

*4. The 6th Air Division will gradually advance its bases of operation to eastern New Guinea and, in cooperation with the Navy, will undertake the following missions....:*

  *a. Destruction of enemy air power in the eastern New Guinea area.*

  *b. Provision of direct air cover for water transport in this area.*

  *c. Direct support, when required, of Army ground operations.*

  *d. Constant reconnaissance of enemy land and sea communication routes in the eastern New Guinea area, and attacks on these lines whenever opportune.*

  *e. Defense of Rabaul.*

  *f. Ferrying of supplies to the front by air, whenever necessary.*

Although the Imperial General Headquarters directive of 25 March and Eighth Area Army's implementing order clearly shifted the weight of the Japanese military effort in the southeast area to New Guinea, actually this shift was difficult to accomplish. By 12 April when the Area Army order was issued, transport by destroyer from New Britain to Finschhafen had already become impossible, and the only means of moving men and supplies to Lae was slow and arduous transport by submarine and small craft. Moreover, the combat effectiveness of Japanese forces already stationed in sectors of New Guinea within range of Allied air power was gradually being worn down even before these forces were engaged in actual fighting.

Under these circumstances only the air forces were capable of taking offensive action on the New Guinea front. In order to deter the build-up of enemy strength and to assist the attempts to reinforce advance positions, Admiral Yamamoto, Commander-in-Chief of the Combined Fleet, promptly ordered the Navy Air forces to launch an all-out offensive directed principally at enemy bases in Papua. In addition to 72

---

114 Extracted from personal memoranda of Col. Sugita, previously cited.

land-based medium bombers, 27 carrier dive-bombers and 86 fighters of the Eleventh Air Fleet, the Third Fleet was ordered to participate in the operation with 54 carrier dive-bombers, 96 fighters and a number of carrier torpedoplanes.[115]

The offensive began under the command of Admiral Yamamoto on 7 April with a powerful strike by 71 dive-bombers and 157 fighters against enemy naval and transport shipping at Guadalcanal, in the Solomons. Air action reports claimed damage to one cruiser, one destroyer and 8 transports, in addition to 28 enemy planes shot down. Japanese losses were 21 planes.

Target of the second attack on 11 April was Allied shipping at Oro Bay, south of Buna. Seventy-two fighters and 22 carrier dive-bombers operating from Rabaul participated in the attack, action reports claiming three transports and one destroyer sunk. Under cover of this attack, two destroyers carrying reinforcements and supplies completed successful runs between Rabaul and Tuluvu.

On 12 April the offensive continued with a heavy raid on Port Moresby by 131 fighters and 43 land-based medium bombers. These blanketed the airstrips with bombs, damaged numerous ground installations, and claimed the sinking of a transport anchored in the harbor. Twenty-eight enemy planes were reported shot down or destroyed, against the loss of seven Japanese aircraft.

The final attack was delivered on 14 April against Milne Bay, with 149 fighters and 37 land medium bombers taking part in the sortie. According to the action reports, ten transports in the bay were either sunk or damaged, and 44 enemy planes were shot down against a Japanese loss of only ten aircraft.[116]

Despite the highly effective results of the air offensive, the Navy Air forces were not capable of continuing sustained attacks on so large a scale. Japanese naval leadership, moreover, suffered a severe blow four days after the Milne Bay attack, when Admiral Yamamoto was killed on an inspection flight from Rabaul to Buin. The Admiral's plane, escorted by nine fighters, was nearing Buin on 18 April when about 24 American fighters suddenly attacked. Admiral Yamamoto's plane crashed in the jungle north of Buin, while a second plane carrying his Chief of Staff, Vice Adm. Matome Ugaki, crash-landed at sea.[117] Admiral Mineichi Koga later was appointed to fill the post of Commander-in-Chief of the Combined Fleet.

On 19 April, five months after its activation, the Eighteenth Army finally transferred its headquarters from Rabaul to Madang in order to enable Lt. Gen. Adachi, Army Commander, to assume personal direction of operations in the New Guinea area. Preparations for the defense of Lae—Salamaua were dangerously behind schedule, and speedy action was imperative to meet the threat of a new northward thrust by General MacArthur's forces.

The next stage of the battle for New Guinea was about to begin.

---

115  Greater East Asia War Summary, op. cit., p. 174.

116  Data on air attacks from 7 to 14 April extracted from (1) Greater East Asia War Summary, op. cit., pp. 174–184. (2) Southeast Area Naval Operations Part II, op. cit., p. 9–10. American Editor's Note: Official Allied sources covering these attacks give the following data, cited for comparison against Japanese claims: 7 April raid on Guadalcanal: participating aircraft, 50 bombers, 48 fighters; 39 shot down; no report of American plane losses or damage to ships. 11 April raid on Oro Bay: 40/45 bombers and fighters; 17 shot down, 16 probables; three ships damaged. 12 April raid on Port Moresby: 45/50 bombers, 50 fighters; 17 shot down, 10 probables; four Allied aircraft destroyed, 14 damaged; fuel ands upply dumps destroyed, buildings damaged. 14 April raid on Milne Bay: 61 bombers, 30 fighters; 15 shot down, 9 probables; one ship sunk, four damaged; fuel dump destroyed.

117  Extracted from the personal diary of Vice Adm. Matome Ugaki, Chief of Staff, Combined Fleet. (Vice Adm. Ugaki sustained severe injuries in the crash-landing.)

# CHAPTER IX

# FIGHTING WITHDRAWAL TO WESTERN NEW GUINEA

## Southeast Area Situation, June 1943

Despite steadily intensified efforts through April and May to strengthen Japanese defenses in New Guinea and the central Solomons in preparation for the next phase of hostilities, the strategic and tactical situation which confronted the Japanese command in the southeast area at the beginning of June 1943 remained distinctly unfavorable.[1] (Plate No. 52)

All along a front of approximately 1,200 miles extending from Lae and Salamaua in Northeast New Guinea, through Tuluvu and Gasmata on New Britain, to Bougainville, New Georgia and Santa Isabel in the Solomons, Japanese forces were thinly spread and on the defensive. Reinforcement and supply of the critical points along this extended front were severely hampered by expanding enemy control of the air and sea.

By June it appeared probable that an Allied attack would not be long delayed. Eighth Area Army intelligence reports indicated that enemy forces, estimated at three to four divisions in eastern New Guinea and three divisions in the southern Solomons, were rapidly being made ready for a new offensive effort. Air bases in the Buna area and on Guadalcanal were being improved and expanded. Enemy planes not only were intensifying their attacks on Lae–Salamaua and on Japanese supply shipping in rear-area ports, but were sowing mines in Japanese-held coastal waters in the Solomons and extending their patrol radius to within close proximity of the Equator.[2] Enemy naval forces were boldly attacking Japanese outposts in the central Solomons, subjecting shore defenses to artillery bombardment.

On the New Guinea flank, the key Japanese positions at Lae and Salamaua, guarding the southern land approach to the Dampier Strait, were expected to be the next major objectives of General MacArthur's forces. These positions already were threatened by reinforced enemy troops in the Wau area directly to the southwest, and increasing enemy activity in this sector indicated the probability of an early attack on the outer defenses of Salamaua. At the same time, infiltration of enemy forces into the Bena Bena and Mt. Hagen areas far to the northwest, where they were developing airfields, created a serious potential menace to Japanese rear bases at Madang, Hansa and Wewak.[3]

In the central Solomons, it was estimated

---

1 This chapter was originally prepared in Japanese by Col. Ichiji Sugita, Imperial Japanese Army. For duty assignments of this officer, cf. n. 1, Chapter VI. All source materials cited in ths chapter are located in G-2 Historical Section Files, GHQ FEC.

2 Enemy air strength in June 1943 was estimated at about 350 planes in the Guadalcanal area, and another 350 in eastern New Guinea. (1) *Nanto Homen Sakusen Kiroku Sono San : Dai Juhachi Gun no Sakusen* 南東方面作戰記錄其の三：第十八軍の作戰 (Southeast Area Operations Record, Part III: Eighteenth Army Operations) 1st Demobilization Bureau, Sep 46. Vol. I, pp. 139–41, 156–7. (2) *Nanto Homen Koku Sakusen Kiroku* 南東方面航空作戰記錄 (Southeast Area Air Operations Record) 1st Demobilization Bureau, Sep 46, p. 14.

3 Aerial reconnaissance of the Bena Bena and Hagen areas in the middle of June revealed the existence of seven large enemy airfields, two of which were still under construction, two medium fields, and three small dispersal strips. Southeast Area Operations Record, Part III, op. cit. Vol. II, pp. 5–6.

that the enemy's next offensive would be directed against Japanese outposts in the New Georgia group. Allied forces were believed likely to attempt initial landings in the vicinity of Wickham, on southern New Georgia, to be followed by later assaults on Munda and Kolombangara. Both in the Solomons and in eastern New Guinea, Eighth Area Army estimated that enemy attack preparations would reach completion by the end of July and that major offensives might be launched in either or both sectors at any time between August and December.[4]

To meet the mounting threat of enemy attack on these widely separated fronts, Eighth Area Army and the Southeast Area Fleet continued to press the reinforcement of Japanese garrisons in the face of steadily increasing transport difficulties.[5] By mid-June, landing barges operating by night along the coastal route from Rabaul had successfully transported to Lae the main strength of the 66th Infantry Regiment, 51st Division, while elements of the 80th Infantry Regiment, 20th Division, were being moved from Madang by similar means. The 65th Brigade, transferred to Rabaul from the Philippines, was meanwhile moved to Tuluvu, on western New Britain, and the 51st Transport Regiment, 51st Division, was dispatched to Manus Island, in the Admiralties, to begin construction of an airfield.

In the Solomons area, concurrent steps had been taken to reinforce the naval garrisons charged with defense of the New Georgia group. The 229th Infantry Regiment, 38th Division, two battalions of the 13th Infantry Regiment, 6th Division, and part of the 10th Independent Mountain Artillery Regiment were moved to New Georgia and Kolombangara, while the 3d battalion of the 23d Infantry Regiment, 6th Division, was dispatched to Rekata, on Santa Isabel.

Although substantial numbers of troops had thus been advanced to the forward areas by dint of slow but persevering effort over a period of months, the fighting effectiveness of these forces was inevitably reduced by logistic difficulties. Large-scale Allied air operations against supply lines severely cut down the amount of rations and forage reaching the front-line forces, necessitating urgent steps to achieve local self-sufficiency.[6] Shortages of food and medical supplies swelled the average proportion of ineffectives due to malnutrition and disease to as high as 40 per cent in front-line combat units.[7]

Owing to new developments in the tactical situation facing the Eighteenth Army in New Guinea, in particular the build-up of enemy strength in the Wau area and the preparation of advance Allied air bases in the vicinity of Bena Bena and Mt. Hagen, Eighth Area Army

---

4  Estimate of enemy situation and intentions given in preceding paragraphs is based on memoranda-notes kept by the writer, at that time Staff Officer (Intelligence Bureau), Imperial General Headquarters. Additional data on the enemy air situation based on reference given in n. 2.

5  In addition to enemy interference, shortage of transport shipping was a major difficulty. Combined shipping available to both Eighth Area Army and Southeast Area Fleet at this time was broken down as follows: Large transports, 15; small transports, 40; powered sailing vessels, 80; fishing boats, 180; powered sampans, 235; large landing barges, 400; collapsible boats, 100. (1) Writer's memoranda-notes; (2) Statement by Capt. Toshikazu Ohmae, Staff Officer (Operations), Southeast Area Fleet.

6  Goals fixed for the end of 1943 were total self-sufficiency in the Solomons and New Britain, at least 50 per cent self-sufficiency for the Madang and Wewak areas in New Guinea, and 25 per cent for other New Guinea areas. *Nanto Homen Sakusen Kiroku Sono Shi: Dai Hachi Homen Gun no Sakusen* 南東方面作戰記錄其の四: 第八方面軍の作戰 (Southeast Area Operations Record, Part IV: Eighth Area Army Operations) 1st Demobilization Bureau, Jul 49, pp. 100–1.

7  Writer's memoranda-notes.

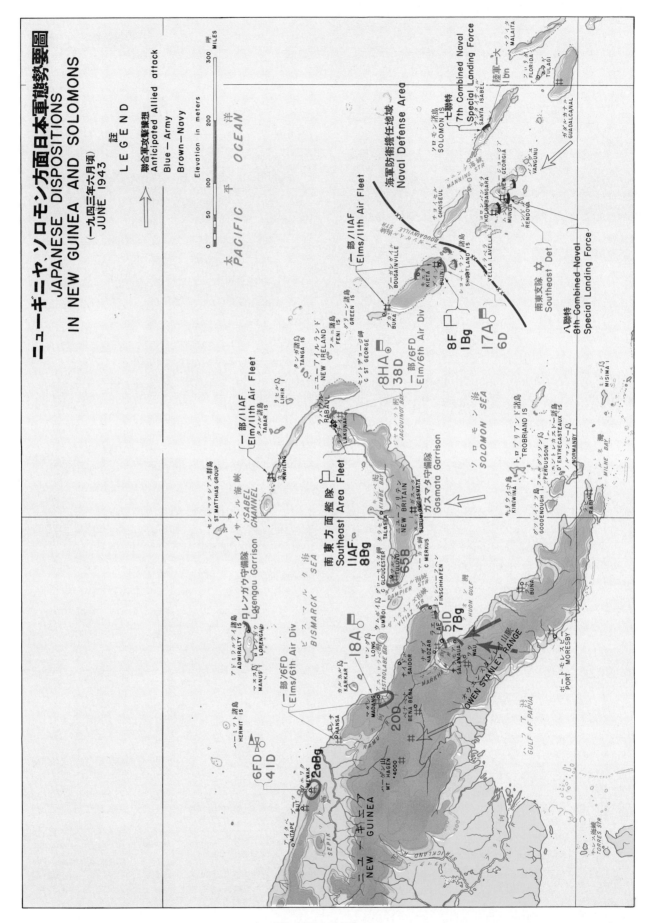

PLATE NO. 52

Japanese Dispositions in New Guinea and Solomons, June 1943

on 20 June made some revisions in its general operations plan of 12 April. By virtue of these revisions, missions allotted to the various forces under Area Army command were newly specified as follows:[8]

>   1. *Eighteenth Army: To effect the rapid completion of strong defensive positions in the Lae–Salamaua area; to prepare for a new offensive against enemy forces in the Wau sector; to plan operations for the capture of enemy airfields in the Bena Bena and Mt. Hagen areas; to hasten completion of airfields at Wewak and Hansa; and to speed up construction of the Madang–Lae road.*
>
>   2. *Seventeenth Army: To accelerate defensive preparations of the 6th Division on Bougainville.*
>
>   3. *65th Brigade: To complete construction of the airfield at Tuluvu on Cape Gloucester.*
>
>   4. *6th Air Division: To begin immediate attacks on the enemy airfields under construction in the Bena Bena and Mt. Hagen areas and on the already existing enemy airfield at Wau; to attack Buna and enemy small craft moving along the New Guinea coast.*[9]

In collaboration with the Eighth Area Army's revised plan, the Southeast Area Fleet continued to allot a portion of the Eleventh Air Fleet's land-based aircraft to support army air operations in New Guinea, particularly against enemy surface transport and amphibious convoys. The major mission of the Eleventh Air Fleet, however, remained to conduct air operations in the Solomons area, operating both from Rabaul and from an advance base at Buin, on southern Bougainville.[10] Ground defense of Munda, Kolombangara and Santa Isabel, in the central Solomons, was also a navy responsibility, the Eighth Fleet exercising operational command over both army and navy units garrisoned there.

The Eighteenth Army command at Madang, in compliance with the new Area Army plan, immediately took steps to speed operational preparations on the Lae–Salamaua front and simultaneously began formulating concrete plans for an attack against the enemy airfields in the Bena Bena and Mt. Hagen areas. On 23 June, when the Deputy Chief of Army General Staff visited Madang to confer with the Eighteenth Army Commander, a preliminary plan envisaging the start of operations against Bena Bena in September had been elaborated, and a request was submitted for the allotment of an additional division, plus supporting troops, to carry out the operation.[11]

Meanwhile, on the Mubo front southwest of Salamaua, Lt. Gen. Hidemitsu Nakano, 51st Division commander, had already initiated a local offensive designed to forestall apparent enemy plans for a move against the Japanese defenses. On 20 June the newly-arrived 66th Infantry Regiment launched a coordinated attack from Mubo toward Guadagasal, but soon ran into difficulties due to the unexpected strength of the enemy positions. Despite heavy casualties, the regiment penetrated the forward positions and, on the night of 20 June, entered the secondary enemy defense line, where severe hand-to-hand combat took place. Again, losses were so heavy due to enemy superiority in automatic weapons that, on 22 June, the

---

   8  (1) Southeast Area Operations Record, Part III, op. cit. Vol. II, pp. 1–3.  (2) Southeast Area Operations Record, Part IV, op. cit., pp. 84–93.
   9  For strength of 6th Air Division, cf. n. 3, Chapter VIII.
   10  At this time the Eleventh Air Fleet had an operational strength of approximately 300 planes of all types. *Nanto Homen Kaigun Sakusen Sono Ni* 南東方面海軍作戰其の二 (Southeast Area Naval Operations Part II) 2d Demobilization Bureau, Feb 47, pp. 14–5, 18–9, 26.
   11  The Eighteenth Army plan envisaged employing the main strength of the 20th Division against Bena Bena, Kainantu and the Mt. Wilhelm area, and elements of the 41st Division against the Mt. Hagen area. Ground operations were to be preceded by air attacks to neutralize enemy air bases, and use of airborne troops was also contemplated. All objectives were to be occupied within two to three months from the start of operations in early September. Southeast Area Operations Record, Part III, op. cit. Vol. II, pp. 10–12.

attack was called off on the verge of success, and the 66th Infantry retired to Mubo. Its short-lived offensive had cost about 200 men.[12]

Without sufficient power to strike a decisive blow at any point on the southeast area front, the Japanese forces could do little but brace themselves to meet impending Allied attack. That attack came a full month in advance of the critical period forecasted by Eighth Area Army headquarters.

## Defense of Salamaua

On 30 June the Allied forces struck with a two-pronged offensive launched simultaneously against Salamaua, in New Guinea, and Rendova Island, in the central Solomons.[13] By striking earlier than the Japanese command had anticipated and at both places simultaneously, the enemy not only achieved tactical surprise but again divided the Japanese effort, particularly forestalling the concentration of 6th Air Division and Eleventh Air Fleet strength at either point of attack.

On the New Guinea front, the first wave of enemy troops, estimated at about 1,000, began landing at Nassau Bay, ten miles south of Salamaua, at 0330 on 30 June.[14] (Plate No. 53) The suddenness of the attack caught the 51st Division forces guarding Salamaua off balance, with their main strength disposed to meet increasing enemy pressure on the Mubo and Bobdubi fronts, to the southwest and west of Salamaua. The only force in the immediate area of the enemy landing was the Nassau Garrison Unit, made up of the 3d Battalion, 102d Infantry Regiment, with a reduced strength of only a few hundred troops.[15]

After offering brief resistance to the first enemy elements put ashore, the Nassau Garrison Unit, on division orders, withdrew northward on 1 July, while the 3d Battalion, 66th Infantry was ordered forward to help stem the enemy advance from the beachhead. Meanwhile the 6th Air Division launched a series of attacks on the beachhead area, destroying or damaging a large number of enemy landing craft, although at the cost of relatively high plane losses.[16] Despite these attacks, the enemy continued to put ashore additional troops and equipment.

Enemy forces now became active on the extreme right flank of the Salamaua defense perimeter, in the Bobdubi sector. To meet this simultaneous threat, Lt. Gen. Nakano ordered forward two battalions under command

---

12 Ibid., pp. 24–5.

13 The Japanese had no knowledge of the simultaneous Allied landings on Kiriwina and Woodlark Islands, in the Trobriand Group, and for several months were unaware that they had been occupied.

14 Southeast Area Operations Record, Part III, op. cit. Vol. II, pp. 27–8.

15 (1) Ibid., pp. 26–30. (2) Order of battle of the Japanese forces in the Lae–Salamaua area at the time of the Nassau Bay landing was as follows:

| | |
|---|---|
| 51st Division | 2d Bn, 21st Infantry Regt. |
|     Division Headquarters | 1st Bn, 80th Infantry Regt. |
|     66th Infantry Regt. | 15th Independent Engr. Regt. |
|     102d Infantry Regt. | 30th Independent Engr. Regt. |
|     115th Infantry Regt. | 5th Independent Hvy Arty Bn. |
|     14th Fld. Artillery Regt. | One antiaircraft battery |
|     51st Engineer Regt. | |
| 7th Naval Base Force | |
|     Sasebo 5th Special Naval Landing Force | |
|     82d Naval Garrison Unit | |

(Statements by Lt. Col. Kengoro, Tanaka, Staff Officer (Operations), Eighteenth Army and Capt. Ohme, previously cited.)

16 Strikes were flown on 1, 3 and 11 July. Southeast Area Operations Record, Part III, op. cit. Vol. II, p. 30.

of Maj. Gen. Chuichi Murotani, 51st Infantry Group Commander.[17] This force, immediately upon its arrival at the front, was engaged in heavy fighting.

The Eighteenth Army command now decided that the decisive battle must be fought at Salamaua since loss of that base would render Lae, to the north, untenable. After returning to Madang on 7 July from an urgent air trip to the Salamaua front, Lt. Gen. Hatazo Adachi ordered the 238th Infantry Regiment of the 41st Division, which had moved up to Madang from Wewak, to advance to Lae via Finschhafen in order to reinforce the 51st Division.[18]

Despite depleted strength and an acute shortage of ammunition, the 51st Division meanwhile fought desperately to defend Salamaua. By 10 July, however, 66th Infantry troops defending Mubo and the left flank coastal sector north of Salus were in serious danger, and Lt. Gen. Nakano decided to tighten his defenses by pulling back to a new semicircular line of positions running from Bobdubi through Komiatum and Mt. Tambu to Boisi, on the coast.[19] This line gave the 51st Division commanding positions along the heights skirting the Salamaua basin and guarding the approach routes from south and west. During the latter part of July, these positions were organized into a main line of resistance, but as the weight of the enemy assault increased, it became doubtful whether even this line could be held.

On about 20 July, 50 large enemy landing craft and four transports anchored in Nassau Bay and off Salus, indicating further reinforcements. The Allied troops began attacking in waves at close intervals, allowing the combat-weary Japanese forces no time to rally between assaults. Mortar and artillery bombardment of the Japanese positions became incessant, and enemy long-range guns began shelling Salamaua itself. In the Bobdubi sector, enemy troops succeeding in taking scattered strongpoints and poured reinforcements into the gaps. The fighting entered a bitter hand-to-hand phase, in which Japanese offensive action was limited to daring night infiltration raids behind the enemy lines.[20]

Alarmed at the unfavorable trend of the fighting, Lt. Gen. Adachi again flew to Salamaua on 2 August and, after studying the situation, ordered the 51st Division to further contract its over-extended front. Under this Army order, the 51st Division commander on 15 August ordered his troops to relinquish the Komiatum—Mt. Tambu—Boisi positions and fall back to a line from Bobdubi to Lokanu.[21] The new dispositions were effected by 23 August, but the line was still too long, the positions had not been previously prepared, and despite the arrival from Madang of the first reinforcements of the 238th Infantry Regiment, troop strength was still inadequate at all points. Nevertheless, Lt. Gen. Nakano ordered a final stand to be made on the new line. In a message to troops on 24 August, he declared:

*The mission of our division is to hold Salamaua without yielding a single foot of ground. Our existing positions constitute the very last line of defense....*[22]

The simultaneous operations on the Sala-

---

17  These units were the 1st Battalion, 80th Infantry (20th Div.), which had recently arrived from Madang, and the 1st Battalion, 66th Infantry. Maj. Gen. Murotani had replaced Maj. Gen. Okabe as 51st Infantry Group Commander.

18  Southeast Area Operations Record, Part III, op. cit. Vol. II, pp. 32-3, 104.

19  Ibid., pp. 37-8.

20  Ibid., pp. 46-7.

21  Ibid., pp. 56-7.

22  Ibid., pp. 67-8.

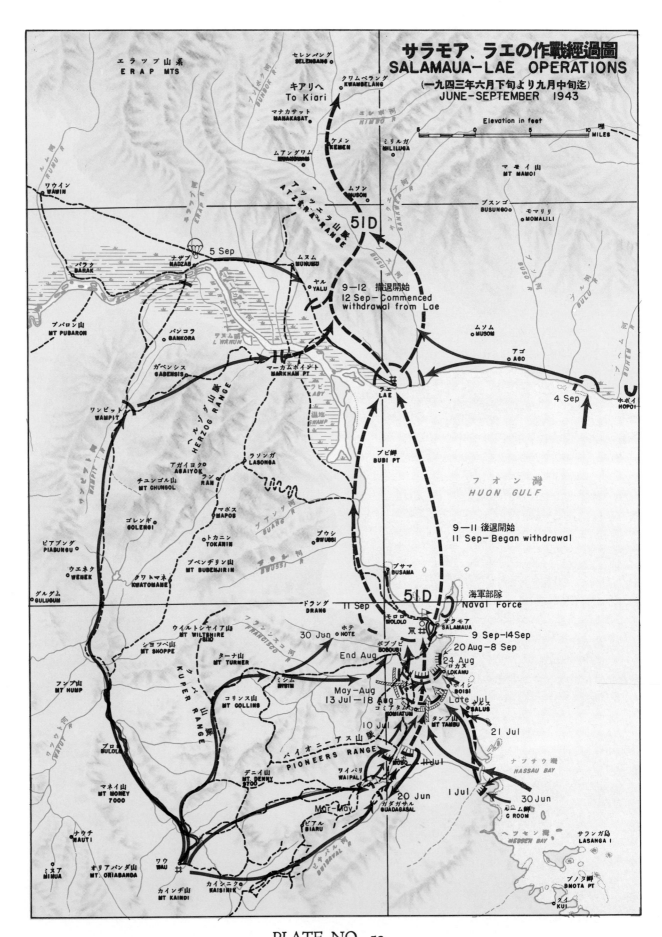

PLATE NO. 53

Salamaua—Lae Operations, June–September 1943

maua front and in the central Solomons had meanwhile strained available Japanese air resources to the limit. Under an Eighth Area Army—Southeast Area Fleet joint agreement reached at Rabaul on 4 July, the 6th Air Division was used temporarily to strengthen the central Solomons area where a chance of a successful counterattack was foreseen. Combat losses and fatigue as a result of incessant activity seriously weakened the 6th Air Division, leaving no margin of strength for employment in the planned Bena Bena—Hagen operations.[23]

In mid-July, therefore, Imperial General Headquarters transferred the 7th Air Division from Ambon, in the Dutch East Indies, to the southeast area, and Fourth Air Army Headquarters was established to command all army air units operating under Eighth Area Army.[24] First elements of the division arrived at Wewak on 25 July and, despite unfamiliarity with the New Guinea terrain, began operating immediately against enemy airfields in the Bena Bena–Hagen areas and the upper Markham Valley. After 1 August, however, in view of the greater urgency of the situation on the Salamaua front, the division shifted its primary effort to that area.[25]

The step-up of the Japanese air effort on the New Guinea front was not long in evoking severe enemy retaliation. On 17 August airfields in the Wewak–But area were subjected to a surprise attack by enemy aircraft, far surpassing in scale and intensity any previous air assaults.[26] Destruction of 100 aircraft in this single attack cut down by more than half the total serviceable plane strength of the Fourth Air Army and rendered the enemy's margin of air superiority so decisive that all phases of the Japanese military effort in New Guinea were severely affected.[27]

On the Salamaua front, the loss of air support immediately resulting from the Wewak raid further compromised the already perilous position of the 51st Division. Enemy bombing and strafing of the Japanese positions intensi-

---

23  In June Imperial General Headquarters and Eighth Area Army were still planning to execute the plan for ground operations against the Bena Bena and Hagen areas. *Daihonyei Rikugun Tosui Kiroku* 大本營陸軍統帥記録 (Imperial General Headquarters Army High Command Record) 1st Demobilization Bureau, Nov 46. p. 153.

24  Composition of the Fourth Air Army on 28 July was as follows:
  6th Air Division (2 light bomber regts., 1 hvy bomber regts., 2 fighter regts., 1 rcn. regt., 1 rcn. sq.
  7th Air Division (1 light bomber regt., 2 heavy bomber regts., 2 fighter regts., 2 rcn. sq.)
  14th Air Brigade (2 fighter regts.)
  1st Parachute Group
  Reconnaissance, photographic and transport elements.
Southeast Area Operations Record, Part III, op. cit. Vol. II, pp. 18–9, 50–2.

25  In a report to Eighth Area Army on 1 August, the Eighteenth Army Commander stated his opinion that the projected Bena Bena—Hagen operations should be treated as secondary to the defense of Lae–Salamaua and the Huon Peninsula area. Finschhafen was to be treated as the most important area. Ibid., pp. 13–8.

26  Although enemy air attacks on the Wewak area were naturally anticipated, the scale and suddenness of the 17 August raid took the Japanese defenses completely by surprise. Defensive precautions were relaxed at the time of the attack. (1) Interrogation of Col. Kazuo Tanikawa, Staff Officer (Operations), Eighth Area Army. (2) Southeast Area Air Operations, op. cit., p. 45.

27  Henceforth the Japanese army air force was obliged to adopt negative strategy and defensive tactics, involving a general retreat to rear-line airfields. Air support of ground operations was severely curtailed, and the schedule of surface transport movement was completely thrown off owing to the impossibility of providing air escort for convoys. (1) Interrogation of Col. Kazuyoshi Obata, Staff Officer (Supply), Eighteenth Army. (2) Southeast Area Operations Record, Part III, op. cit. Vol. II, pp. 249–50.

small security detachment, which was swiftly overwhelmed and annihilated. Meanwhile, naval air units immediately launched repeated attacks on the enemy beachhead and invasion shipping with 106 aircraft, but encountered such strong opposing air cover that plane losses became prohibitive. Additional enemy troops were put ashore on 1 July, and on 2 July American heavy artillery emplaced on Rendova began shelling Munda airfield, on New Georgia.

In complete control of Rendova, the enemy proceeded to dispatch small amphibious forces to the nearby islands of Roviana and Aumbaaumba, lying in the narrow channel between Rendova and Munda, and followed up almost immediately by moving an advance force across Roviana Lagoon to land on the mainland of New Georgia. On 2 July, realizing that a major enemy offensive was in the making, the Southeast Detachment and the 8th Combined Special Naval Landing Force concluded a joint local agreement placing all army and navy ground forces defending Munda under operational command of Maj. Gen. Minoru Sasaki, Southeast Detachment commander.[37]

While Maj. Gen. Sasaki took initial steps to bolster Munda's defenses,[38] another step in the Allied invasion plan unfolded. On 4 July enemy marines landed at Rice Anchorage on the northwest coast of New Georgia, about 15 miles from Munda, and before resistance could be organized, drove swiftly to the vicinity of Bairoko on 10 July.[39] With Munda thus threatened from two directions, Maj.Gen.Sasaki decided to make the main defensive effort on the right flank. The 2d Battalion, 45th Infantry Regiment, newly arrived from Bougainville, was ordered to Bairoko to fight a holding action against the marines. Meanwhile, the 13th Infantry was moved forward from Kolombangara to Munda to bolster the main line already defended by elements of the 229th Infantry.

On 15 July the Southeast Detachment forces launched a coordinated attack on the Munda front in an attempt to turn the enemy flank, but repeated assaults failed to make headway, and the attack ended in failure. Thereafter, tightening enemy control of the air and sea made it increasingly difficult to move in reinforcements even from nearby Kolombangara,[40] and the fighting power of the Japanese forces steadily declined under heavy attack by American aircraft, artillery and armor. By 31 July, the units defending Munda airfield were standing on their final defense line.

In view of the evident hopelessness of continued resistance on New Georgia, Maj.Gen. Sasaki on 7 August ordered the gradual withdrawal of the forces defending Munda to Kolombangara. The 13th Infantry, covered by rear-guard actions fought first at Munda and then on Baanga Island,[41] successfully effected its withdrawal. On the left flank around Bairoko, fighting continued until 19 August, when the 2d Battalion, 45th Infantry finally

---

37 Southeast Area Operations Record, Part II, op. cit. Vol. II, pp. 116–7.

38 On 4 July the 3d Battalion, 229th Infantry Regiment, was moved from Kolombangara to Munda. This unit, which as the Kenmotsu Battalion, had been virtually annihilated in the Buna campaign, had been reconstituted and refitted in Kolombangara.

39 Southeast Area Operations Record, Part II, op. cit. Vol. II, pp. 118–9.

40 Despite these difficulties, the Eighth Fleet, by an all-out and costly effort, succeeded in moving several reinforcement groups to the New Georgia area. The total troop strength transported amounted to about five infantry battalions. Ibid., pp. 104–6.

41 An aggressive rear-guard action was fought on Baanga Island by the 3d Battalion, 23d Infantry Regiment, between 11 and 22 August. The battalion then retired to Arundel Island, where it continued to resist the enemy advance until ordered to evacuate in mid-September. Ibid., pp. 131–5.

evacuated to Kolombangara, thence moving on to Gizo Island.[42]

The main strength of the New Georgia defenders was now added to the forces available for the defense of Kolombangara, and Maj. Gen. Sasaki began an immediate reorganization of his troops with a view to launching an early counteroffensive. American troops, however, were already on Arundel Island, just across a narrow, mile-wide channel from the main Japanese base at Vila, on southern Kolombangara. While fighting continued there and on Baanga, an enemy amphibious force on 15 August suddenly seized Vella Lavella Island, 17 miles northwest of Kolombangara,[43] thus threatening the Japanese defenses from a new direction.

In view of these unfavorable developments in the tactical situation as well as critical supply difficulties, Eighth Fleet Headquarters on 15 September ordered the evacuation of all Army and Navy forces from Kolombangara.[44] The withdrawal operation was begun immediately, with barges and destroyers serving as the chief means of transport. American fleet units, which were then actively patrolling the waters west of the island, subjected the movement to extreme danger, but despite this menace the bulk of the Southeast Detachment and naval forces on Kolombangara were successfully evacuated to Bougainville and Rabaul.[45]

## Evacuation of Lae and Ramu Valley Operations

At the same time that the Japanese forces were preparing to withdraw from Kolombangara, marking the end of the campaign in the central Solomons, the 51st Division and attached forces on the distant New Guinea front were pulling out of the besieged Lae area on the first stage of a difficult and costly retreat toward the north coast of the Huon Peninsula.

The Eighteenth Army plan had directed the 51st Division commander to route the withdrawal along the southern slopes of the Finisterre range, through Kaiapit, and into the upper Ramu Valley.[46] To cover the withdrawal, a force composed of the 78th Infantry Regiment (reinf.), under the command of Maj. Gen. Masutaro Nakai, 20th Infantry Group Commander, was ordered to launch operations up the Ramu Valley and into the upper Markham Valley.[47] (Plate No. 56)

However, the situation west of Lae and the open terrain made it impossible to use the Ramu Valley withdrawal route. Enemy forces occupied both banks of the Markham River, and the units which had landed at Nadzab on and after 5 September were squarely astride the road of escape. Lt. Gen. Nakano therefore elected to use a steep mountain trail, reconnoitered in April, which led north from Lae, up the Busu

---

42 Ibid., p. 136.

43 Ibid., pp. 136–7.

44 Ibid., p. 144.

45 The Southeast Detachment was subsequently dissolved, and the units which had been attached from the 6th and 38th Divisions rejoined their parent organizations. Ibid., pp. 153–4.

46 This route, leading over the low saddle of the Markham–Ramu divide, is the natural route of access from the Lae–Salamaua area to Madang. The area in the divide is the largest area free of forest cover on the New Guinea mainland, and is passable for all types of transport throughout the year, with the exception of certain small localities which become boggy during the rainy season. The highest point along this terrain corridor is not more than 900 feet above sea level.

47 Composition of the Nakai Detachment was as follows: 20th Infantry Group headquarters; 78th Infantry Regiment (less elements); 1st Battalion, 26th Field Artillery Regiment; miscellaneous service units. Southeast Area Operations Record, Part III, op. cit. Vol II, pp. 158, 275–6.

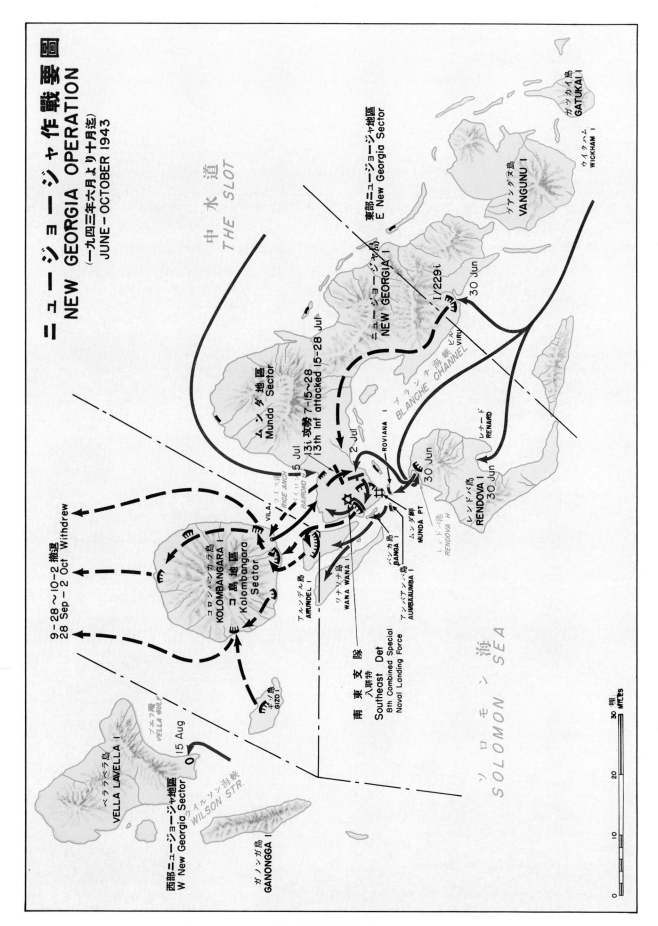

PLATE NO. 55

New Georgia Operation, June—October 1943

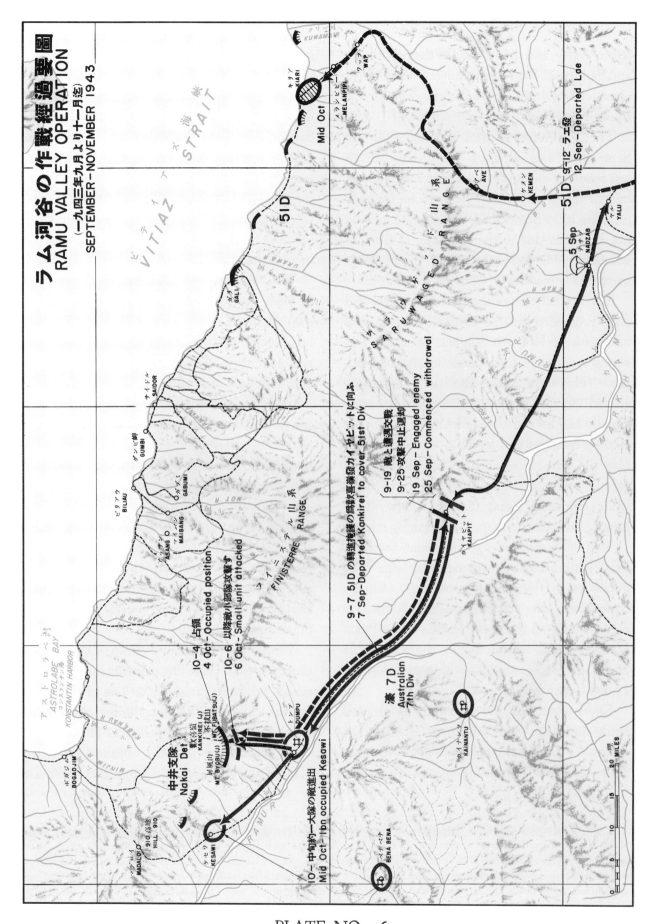

PLATE NO. 56

Ramu Valley Operation, September—November 1943

major Japanese strategic outposts in the Central Solomons, forcibly impressed upon the High Command that the southeast area was not being subjected to a more harassing or secondary attack. General MacArthur's objective was now recognized as the complete disintegration of the Japanese position south of the Mandated Islands.

Developments on other sectors of the Pacific front also influenced the shift in Imperial General Headquarters strategy. In the North Pacific, Japanese troops had been forced to evacuate Kiska in July. In the Central Pacific, a powerful attack had been carried out by American carrier aviation against Marcus Island, 1,100 miles from Tokyo. Wake, Kwajalein, Majuro and other key Japanese-held islands in the Marshalls were also struck, and American shipping was moving through the water of the Ellis Archipelago. Thus, a new enemy thrust appeared imminent in the Central Pacific parallelling the MacArthur drive from the southeast area.

In view of these major threats and Japan's declining military and naval strength, solidification of the inner defenses of the Empire had become imperative. Imperial General Headquarters, in particular, was aware that the national strength was no longer adequate to conduct operations in the southeast area on a large scale, and that Japanese troops must in the future avoid the terrific drain involved in constantly pitting weak ground contingents against stronger, better-equipped and more adequately supplied enemy forces. The swift build-up of decisive enemy superiority in the air also underlined the folly of such operations.

Moved by these considerations, the Army Section of Imperial General Headquarters had for some time urged that a strategic line delineating Japan's " absolute zone of national defense " be fixed, behind the periphery of which air and ground strength could be replenished and marshalled for decisive battle. If this meant excluding the southeast area bastion of Rabaul, now almost in the front line, the Army stood ready to draw the perimeter line west of that locality.[53]

Of such fundamental importance was the decision on this issue that, on 30 September, the fourth Imperial Conference held since the start of the war was convoked to sanction the agreement finally reached between the Army and Navy.[54] Under this agreement, the perimeter delimiting the absolute defense zone was drawn from Western New Guinea to the Mariana Islands via the Carolines. (Plate No. 57) This perimeter was to be strongly manned and fortified so as to deny the enemy further gains and to provide a bastion behind which forces could be gathered for offensive blows. Defense preparations behind that line were to be completed by the spring of 1944.[55]

The positions held by the Eighth Area Army and Southeast Area Fleet in Northeast New Guinea, the Bismarcks, and the Northern Solomons now constituted the outpost line in the southeast area, while the Second Area Army was to be transferred to New Guinea to take charge of operations along and behind the new national defense zone in Western New Guinea. Imperial General Headquarters clearly recognized the necessity of vigorously maintaining the outposts and holding the enemy at bay

---

53 (1) Statement by Lt. Col. Nobutake Takayama, Staff Officer (Operations), Imperial General Headquarters, Army Section. (2) The problem of the national defense zone had been under discussion in the Army Section of Imperial General Headquarters for some time. (Statement by Col. Sei Matsutani, Chief, 20th Group (Coordination), Imperial General Headquarters, Army Sectiion.)

54 (1) Imperial General Headquarters Army High Command Record, op. cit., pp. 174–5. (2) Statement by Col. Takushiro Hattori, Secretary to the War Minister.

55 Details of the Army–Navy Central Agreement of 30 September, together with the implementation thereof by Second Area Army, are dealt with in Chapter X.

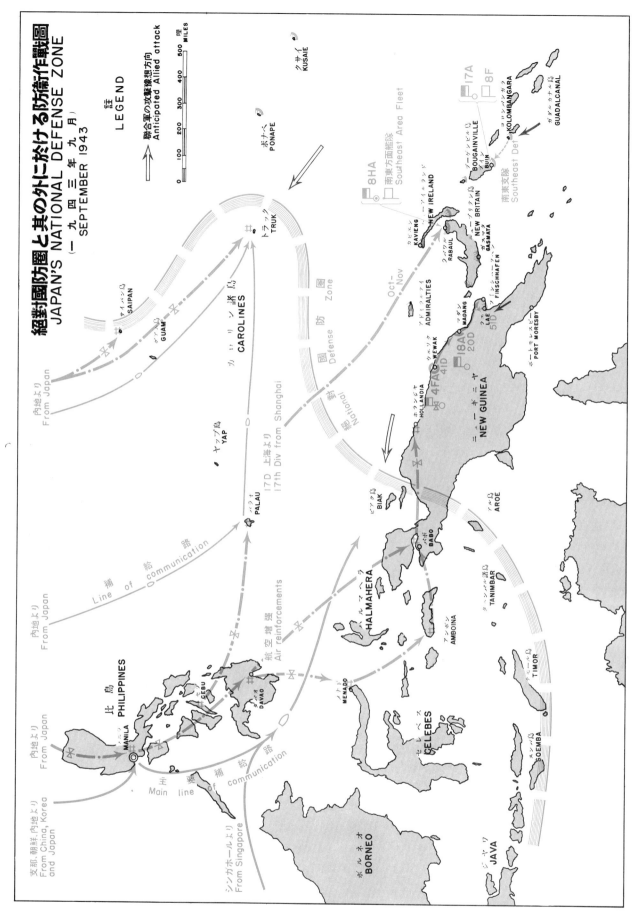

PLATE NO. 57

Japan's National Defense Zone, September 1943

positions to the west were being prepared. With this in mind, the 17th Division was dispatched from Shanghai to Rabaul to reinforce the troops manning the forward wall.

On the basis of Imperial General Headquarters directives, Eighth Area Army revised its operational plans. General Imamura and his staff were of the opinion that the soundest tactics to accomplish the Area Army's mission of strategic delay were to conduct determined counterattacks against all Allied attempts to pierce the outpost line. The major critical areas to be defended were the Dampier Strait region, particularly Finschhafen and Cape Gloucester, and Bougainville in the northern Solomons.

Within the framework of the Area Army plans, instructions were issued to the subordinate commands. The Eighteenth Army in New Guinea was ordered to occupy and defend a line along the Finisterre Range, with emphasis on the Dampier Strait coast near Finschhafen. In order to cover the right flank and assist in the main mission of guarding the west coast of the Dampier Strait, a strong force was to operate in the Ramu Valley, and all enemy attempts to cross the Finisterre Range were to be repelled. The 51st Division was to retire west of Madang and continue its reorganization.

On New Britain, the Matsuda Detachment (65th Brigade, reinf.) operating under Area Army control was charged with strengthening the eastern defenses of the Dampier Strait, particularyl Umboi Island and Cape Gloucester.[56] The airfields at Gasmata, on southern New Britain, and on Los Negros Island in the Admiralties were garrisoned with smaller detachments, whose mission was to destroy enemy landings aimed at seizing the airfields.

On the left flank in the Solomons, the Seventeenth Army was being strengthened for the defense of Bougainville. The chief units stationed there were the 6th Division and the 4th South Seas Garrison Unit,[57] while part of the 17th Division, was to come under Seventeenth Army control upon its arrival from China. The 38th Division was in the Rabaul area and on New Ireland.[58]

At this time the Fourth Air Army, consisting mainly of the 6th and 7th Air Divisions, was operating from widely scattered bases, mainly in Eighteenth Army territory. It was now ordered to operate principally from bases between Madang and Hollandia in support of planned operations. Eighteenth Army and the 65th Brigade were to have priority in calling for air support, and the primary mission was to destroy all enemy landing attempts in the Dampier Strait region.[59]

In support of Army operations along the outpost line, the Southeast Area Fleet planned to utilize available surface and air strength to disrupt enemy convoys en route to landing areas, while supply to the Dampier Strait region was to be assured by the use of submarines. Naval ground forces were also to be disposed at important points along supply and transport routes to insure the smooth functioning of shore activities such as communications, stevedoring, repair, and base defense.[60]

Even before detailed plans for the new dis-

---

56 Southeast Area Operations Record, Part III, op. cit. Vol. II, pp. 123–4.
57 The 4th South Seas Garrison Unit consisted of Garrison Headquarters, three infantry battalions, a field artillery battery, one signal company, and a tank company.
58 (1) Southeast Area Operations Record, Part III, op. cit. Vol. II, pp. 123–4. (2) Southeast Area Operations Record, Part IV, op. cit., p. 200.
59 Ibid., pp. 156–8.
60 In addition to these activities the Combined Fleet was watching for a propitious moment to stage a showdown battle with the American Fleet outside the perimeter of the defense line in the Southeast Area. Such an operation was planned, but the opportunity never arose to set the plan in motion. Meanwhile, Japanese naval air strength was slowly whittled down. (Interrogation of Vice Adm. Shigeru Fukudome, Chief of Staff, Combined Fleet.)

positions were completed, however, the enemy made an initial breach in the outpost line by an assault on the Finschhafen area.

### Dampier Strait Defense: Finschhafen

Ever since the initial Japanese advance into eastern New Guinea, the 60 mile wide Dampier Strait lying between the western tip of New Britain and the Huon Peninsula had been of vital importance to Japanese sea communications between the main southeast area base at Rabaul and the New Guinea fighting front. The majority of troop and supply convoys dispatched to Buna and subsequently to the Lae–Salamaua area had moved via this narrow sea passage, and as General MacArthur's forces pressed northward, it was apparent that control of the strait would be a major Allied strategic objective in order to pave the way for further amphibious moves toward Western New Guinea.

Finschhafen, commanding the strait from the west, was the key point in the Japanese scheme of defense. Prior to the loss of Lae and Salamaua, troops moving forward from Rabaul and Madang had usually been routed through Finschhafen, which served as a stopping point and staging area. One of the best developed localities in New Guinea, the town controlled an area containing two excellent anchorages: Finschhafen itself and Langemak Bay.

While the battle for Salamaua was still in progress, the Eighteenth Army commander foresaw the danger of an attack on the Finschhafen area and took steps to reinforce the weak garrison forces, which then consisted only of a small number of naval landing troops, some army shipping units, and groups of replacements destined for Lae. Army units in the area were under command of Maj. Gen. Eizo Yamada, 1st Shipping Group commander.

To strengthen these inadequate forces, Eighteenth Army on 7 August ordered the main strength of the 80th Infantry Regiment,[61] 20th Division, and one battalion of the 26th Field Artillery Regiment to proceed from Madang to Finschhafen. Subsequently, on 26 August, the 2d Battalion of the 238th Infantry Regiment, 41st Division, which had already advanced to Finschhafen on its way to Lae to reinforce the 51st Division, was ordered to remain at Finschhafen under Maj. Gen. Yamada's command.[62] The Army forces in the area were assigned the mission of reconnoitering and organizing the ground in preparation for a possible enemy landing.[63] Defensive organization of the coastal areas around the mouths of the Mongi, Logaweng, and Bubui Rivers and on Point Arndt was undertaken by Army units, while the naval garrison of about 500 men was given responsibility for the defense of Finschhafen proper.[64]

The enemy attack against Lae on 4–5 September removed all doubt that Finschhafen would soon be in the front line of combat. Eighteenth Army therefore immediately took additional steps to make the area as strong as possible within the limits of the supply and manpower situation. By this time the 20th Division had been relieved of its mission on the Madang–Lae road construction project and was ordered forward to reinforce Finschhafen. On 10 September the main body of the division[65] under command of Lt. Gen. Shigeru Katagiri left the Bogadjim area in eighteen

---

61 Less 1st Battalion and 5th Company.
62 Southeast Area Operations Record, Part III, op. cit. Vol. II, p. 117.
63 Ibid., pp. 112–4.
64 Ibid., pp. 163–4.
65 Elements making up the main body of the 20th Division were: Division headquarters; 79th Infantry Regi-

march serials to move by overland routes to Finschhafen. By 21 September, however, the division, without horses or vehicles to facilitate its movement over the difficult trail, had only reached Gali and still had almost 100 miles to cover. The division actually was not scheduled to reach Finschhafen until 10 October. In the event of a prior enemy attack, Eighteenth Army hoped that the Yamada force, which had ample time to organize the ground, would be able to hold the Satelberg area, so that when the 20th Division main body arrived, it could be assembled for an immediate counterattack.

At dawn 22 September, a large enemy convoy appeared off Point Arndt and after heavy air preparation commenced troop landings. Maj. Gen. Yamada immediately issued orders for the concentration of his main force on Satelberg Hill to prepare for counteroffensive action. The 3d Battalion, 80th Infantry was dispatched as a sortie force to attack the enemy beach and reconnoiter the front in preparation for a full-scale counterattack.

At this critical moment much depended on the rapidity with which Japanese air units could mount large-scale attacks against the enemy amphibious forces. The 7th Air Division, though charged with this responsibility, was under orders to fly cover for a convoy in-bound to Wewak on 23 September[66] and hestitated to go out in force against Finschhafen, leaving the convoy unprotected. Fourth Air Army quickly ended this indecision by ordering the division to attack the Finschhafen landings, but bad weather prevented missions on the 23d and 24th. Meanwhile, however, planes of the Eleventh Air Fleet, flying from Rabaul, were able to take off and conducted heavy strikes on 22, 24 and 26 September against enemy shipping in the Finschhafen area.

Upon receiving confirmation of the enemy landing in the Point Arndt area, Lt. Gen. Adachi, aware that the success or failure of the defense of Finschhafen area would decisively influence the fate of the Dampier Strait region, ordered the Yamada force to launch an immediate attack against the enemy beachhead. (Plate No. 58) Meanwhile, the 20th Division began a race with time across the mountain trails from Gali to Finschhafen.

At Finschhafen Maj. Gen. Yamada and his small force faced a difficult situation. A force of approximately 1,000 enemy troops was approaching the north bank of the Mape River,[67] and at Point Arndt a follow-up convoy was landing an estimated 5,000–6,000 additional troops with tanks and heavy artillery.[68] On 26 September, another force of approximately 500 enemy troops, which had advanced along the coast from the Hopoi landing area east of Lae, appeared in the vicinity of Cape Cretin, six miles south of Finschhafen. Air support of the Yamada Force was meanwhile limited to two Army aircraft a day. The enemy had already overrun the Finschhafen airstrip and was preparing it for use.[69]

Pursuant to Eighteenth Army orders, the Yamada Force on 26 September launched a

---

ment; 26th Field Artillery Regiment (less two battalions); 20th Engineer Regiment; 33d Independent Engineer Regiment; 20th Division special troops. The Nakai Detachment was operating in the Finisterre Mountains under Eighteenth Army control. (Statement by Lt. Col. Kengoro Tanaka, previously cited.)

66 *Dai Shichi Hiko Shidan Kimitsu Sakusen Nisshi Dai Ni-go* 第七飛行師團機密作戰日誌第二號 (Top Secret Operations Diary No. 2, 7th Air Division) Jul—Dec 43, pp. 59–60.

67 Southeast Area Operations Record, Part III, op. cit. Vol. II, p. 165.

68 The Navy stationed four submarines off Cape Cretin in order to intercept the enemy reinforcement convoys, but the Allied ships succeeded in running the blockade. Southeast Area Naval Operations, Part II, op. cit., pp. 57–8.

69 Southeast Area Operations Record, Part III, op. cit. Vol. II, p. 167.

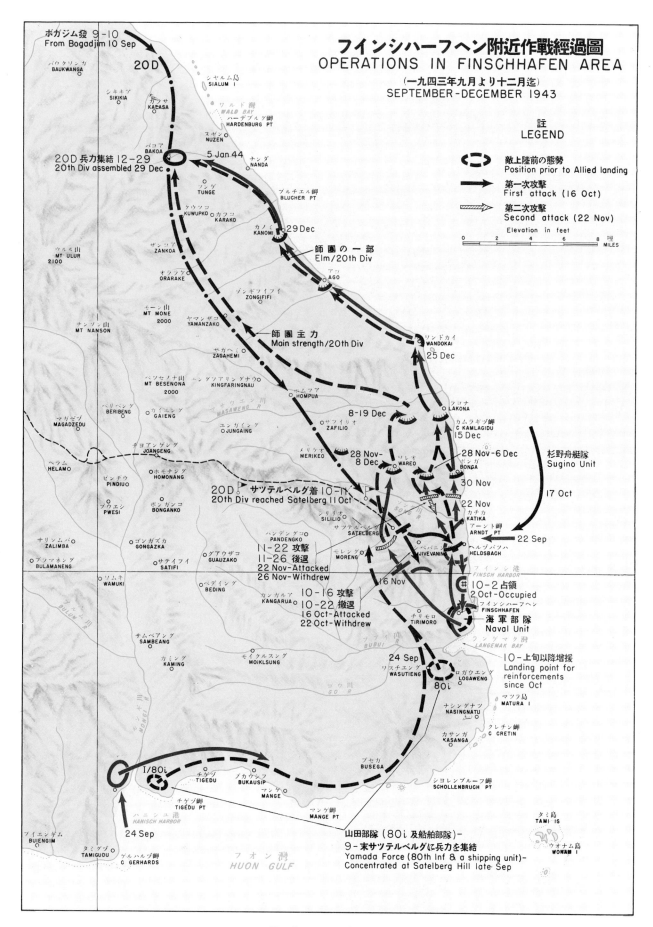

PLATE NO. 58

Operations in Finschhafen Area, September—December 1943

coordinated attack from the Satelberg Hill area in the direction of Heldsbach Plantation, the 80th Infantry Regiment in the assault. The objective was to reduce the enemy beachhead before further reinforcements could be brought ashore. A heavy engagement ensued in the area between Satelberg and the sea. Meanwhile, in Finschhafen, the naval garrison of about 300 men and a company of the 2d Battalion, 238th Infantry, comprising the defense force of the town, were surrounded on 27 September. This force withstood five days of ferocious enemy attack but was finally overwhelmed on 2 October. With the fall of Finschhafen, the Allied force could now turn its full attention to the main strength of the Yamada force operating east of Satelberg. The attack of the 80th Infantry had been halted short of its goal, and the regiment, now on the defensive, was engaged in heavy fighting in the vicinity of Satelberg Hill. During this period the enemy began using the Finschhafen airstrip.

On the morning of 11 October, the command post of the 20th Division was established on Satelberg Hill, and Lt. Gen. Shigeru Katagiri took command of all forces in the Finschhafen area. After four days spent in assembling the newly arrived troops, the division launched a surprise attack early in the morning of 16 October through Jivevaning in the direction of Katika. After heavy fighting, the enemy force was cut squarely in two, and 20th Division units reached the sea at Katika the next day. On the same day, 17 October, the Sugino Boat Unit landed in the enemy rear at Point Arndt and threw the opposing forces into disorder.[70]

The situation was now highly favorable. As a result of the successes of 16–17 October, the 79th Infantry Regiment swarmed along the coast in full force, overrunning enemy positions and capturing large quantities of weapons and ammunition, as well as trucks fully loaded with rations and medical supplies. Prospects for the early recapture of Finschhafen were momentarily bright. However, in the Heldsbach area, the enemy line suddenly stiffened, and the regiment was forced to deploy. On 20 October the Allied force was reinforced from the sea,[71] and a new and ferocious battle ensued around Point Arndt. Since neither side seemed to be able to break the deadlock, the fighting temporarily slackened while both sides prepared new blows.

Because of the supply situation, Lt. Gen. Katagiri recognized the importance of quickly reorganizing his units and resuming the attack in order to gain a quick decision. On 31 October the Army Commander, Lt. Gen. Adachi, arrived at the 20th Division command post and studied the situation. Enemy strength was estimated at one division, confronting the Japanese forces along a winding front extending from the north bank of the Song River south to Logaweng Hill via Jivevaning.[72] Lt. Gen. Adachi ordered a renewal of the attack with the objective of seizing the mouth of the Song River in a series of limited objective operations. In order to prevent enemy landings in the Japanese rear, the 2d Battalion, 79th Infantry was dispatched north on 6 November under orders to reconnoiter as far as Lakona and strengthen coastal defenses. Meanwhile, although the division was on one-third rations and short of

---

70 The Sugino Boat Unit, composed of the 10th Company (155 men) of the 79th Infantry, carried out counter-landings with four landing barges. This unit, catching the enemy completely by surprise, destroyed three antiaircraft guns, four artillery pieces, ten machine guns, two automatic cannons, twenty automatic rifles, and two ammunition dumps. Four hundred casualties were inflicted on the Allied force. Southeast Area Operations Record, Part III, op. cit. Vol. II, p. 175.

71 Ibid., p. 176.
72 Ibid., p. 180.

ammunition,[73] Lt. Gen. Katagiri set the date for the new attack at 23 November.

While the 20th Division was preparing its new assault, the Allied forces were pouring ashore reinforcements and supplies.[74] On 16 November, a full week before the scheduled attack date, a coordinated enemy drive supported by a large-scale air strike was launched against the Satelberg position. The struggle that followed was one of the heaviest battles fought in the southeast area. Units of the 6th Air Division turned out in force and gave unusually heavy support to the ground effort.[75] Beginning on 23 November, heavy meeting engagements developed north of the Song River as the Allied force drove toward Bonga and Wareo. On Satelberg Hill, the 80th Infantry, which had been subjected to ten days of sustained air and ground attack, terminated its heroic defense on 26 November and withdrew beyond Wareo on division order. In spite of its failing troop strength and scanty store of provisions, the 20th Division on 30 November launched a counterattack to ease the pressure on Wareo and Bonga. This desperate attempt failed, and on 8 December, under renewed enemy assaults, Wareo fell. A week later the enemy entered Lakona after shattering the determined resistance of the 79th Infantry.

It was now clear that the Finschhafen base was irretrievably lost and that the 20th Division lacked sufficient combat strength to renew the offensive.[76] Accordingly, on 17 December, pursuant to directives from General Imamura at Rabaul, Eighteenth Army ordered the division to withdraw behind the Mesaweng River, adopt delaying tactics, and hold out in the Sio area.[77] The retirement behind the Mesaweng began on the night of 19 December. Along the coast, the 2d Battalion, 79th Infantry, conducted delaying actions on successive positions all the way from Wandokai to Kalasa. The main body of the 20th Division closed into Kalasa on 29 December.[78] Thus, in two and a half months, the opening battle in the defense of the Dampier Strait had resulted in disaster.

## Bougainville

While the 20th Division was still conducting its gallant but unsuccessful defense of the

---

73 (1) Ammunition levels of the 20th Division at this time were as follows: Type 94 mountain guns, 135 rounds; Type 41 mountain guns, 78 rounds; Infantry guns, 36 rounds; Mortars, 102 rounds; Demolitions, 436 kilograms. (2) After reaching Satelberg Hill the 20th Division received practically no supplies from rear areas. From 1 October the ration was about 6 shaku (about 1/5 pint) of staple food per day. Southeast Area Operations Record, Part III, op. cit. Vol. II, pp. 185-7, 191-2. (3) "We have been without rations for a month....We have eaten bananas, stems and roots, bamboo, grass, ferns, and, in fact, everything edible up to the leaves of the trees." Diary of Officer (rank not given) Kobayashi, 80th Infantry Regiment Headquarters. ATIS Current Translations, No. 106, 20 Mar 44. pp. 35-6.

74 The enemy had the following assault shipping at Finschhafen: 6 November, 30 transports; 12 November, 72 transports; 15 November, 3 transports and 22 landing barges. Southeast Area Operations Record, Part III, op. cit. Vol. II, pp. . 185-6.

75 On 23 November enemy positions in the vicinity of Jivevaning were bombed by 44 aircraft, and on 26 November, 47 planes hit enemy positions in and around Finschhafen. The 7th Air Division had already returned to Ambon early in November to aid in the establishment of the new national defense line. (1) Ibid., pp. 199-200. (2) Southeast Area Air Operation Record, op. cit., p. 60.

76 The total casualties of the 20th Division were 5,761 or 45 per cent of the total strength. The units hardest hit were the 80th Infantry (59 per cent losses) and the 20th Engineers (66 per cent losses). Southeast Area Operations Record, Part III, op. cit. Vol. II, pp. 219-20.

77 Ibid., pp. 214-6.

78 The advance guard of the enemy forces advanced to the Sio vicinity by 6 January 1944. Action at this time, however, was limited to light skirmishes. Ibid., pp. 218-9, 323.

west coast of the Dampier Strait, the enemy forces in the Solomons again unleashed a new assault. This time the target was Bougainville, left flank strongpoint of the Eighth Area Army line and strategic key to Rabaul.

Following determination of the new national defense zone at the end of September, Eighth Area Army had notified Lt. Gen. Haruyoshi Hyakutake, Seventeenth Army Commander, on 7 October that the army's primary mission was to organize and strengthen the defenses of Bougainville. Under this directive Lt. Gen. Hyakutake immediately concluded a local operational agreement with the Eighth Fleet, formulating joint plans to meet an eventual enemy landing in the Bougainville area.

Execution of these plans had barely gotten under way when, on 27 October, enemy forces landed on Mono Island, due south of Bougainville, and five days later followed up with the main landing effort in the Cape Torokina area, on the northern side of Empress Augusta Bay. (Plate No. 59) Since the terrain on the western side of Bougainville was so low and damp as to render it relatively unsuitable for attack operations, Seventeenth Army had not anticipated a landing in the Empress Augusta Bay area, and the only force there was a small observation unit.[79] This was speedily overwhelmed, and the enemy began a slow advance inland.

The command post of the Seventeenth Army and the bulk of the forces on Bougainville were at this time in the Erventa area, on the southeast tip of the island.[80] Here, they were disposed to cover the shores of Tonolei Harbor and the Buin area, where the Navy had a base and also the largest operational airfield in the Solomons.

Confronted by the enemy landing on Empress Augusta Bay, Lt. Gen. Hyakutake immediately dispatched the 23d Infantry Regiment (less 1st Battalion) of the 6th Division to the Torokina area by overland routes.[81] Meanwhile, the 2d Battalion of the 54th Infantry Regiment, 17th Division, which had arrived at Rabaul, was ordered to undertake an amphibious operation directly against the enemy beachhead. The Eleventh Air Fleet, newly reinforced by 173 carrier-based aircraft, prepared to render strong support to these operations.

While the 23d Infantry moved overland toward Empress Augusta Bay, six destroyers carrying the 2d Battalion, 54th Infantry sailed from Rabaul on 2 November, covered by a naval support force of four cruisers and six

---

79 Eighth Area Army also did not regard an enemy landing in the Empress Augusta Bay area as likely. "The first real surprise maneuver after I arrived at Rabaul occurred when the enemy landed on Cape Torokina ....Because we thought the poor topographical features of this area would hamper enemy landing operations, we did not anticipate a landing....and were not adequately prepared." (Interrogation of Lt. Col. Matsuichi Iino, Staff Officer Intelligence, Eighth Army Army.)

80 Order of battle of the Japanese forces on Bougainville on 1 November 1943 was as follows:
        Headquarters, Seventeenth Army
           6th Division
           Elements of 17th Division
           4th South Seas Garrison Unit
           15th Antiaircraft Artillery Group.
           17th Signal Unit
           2d Shipping Group
           Miscellaneous elements.
Southeast Area Operations Record, Part II, op. cit. Vol. II, pp. 160–4.

81 This force was later placed under command of Maj. Gen. Shun Iwasa, 6th Infantry Group commander, and designated as the Iwasa Detachment. Ibid., p. 172–3.

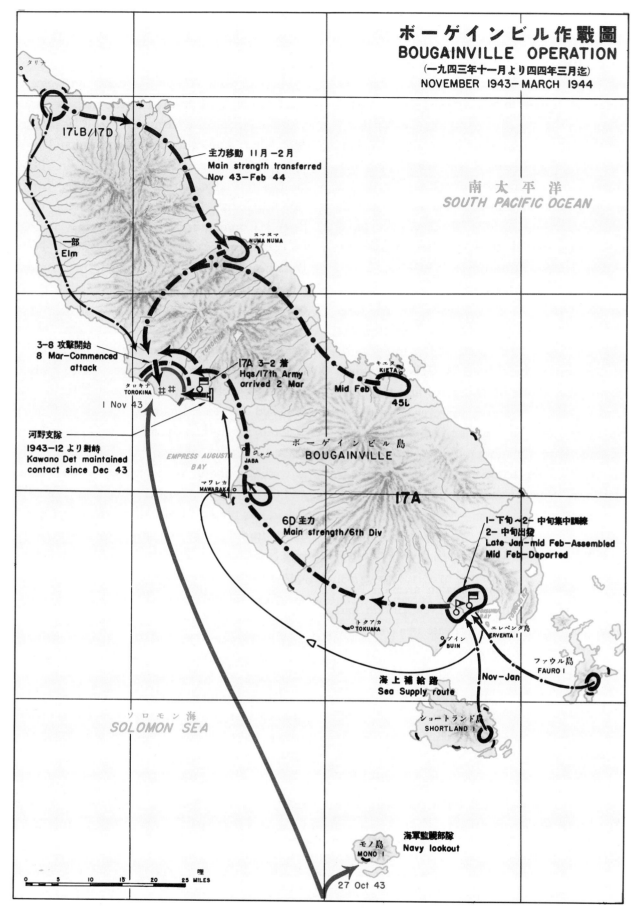

PLATE NO. 59

Bougainville Operation, November 1943—March 1944

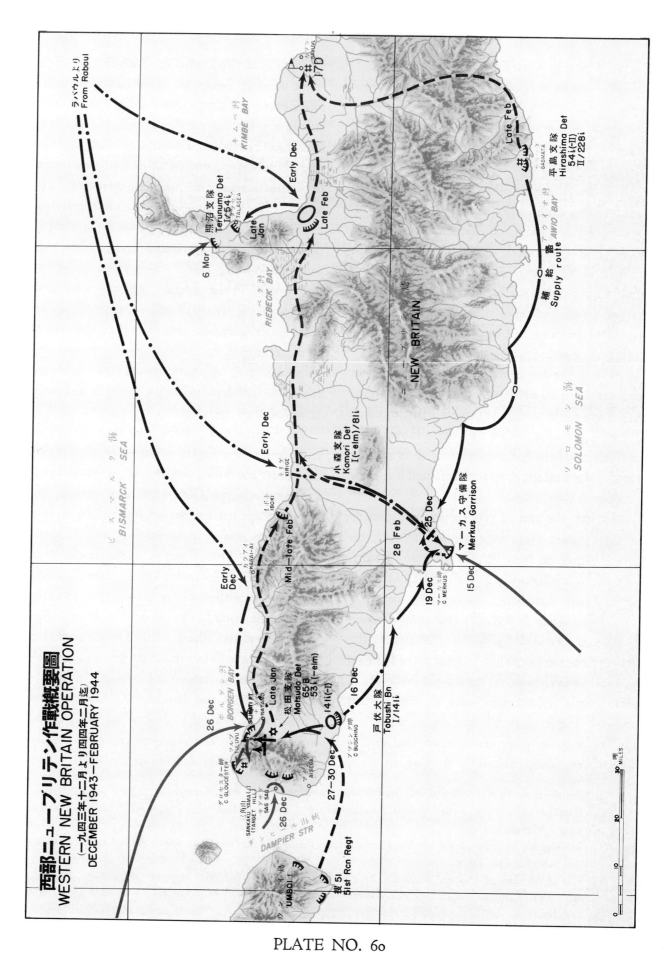

PLATE NO. 60

Western New Britain Operation, December 1943—February 1944

December. On the following day, the Tobushi Detachment joined the Komori Detachment and futile attacks were repeated for several days but all efforts failen.[91] Throughout this period naval air units operating from Rabaul and 6th Air Division planes operating chiefly from Wewak carried out repeated assaults on the enemy ships supplying the beachhead, inflicting substantial damage but also taking severe aircraft losses.[92]

Lt. Gen. Sakai estimated that the Merkus landing was not the main enemy effort, and his suspicions were soon confirmed. From 19 December, Japanese positions on and around Cape Gloucester were subjected to heavy daily air attack. On 20 December a reconnaissance plane reported that a large concentration of enemy transports and other assault shipping had rendezvoused in Buna Bay.[93] On 25 December a large enemy convoy was spotted moving northward in the direction of Dampier Strait, and the Matsuda Detachment, responsible for the defense of the Borgen Bay–Cape Gloucester area, was promptly ordered to battle stations.

At 0400 on 26 December, the defenses skirting the Cape Gloucester airfield and Silimati Point were subjected to devastating naval gunfire preparation, followed by heavy air strikes. At dawn enemy troops began landing at two points, the main force going ashore northwest of Silimati Point and a smaller contingent debarking just west of the Cape Gloucester airfield.[94] Maj. Gen. Matsuda immediately began deploying his troops to meet this two-pronged attack, which obviously aimed at the encirclement of the airfield area. The nearest unit to the main enemy landing was the 2d Battalion, 53d Infantry, which was quickly ordered to launch local counterattacks pending commitment of the main strength.

On the morning of the landings, naval air units from Rabaul and Kavieng went out in strength to smash the enemy convoy. Hostile fighter cover over the beach was so thick, however, that the results were inconclusive, although the enemy's transport formation was dispersed. At twilight the attack was renewed with more success, several enemy ships receiving heavy damage. Japanese plane losses in these and subsequent operations, the air raids on Rabaul, and the appearance of enemy carriers in the waters east of New Ireland prevented the naval air units from carrying out further operations. By early January the number of serviceable aircraft other than fighters available for operations in New Britain had dwindled to approximately 24 navy medium attack planes, 8 navy bombers, and no army bombers at all.[95] From this time on, the air forces were unable to influence the decision at Cape Gloucester.

On the ground, fierce fighting took place for possession of the airfield area. Every effort to defend this locality by the Matsuda Detachment proved unavailing, and the detachment was forced to evacuate its positions. Maj. Gen. Matsuda, however, had concentrated his main strength, consisting of the 53d, 141st Infantry and the 51st Reconnaissance Regiments, on commanding ground in the area southwest of

---

91 Southeast Area Operations Record, Part IV, op. cit. Supplement I, pp. 22–7.

92 (1) Southeast Area Operations Record, Part IV, op. cit., pp. 198, 212–3. (2) Southeast Area Naval Operations Part III, op. cit., pp. 43–5.

93 Enemy shipping at Buna was reported as follows: Medium-sized transports, 17; small transports, 60; landing craft, 20; destroyers, 8. ATIS Current Translations No. 106, p. 28.

94 Southeast Area Operations Record, Part IV, op. cit., p. 207.

95 (1) Ibid., pp. 212–4. (2) *Nanto Homen Kaigun Koku Keikano Gaikyo Sono Go* 南東方面海軍航空經過の概要其の五 (Outline of Southeast Area Naval Air Operations, Part V) 2d Demobilization Bureau, Nov 42, pp, 15–7.

Silimati Point. Though cut off from all reinforcement and supply, and with no air support and very little artillery, this force on 3 January launched a counterattack against the American troops advancing along the west side of Borgen Bay and succeeded in carrying *Sankaku Yama* (三角山)[96], a prominent terrain feature overlooking the beach. The hill could not be held, however, due to mass enemy artillery fire and air bombardment, and the Japanese were forced to withdraw.

After this effort failed, further resistance on Cape Gloucester was out of the question, and the Matsuda Detachment was ordered to withdraw toward Talasea. The Dampier Strait campaign thus ended on 24 January 1944, four months after the opening gun was fired at Finschhafen. The main route of advance to the national defense zone in Western New Guinea now lay open to the enemy.

### Saidor

The situation which confronted the Eighteenth Army at the end of 1943 was dark indeed. The bitter loss of Finschhafen, coupled with the enemy advance into the upper Ramu Valley, had virtually opened the way for an assault by General MacArthur's forces on the nerve-center of Eighteenth Army resistance at Madang. Sio, on the northeast corner of the Huon Peninsula, remained the only important defense area between Madang and the enemy forces driving up the coast from Finschhafen. At Sio, the 20th Division, seriously weakened in the Finschhafen campaign, was reassembling its depleted forces in preparation for a new defensive stand. The 51st Division, consisting largely of combat ineffectives, was also spread out to the west of Sio waiting to move to the rear for rest and refitting.[97]

The enemy was not slow to take advantage of this favorable situation. Striking swiftly in a new amphibious operation, Allied forces on the morning of 2 January 1944 landed in the vicinity of Saidor, about halfway between Sio and Madang.[98] (Plate No. 61) Eighteenth Army, although it had feared a new enemy landing somewhere along the north coast of the Huon Peninsula,[99] was powerless to combat it. With the enemy at Saidor, the Army strength was split squarely in two, and two exhausted, understrength divisions were cut off in the Sio Area.

In view of the weakened condition of these units and the urgent need of bolstering the thin defenses of Madang, the Eighth Area Army headquarters in Rabaul relieved the Eighteenth Army Commander of further defense of the Sio area. The Eighteenth Army Commander, who was then at Sio, directed their withdrawal past the Saidor beachhead to the Madang area.

In compliance with the Area Army order, Lt. Gen. Adachi ordered the 20th and 51st Divisions to proceed as quickly as possible to the Madang area, and placed Lt. Gen. Nakano, Commander of the 51st Division, in over-all command of forces east of Biliau, including the 20th Division. Lt. Gen. Adachi then left by submarine for Madang, where the 41st Divi-

---

96  This hill had been subjected to extremely severe air bombardment and naval gunfire concentrations prior to the enemy landings. It was known to the Americans as Target Hill.

97  In late December, owing to the seriousness of the situation, Lt. Gen. Adachi made a trip from Madang to Kiari to direct the dispositions of the 20th and 51st Divisions for the defense of the Sio area. Southeast Area Operations Record, Part III, op. cit. Vol. II, pp. 240–1, 247–8.

98  Ibid., pp. 247–8, 301, 323.

99  The usual signs of a forthcoming enemy attack were all present during the latter part of December. Enemy PT boats were active along the coast, and there was a marked acceleration of air activity against Madang and the Saidor area. Ibid., pp. 305–7.

sion, moved forward from Wewak, was charged with organizing the defenses of the area pending the arrival of the 20th and 51st Divisions.

Lt. Gen. Nakano now tackled the problem of evacuating his command past the expanding enemy beachhead at Saidor. Two routes were chosen, one running fairly close to the coast and the other following the ridge-line of the Finisterre foothills.[100] The 20th Division was to take the coastal route, while the 51st, together with some naval units, was to use the route inland. The first echelon was scheduled to reach Madang on 8 February, and the entire movement was to be completed by 23 February. To divert the enemy at Saidor, eight infantry companies of the Nakai Detachment, under command of Maj. Gen. Nakai, were ordered to withdraw from the Ramu Valley front and advance down the coast from Bogadjim to threaten the enemy beachhead.[101]

The first echelon of the retiring 20th and 51st Division forces left Sio on schedule. At the last minute, however, the withdrawal plan was modified in favor of moving both divisions via the inland route. The covering operations of the Nakai force were carried out according to plan, and the Nakano group successfully negotiated the withdrawal without encountering enemy resistance. Illness and starvation, however, took a heavy toll along the difficult retreat route. At the end of December the combined strength of the 20th and 51st Divisions had been about 14,000. About 9,300 finally reached Madang by 1 March.[102]

While the withdrawal was in progress, the Nakai force engaged the American troops attempting to break out of the Saidor beachhead toward Madang. To forestall this move, the force deployed along the Mot River line in the vicinity of Maibang and Gabumi. Enemy attempts to cross the river, although supported by heavy artillery fire, were repelled, and the line held firm until 21 February when the force, having completed its mission of covering the Nakano group withdrawal, retired to Bogadjim.

On the Ramu front, the enemy had meanwhile taken advantage of the reduced strength of the Japanese forces. On 19 January, the Australians, operating out of Dumpu, launched a determined attack on the Kankirei positions and, by 27 January, took possession of Kankirei. The Nakai force returned to Bogadjim just in time to bolster the line and prevent the Australians from advancing down the Mintjim River to join forces with the Americans coming up from Saidor.

Madang was now extremely vulnerable to attack. The condition of Japanese air units did not permit reliance on air power to protect the town. The 41st Division and the Nakai Detachment were forced to spread their strength from the Finisterre Mountains to Astrolabe Bay and up through Alexishafen to Mugil. Thus dispersed, these units could not give adequate protection to Madang, even with the cooperation of the Ninth Fleet.[103] The tide of

---

100 The mountain route had already been reconnoitered and afforded the least danger of encountering enemy patrols. Southeast Area Operations Record, Part III, op. cit. Vol. II, pp. 335-9.

101 Seven infantry companies of the Nakai Detachment remained in the Kankirei area on the Ramu front, under command of the 78th Infantry Regiment commander. Since mid-October the detachment had successfully checked all enemy attempts to penetrate the Ramu Valley line toward Madang. In early December the enemy attempted to flank the Japanese line on the right by sending an Australian infantry battalion to Kesawi, but on 8 December the Nakai Detachment attacked and drove the enemy back to Dumpu. Ibid., pp. 289-96, 343-4, 360-2.

102 Ibid., p. 355.

103 The Ninth Fleet was activated in the eastern New Guinea area and consisted of a small number of surface vessels. Southeast Area Naval operations, Part III, op. cit., p. 38.

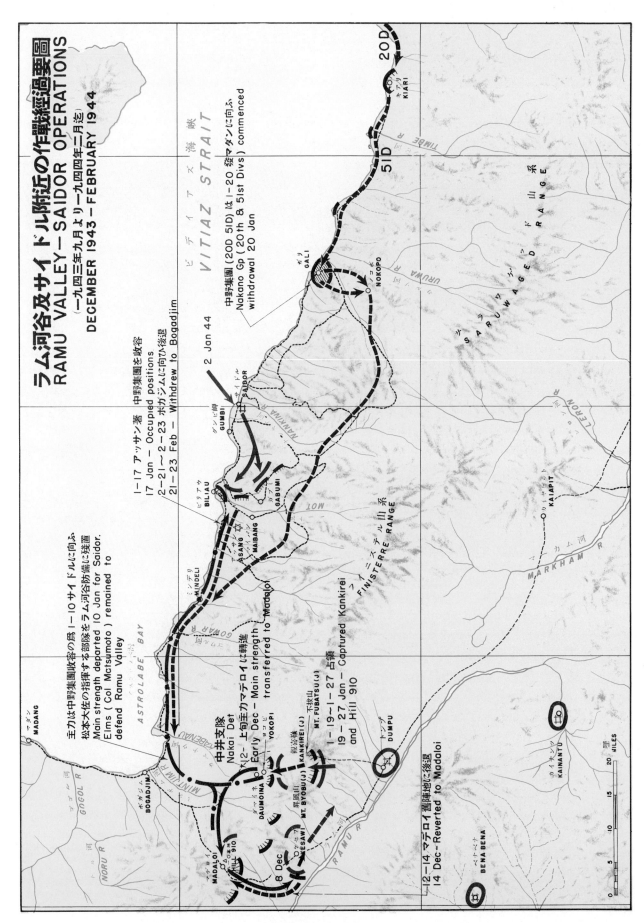

PLATE NO. 61

Ramu Valley and Saidor Operations, December 1943—February 1944.

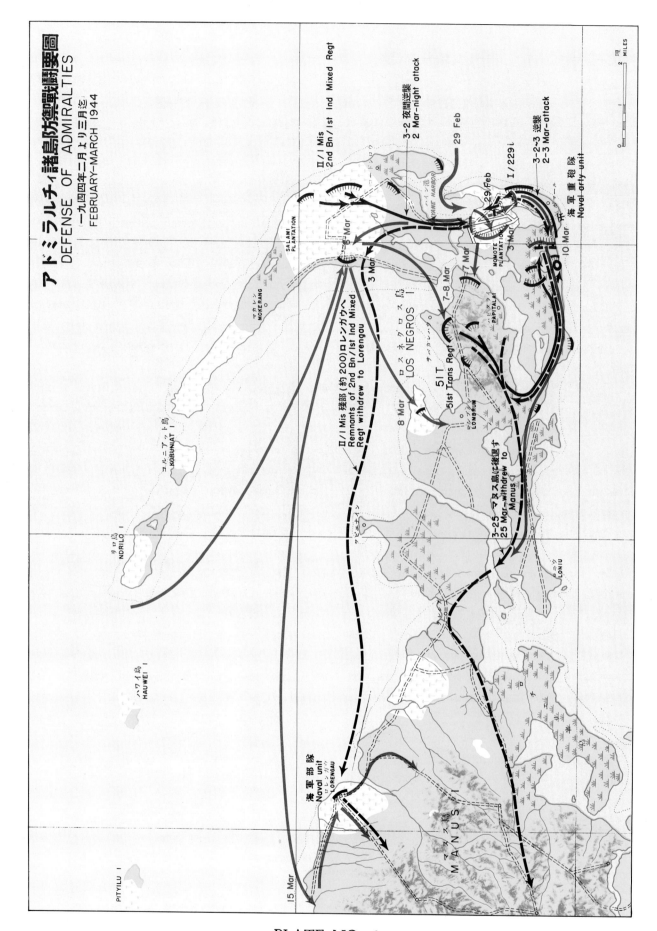

PLATE NO. 62

Defense of Admiralties, February—March 1944

battle was moving irresistibly westward.[104]

## Isolation of Rabaul

While the enemy forged steadily ahead in New Guinea, his forces already entrenched on Bougainville in the Solomons and on western New Britain itself created an ever-growing threat to the heart of Japanese power in the southeast area at Rabaul. On Bougainville, Seventeenth Army succeeded in containing the enemy beachhead but was unable to oust the invading forces. In the Cape Gloucester area on New Britain, the situation was still darker. The Matsuda Detachment, down to less than half its combat strength, faced starvation or annihilation at the hands of the superior enemy. On 23 January 1944, Eighth Area Army ordered Lt. Gen. Sakai to move immediately all troops from western New Britain to the area east of Talasea, thereby easing the supply problem and tightening the defenses of Rabaul.

In February the southeast area situation underwent a further radical change as a result of new developments in the Central Pacific. On 1 February an American amphibious force invaded Kwajalein Atoll in the Marshall Islands. On 17 February a large enemy carrier task force struck at Truk, the main Japanese naval base in the mandated islands, and on the following day American troops landed on Eniwetok, westernmost island in the Marshall group.

To cope with this new threat, naval air units at Rabaul were immediately ordered to Truk, leaving Rabaul virtually without air protection and depriving the forces throughout eastern New Guinea, New Britain and the Solomons of most of their air support. Sea traffic practically ceased. Eighth Area Army, increasingly concerned over the defense of Rabaul and nearby New Ireland, ordered the 17th Division on 23 February to move immediately to reinforce eastern New Britain.[105]

The buttressing of Rabaul's shrinking defenses had barely gotten under way when the Allied forces, in a new surprise move of consummate boldness and far-reaching strategic consequences, invaded Los Negros Island, in the Admiralties, 365 airline miles northwest of Rabaul and 250 miles farther into Japanese-held territory than the deepest previous penetration by amphibious forces.

Strategically located on the main supply route to Rabaul, the Admiralties also served as an intermediate air stop between Rabaul and rear bases in Northeast New Guinea. By June 1943, the 51st Transport Regiment had almost completed one airfield at Lorengau, on Manus Island, and was beginning construction of a second at Momote Plantation, on Los Negros. Thereafter the strategic importance of the islands increased steadily as the Eighteenth Army front line on New Guinea was pushed rapidly backward and western New Britain fell under enemy control.

Eighth Area Army, in December 1943, ordered one infantry regiment reinforced by one field artillery battalion, to strengthen the defense of the Admiralties. However, the ships carrying the first echelon of these forces were attacked en route and forced to turn back, with the result that reinforcement attempts from Palau were given up. On 23 January Eighth

---

104 General Adachi's plans in early March were as follows: The 41st Division was to station itself in the sector between Madang and Mugil and prepare to attack any enemy landing in the sector east of Hansa. The Nakai Detachment was to be relieved and rejoin the 20th Division. The 20th Division was to recuperate at Hansa and prepare the defenses of that area. The 51st Division, when relieved by the 20th at Hansa, would immediately leave for Wewak to reorganize and defend that area. Southeast Area Operations Record, Part III, op. cit. Vol. III, pp. 6–7.

105 Southeast Area Operations Record, Part IV, op. cit., pp. 224–5.

Area Army ordered the 2d Battalion of the 1st Independent Mixed Infantry Regiment, stationed on New Ireland, to move to the Admiralties. A week later the 1st Battalion, 229th Infantry, 38th Division was also ordered to proceed from Rabaul to the Admiralties by destroyer. All forces in the Admiralties were placed under Col. Yoshio Ezaki, 51st Transport Regiment commander, with the missions of securing Los Negros Island with its vital airfield[106], and preventing the enemy from seizing and establishing airfields on Manus, Pak and Pityilu Islands. Fourth Air Army was to cooperate in the defense of the Admiralties in the event of an enemy landing attempt.[107]

Defensive dispositions in the Admiralties were based on the anticipation that an Allied landing attempt would probably be made somewhere along the shore of Seeadler Harbor or the eastern and southern shores of Los Negros. (Plate No. 62) Hyane Harbor, on the opposite side of Los Negros, was not considered a likely landing point due to its smallness and the danger to which enemy movement through the narrow harbor entrance would be subject.

The defending forces on Los Negros were taken by surprise, therefore, when a small enemy invasion force, following a brief but intense naval gunfire preparation, began landing on beaches inside Hyane Harbor at 0815 on 29 February, striking directly at Momote airfield. The 1st Battalion, 229th Infantry, defending the airfield sector, was slow in reacting but, on the night of the 29th, launched a counterattack which failed to dislodge the beachhead. On 1 March a small number of aircraft attempted to support the ground defense but were driven off by strong enemy air cover. A further counterattack launched at 1700 the same day was broken up by enemy artillery.

On 2 March the enemy, heavily reinforced from the sea, gained possession of the airfield. Col. Ezaki now planned a pincers attack from north and south, ordering the 2d Battalion, 1st Independent Mixed Infantry Regiment to attack from the Salami Plantation area, north of Hyane Harbor, in conjunction with a further counterattack by the 1st Battalion, 229th Infantry, from the southern sector. The attack was to be launched on the night of 2 March, but due to heavy enemy air and naval bombardments, it had to be delayed.

Spearheaded by the 2d Battalion, 1st Independent Mixed Infantry, the main attack was finally launched on the night of 2 March. The enemy positions were successfully infiltrated, but the curtain of mortar and artillery fire encountered by the 2d Battalion was so intense that its gains could not be exploited. This abortive attack was the last large-scale effort which the Japanese forces in the Admiralties were able to mount. Thereafter, although fighting continued on Los Negros until 12 March and on Manus until early April, the outcome was inevitable.

With the Admiralties in their grasp, the Allied forces were now squarely astride Eighth Area Army supply lines and in a position to isolate the large numbers of Japanese troops remaining on New Britain, New Ireland and

---

106 The Hyane (Momote) airfield had been used mainly as a staging field for air units moving to advanced bases in the southeast area. At the time of the Allied landing, there were no operational aircraft on the field. (Statement by Lt. Col. Ohta, previously cited.)

107 In anticipation of new Allied attacks in late February, Imperial General Headquarters had ordered the transfer of five air regiments to New Guinea from the southwest area. These reinforcements had arrived in the theater prior to the enemy landing in the Admiralties, but adverse weather at the time of the landing prevented effective operations against the enemy invasion force. (1) (Statement by Lt. Col. Koji Tanaka, Staff Officer (Air), Imperial General Headquarters, Army Section. (2) Southeast Area Air Operation, op. cit., pp. 62–7.

in the Solomons. They were also in possession of a base from which further amphibious attacks might be mounted against Japanese rear bases on New Guinea. Such an attack was anticipated about the end of April.[108]

## Bougainville Counteroffensive

While the Admiralties campaign was in progress far to the rear, the Seventeenth Army on Bougainville launched the last large-scale Japanese offensive effort in the Solomons in an attempt to wipe out the enemy beachhead in the Empress Augusta Bay area.

On 21 January, the Eighth Area Army Commander, General Imamura, flew from Rabaul to Bougainville and set early March as the target day for the attack. In mid-February Lt. Gen. Hyakutake, Seventeenth Army Commander, conducted extensive reconnaissance of the enemy front and began assembling his forces for the offensive. Training had been tough and thorough, and the assault units moved overland from their bases confident of victory.

By 25 February the troops were in forward assembly areas, and on 2 March Lt. Gen. Hyakutake arrived from Erventa to take command.[109] The plan of maneuver called for the main effort to be made on the right across the Laruma River by a force consisting of the 6th Infantry Group (one battalion, 13th Infantry and 23d Infantry Regiment) and the 45th Infantry Regiment. A secondary attack was to be launched across the Torokina River by the 13th Infantry (less one battalion). The 6th Field Artillery Regiment, reinforced by a battalion from the 4th Heavy Artillery Regiment, was to support the attack.

On the night of 6 March the Japanese units deployed in great secrecy along a line 500 to 800 yards from the enemy positions. During the following day enemy outposts were driven in, and final preparations were completed. At 0415 on 8 March, under cover of a heavy artillery barrage, the attack was launched. Operations proceeded smoothly, and by the morning of 12 March several deep wedges had been driven into the American positions and the main line of resistance penetrated. The enemy, however, counterattacked with powerful armored forces, inflicting heavy casualties and making it impossible to fully exploit the initial success.[110]

Lt. Gen. Hyakutake planned to renew the offensive on 15 March with the direction of the attack slightly altered. Since reinforcements were badly needed, the 2d Battalion of the 4th South Seas Garrison Unit, which was following from Erventa, and the 6th Cavalry Regiment (dismounted), which had been guarding rear areas, were both ordered into the line. These dispositions completed, the attack was resumed on 16 March. On 17–18 March the enemy lashed back in another fierce counterattack. The Japanese front-line units, decimated by disease and casualties and without air support, were gradually forced to withdraw to their initial positions.

On 26 March the Torokina campaign was

---

108 (1) Southeast Area Operations Record, Part III, op. cit. Vol. III, p. 5. (2) "The taking of the Lae–Salamaua area was the turning point of the New Guinea campaign, but the final step was the taking of the Admiralty Islands.... Two large airfields fell to the Allies, and Japanese supply lines (to Rabaul) were cut off. Also from these islands the Allies were able to isolate the individual Japanese positions along the New Guinea coast and to prevent any large-scale withdrawal." Interrogation of Col. Shigeru Sugiyama, Senior Staff Officer, Eighteenth Army.

109 There were no important changes in the order of battle of the forces on Bougainville after the original Allied landing on 1 November. These forces were assembled in February from Erventa on southern Bougainville, Kieta on the east coast, and from the northern tip of the island. Southeast Area Operations, Record, Part II, op. cit. Vol. II, pp. 189–29.

110 Ibid., pp. 196–202

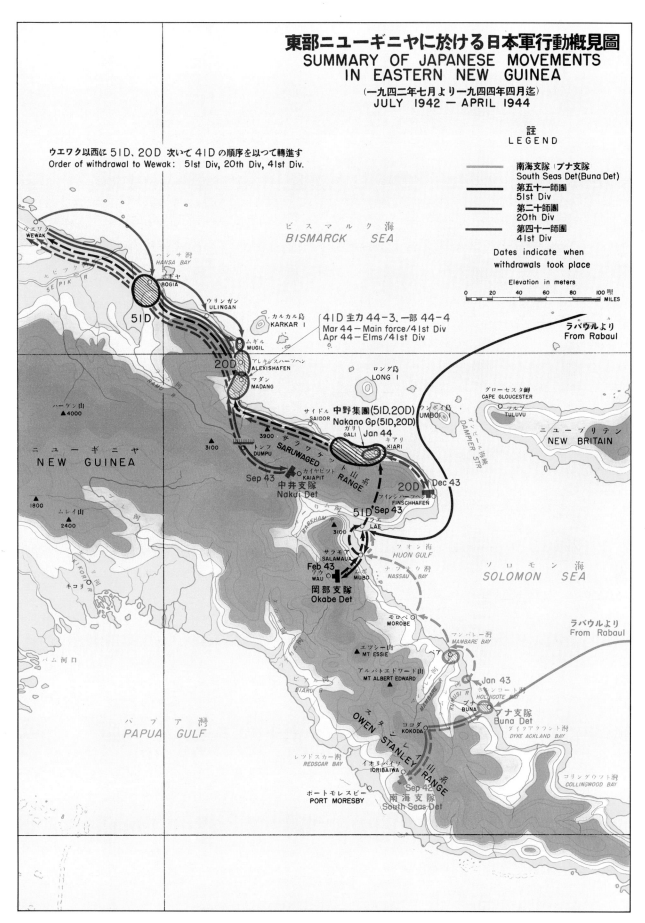

## PLATE NO. 63

Summary of Japanese Movements in Eastern New Guinea, July 1942—April 1944

brought to a final halt by Eighth Area Army order, and Seventeenth Army began to redeploy its units for a holdout campaign. The 6th Division was assigned to the western and southern parts of Bougainville, while the 17th Infantry Group was deployed along the east coast. The main body of each unit was pulled from the line behind covering rearguard actions and proceeded to the Erventa and Kieta areas, respectively. By this time sea and air communications with Rabaul were almost completely cut off, and it became necessary for the units on Bougainville to maintain their fighting strength by becoming totally self-sufficient. So successfully was this accomplished that the defenders were still in possession of the greater part of the island at the conclusion of hostilities.

### Southeast Area Situation, March 1944

The nine-month period which ended on 30 March 1944 had witnessed a serious disintegration of the Japanese position in the southeast area under the impact of swift, overpowering blows by the Allied forces, severe attrition of ground, sea and air strength, and insurmountable logistical difficulties.

The vast quantities of men and material which were poured into the area had not sufficed to turn the tide.[111] Despite the utmost efforts of the area army and navy commands, and despite the fortitude and endurance of the officers and men of both fighting services under conditions of severe hardship, Lae and Salamaua, New Georgia, Kolombangara, the Ramu Valley, Finschhafen, Cape Gloucester, Saidor, the Admiralties and part of Bougainville had been wrested from Japanese hands in a series of disheartening reverses.

Moreover, as a consequence of the Allied seizure of the Admiralties, approximately 175,000 army troops and naval personnel in eastern New Britain and Rabaul, on Bougainville and New Ireland had been by-passed and isolated in the wake of battle, and were henceforth unable to make any significant contribution to the war effort.[112]

On the New Guinea front, Eighteenth Army, also cut completely off from Rabaul by the loss of the Admiralties and Cape Gloucester, was forced to move the center of its resistance to the west. The seizure of the Admiralties also created the grave possibility that the next Allied attack in New Guinea might be much more ambitious than a move merely against the area east of Hansa Bay. Consequently, Eighteenth Army was forced to begin full-scale defensive preparations at Wewak and Aitape as well. (Plate No. 63)

On 10 March Lt. Gen. Adachi ordered the main body of the 41st Division to proceed to Hansa Bay and begin organizing the defenses of that area. The Eighteenth Army command post itself displaced from Madang to Hansa Bay on 17 March, and only the Shoge Detachment,

---

111 Total army and navy forces dispatched to the southeast area from the initial invasion of the Bismarcks up to March 1944 aggregated roughly 300,000. The Army alone supplied 1,800 aircraft and 2,000 pilots. Both the Army and Navy sent the largest consignments of newly manufactured planes to the southeast area. Arms and ammunition enough to equip six combat divisions passed through or were stocked at Rabaul. The Navy lost 50 combat ships and 300,000 tons of transport shipping. (Statements by Col. Kumao Imoto, Staff Officer (Operations), Eighth Area Army ; and Capt. Ohmae, Col. Takayama and Lt. Col. Tanaka, previously cited.)

112 Distribution of isolated Eighth Area Army troops was as follows: Rabaul area, 56,512; Bougainville, 31,024; New Ireland, 8,082. Total Army troops, 95,618. In addition there were 12,416 military labor personnel, mostly in the Rabaul area, and about 53,000 naval shore personnel throughout the Army area. Grand total, about 161,000 (1) Southeast Area Operations Record, Part IV, op. cit., pp. 324–5 (2) Statistics complied by 2d Demobilization Bureau, Nov 50.

consisting of elements of the 41st Division, was left in Madang as an outpost guard.¹¹³ The 51st Division was ordered to proceed with all possible haste to Wewak to organize ground defenses, while the 20th Division was given the same mission in the Aitape area.

In view of the fact that the outpost line in the southeast area had now been seriously breached, Imperial General Headquarters hastened action to organize and strengthen the new national defense zone in western New Guinea.

On 25 March the Second Area Army, under command of General Korechika Anami, with headquarters at Davao on Mindanao, absorbed the Fourth Air Army and the Eighteenth Army in the New Guinea area. Responsibility for operations west of the 147th meridian was transferred to the Second Area Army, while east of this line the isolated Eighth Area Army, supported by the Southeast Area Fleet, was to continue the defense of eastern New Britain, New Ireland, and Bougainville.

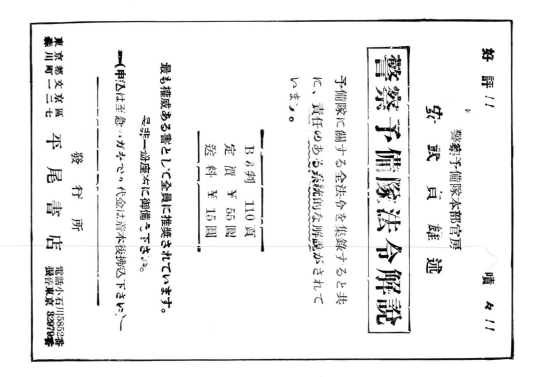

---

113 Southeast Area Operations Record, Part III, op. cit. Vol. III, pp. 21-2.

# CHAPTER X

# WESTERN NEW GUINEA OPERATIONS

## Strategic Planning

By April of 1944, under the impact of General MacArthur's two-pronged offensive against the Japanese forward line in the southeast area and the parallel enemy thrust into the outer defense rampart of the Central Pacific mandated islands, the operational center of gravity in the Pacific theater of war was moving relentlessly closer to the line which the Army-Navy Central Agreement of 30 September 1943 had defined as the boundary of Japan's "absolute zone of national defense."[1]

In drafting this agreement, Army and Navy strategists recognized that the continuous attenuation of Japan's fighting potential made it unwise, if not impossible, to attempt a decisive defense of the existing Pacific front line under the increasing weight of Allied offensives. Therefore, the mission of the forces in Northeast New Guinea, the Bismarcks, Solomons, Marshall and Gilbert Islands was limited to one of strategic delay, and plans were laid to build a main line of resistance along a restricted perimeter from the Marianas and Carolines to Western New Guinea and the Banda and Flores Seas.[2] These were to be flanked by the Bonins and Kuriles to the north and the Sundas to the west.

The essential points of the Army-Navy Central Agreement embodying this vital revision of Central and South Pacific war strategy were as follows:[3]

*1. Key points in the southeast area, extending from Eastern New Guinea to the Solomon Islands, will be held as long as possible by destroying enemy forces whenever they attack.[4]*

*2. With a view to the rapid completion of counteroffensive preparations, the following missions will be accomplished by the spring of 1944:*

*a. Defenses will be strengthened, and tactical bases developed, in the areas of the Marianas and Caroline Islands, Western New Guinea, and the Banda and Flores Seas.*

*b. Bases will be developed in the Philippines area for strategic and logistic support.*

*c. Ground, sea, and air strength will be built up in preparation for counteroffensive action.*

*3. In the event of an enemy approach toward the areas mentioned in paragraph 2a, powerful com-*

---

1 This chapter was originally prepared in Japanese by Capt. Atsushi Oi, Imperial Japanese Navy. Duty assignments of this officer were as follows: Personnel Bureau, Navy Ministry, Jan 41—Mar 43; Executive Officer, 21st Base Force (Soerabaja), 23 Mar-25 Jun 43; Staff Officer 1st Bureau (Operations), Imperial General Headquarters, Navy Section, 2 Jul-14 Nov 43; Staff Officer (Operations), General Escort Command, 15 Nov 43-21 Aug 45. All source materials cited in this chapter are located in G-2 Historical Section Files, GHQ FEC.

2 Cf. Plate No. 57, Chapter IX.

3 *Daikaishi Dai Nihyakuhachiju-go Bessatsu: Chunambu Taiheiyo Homen Riku-kaigun Chuo Kyotei* 大海指第二百八十號別冊中南部大平洋方面陸海軍中央協定 (Imperial General Headquarters Navy Directive No. 280, Annex: Army-Navy Central Agreement re Central and South Pacific) 30 Sep 43.

4 Cf. Chapter IX for discussion of Eighth Area Army and Southeast Area Fleet plans implementing the Central Agreement provisions for Eastern New Guinea, New Britain and the Solomons.

*ponents of all arms will be concentrated against his main attacking front, and every means will be employed to destroy his forces by counteroffensive action before the attack is launched.*

*4. After the middle of 1944, if conditions permit, offensive operations will be undertaken from the area including Western New Guinea and the Banda and Flores Seas. Separate study will be made to determine the front on which such operations should be launched, and necessary preparations will be carried out accordingly.*

The deadline fixed by Imperial General Headquarters for the completion of preparations along the new defense perimeter was based upon the estimate that full-scale Allied offensive operations against either the Western New Guinea or Marianas-Carolines sectors of the line, or possibly against both sectors simultaneously, would develop by the spring and summer of 1944. Although a six months' period was thus allowed for execution of the program, its actual start was somewhat delayed. Moreover, the scope of preparations envisaged was so vast that it was problematical whether the nation's material and technical resources would be equal to the task.

Primary emphasis in these preparations was placed upon the development of air power. After the bitter lessons taught by the southeast area campaigns of 1942–43, Army and Navy strategists unanimously agreed that the air forces must be the pivotal factor in future operations, whether defensive or offensive. To successfully defend the new "absolute defense zone" against the steadily mounting enemy air strength, they believed it imperative to have 55,000 planes produced annually. At the same time a large number of air bases, echeloned in depth and mutually supporting, had to be built and equipped over the widely dispersed areas of the new defense zone.

To meet the first of these requirements was impossible considering the current production level and the overall natural resources.[5] Therefore, at the Imperial conference of 30 September 1943, a compromise was reached which set a production goal of 40,000 planes for the fiscal year 1944, a goal still thought extremely difficult to attain. The airfield construction program was equally ambitious. In the area embracing Western New Guinea, the Moluccas, Celebes, and the islands of the Banda and Flores Seas, where the existing number of fields totalled only 27, plans were laid for the construction of 96 entirely new airfields and the completion of 7 others already partially built, bringing the total number of airfields planned for the area to 120.[6] This program was to be completed by the spring or, at latest, by the summer of 1944.

Although the central strategic concept of the new defense zone was one of powerful air forces rapidly deployable to prepared bases in any threatened sector, it was also obviously essential to build up adequate ground defenses to protect these bases from attack. In the early stages of the war, troops and materiel had been thrown into the exterior perimeter of advance, and development of a reliable inner defense system had been neglected. The powerful Allied offensives of 1943 in the southeast area aggravated this situation by drawing off and consuming a large portion of Japanese war strength, with the result that rear-area defenses in Western New Guinea, the Carolines and Marianas remained seriously weak and, at some points, non-existent. The

---

5 Monthly aircraft production figures for the period August–October 1943 were as follows: August, 1,360; September, 1,470; October, 1,620. *Dai Toa Senso Shusen ni kansuru Shiryo* 大東亞戰爭終戰に關する資料 (Data Bearing on the Termination of the Greater East Asia War) Ministry of Commerce and Industry, 14 Aug 45, p. 22.

6 Imperial General Headquarters Navy Directive No. 280, Annex, op. cit.

plan to forge these areas into a main line of resistance consequently necessitated the movement of substantial troop reinforcements and a large volume of supplies.

In view of Russia's continued neutrality and a relatively quiet situation on the China front, Imperial General Headquarters decided to redeploy a number of troop units from the Continent to the areas along the new Pacific defense line. The transportation of these units, however, presented a difficult problem because of the serious depletion of ship bottoms. The Army and Navy pressed for the allocation of additional non-military shipping to military use, but the tonnage demanded was far in excess of what could be spared without impairing the movement of raw materials urgently required for the war production program. The compromise figure of 250,000 tons finally agreed upon at the Imperial conference of 30 September was barely enough to compensate for losses of military shipping in current operations.[7] However, it was considered the maximum that could be drawn from the non-military shipping pool, which itself was below existing requirements.[8]

The critical shipping situation and the difficulties of procuring defense equipment greatly retarded the reinforcement of the new defense zone. At the end of 1943 the Marianas and Carolines, forming a vital sector of the perimeter line, were still garrisoned only by skeleton naval base forces.

Western New Guinea, lying directly astride the axis of General MacArthur's advance, also was weakly held by scattered naval base units and Army line of communications troops. The only sector adequately manned was the southern flank of the line in the Banda and Flores Seas area, the defenses of which had been comparatively well organized by the Nineteenth Army, with headquarters at Ambon.[9]

## Western New Guinea Defenses

To provide for the defense of Western New Guinea, Imperial General Headquarters had decided at the end of October to transfer from Manchuria the headquarters of the Second Army, commanded by Lt. Gen. Fusataro Teshima, and to assign to it two first-line divisions, the 3d and 36th, then stationed in China. At the same time, it was decided to relieve the Second Area Army headquarters of its current duties in Manchuria and to place it in command of both the Second and Nineteenth Armies, thus unifying the direction of Army forces in the Western New Guinea and Banda-Flores Sea sectors.[10] General Korechika Anami, Second Area Army commander, provisionally established his headquarters at Davao, in the southern Philippines, on 23 Novem-

---

7 *Gun Hoyu Sempaku Hendo ni kansuru Shuyo Jiko* 軍保有船舶變動ニ關スル主要事項 (Principal Matters Pertaining to Changes in Military Shipping) Shipping Division, General Maritime Bureau, Ministry of Transportation, 20 Jan 46.

8 As of 1 September 1943, 2,497,000 gross tons of shipping were available for non-military use, while 2,844,000 gross tons were allocated to the Army and Navy. Even prior to the planning of the huge aircraft production program, the minimum estimated tonnage requirement for non-military use was 3,000,000 tons. Ibid.

9 Main combat strength of the Nineteenth Army consisted of the 5th Division on the Aroe, Kai, and Tanimbar Islands, and the 48th Division on Timor. These were reinforced in February 1944 by the 46th Division (less 145th Infantry Regt.), which was stationed on Soemba Island, west of Timor. (Interrogation of Col. Kazuo Horiba, Staff Officer (Operations), Second Area Army.)

10 *Daihonyei Rikugun Tosui Kiroku* 大本營陸軍統帥記錄 (Imperial General Headquarters Army High Command Record) 1st Demobilization Bureau, Nov 46, pp. 186–7.

ber,"[11] and on 1 December assumed operational command of the Second and Nineteenth Armies, the 7th Air Division, and the 1st Field Base Unit.[12] Also by 1 December, Second Army headquarters had moved to Manokwari, Dutch New Guinea, where Lt. Gen. Teshima took command of forces in the assigned Army area.

The operational zone assigned to the Second Area Army extended on the west to the Makassar and Lombok Straits, on the north to five degrees N. Latitude, and on the east to the 140th meridian, which marked the boundary with the Eighth Area Army. (Plate No. 64) Within this zone, the Area Army was to exercise direct command over the northern Moluccas, northeastern Celebes, and Talaud Islands. The Nineteenth Army remained charged with operations in the Banda-Flores Seas area, and the Second Army was assigned responsibility for all of Dutch New Guinea west of the 140th meridian.[13]

Imperial General Headquarters instructed General Anami that the main defensive effort of the Area Army should be made in Western New Guinea. However, when the new command dispositions went into effect on 1 December, the situation of the Second Army was hardly favorable for the establishment of strong defenses in this area. The 36th Division was still en route from China, while the 3d Division, operating on the Central China front, had not yet been released for shipment, with the result that there was not a single ground combat unit in the entire Army zone. Nor could Nineteenth Army furnish reinforcements since its two (later three) combat divisions were scattered over the many islands of the Banda and Flores Seas, then still considered a vital sector of the defense zone.[14] Moreover, the shortage of shipping and the menace of Allied air and submarine attacks militated against the ready transfer of units from the Nineteenth Army area to Western New Guinea.

Although the situation improved with the arrival of the main elements of the 36th Division[15] on 25 December, Second Army troop strength was still inadequate to assure the defense of its broad operational zone. Pending final formulation of an over-all defense plan for Western New Guinea, Lt. Gen. Teshima stationed the 36th Division (less 222d Infantry) at

---

11 The provisional Area Army headquarters was set up at Davao in accordance with a directive by Imperial General Headquarters. It was not until 26 April, following the Hollandia landing, that the headquarters finally advanced into the Area Army's operational zone, establishing itself at Menado, in the northern Celebes. *Gohoku Sakusen Kiroku* 濠北作戰記錄 (North of Australia Operations Record) 1st Demobilization Bureau, Jul 46, pp. 13–4, 107.

12 The 1st Field Base Unit was activated in Japan in October and assigned to Second Area Army to control all service and rear-echelon units in the Area Army zone of direct command. The 2d Field Base Unit, activated simultaneously, was assigned to Second Army to perform the same mission in Western New Guinea. Both were commanded by major generals and were the only headquarters of this type in the southern area. (Statement by Lt. Col. Kotaro Katogawa, Staff Officer (Operations), Second Area Army.)

13 North of Australia Operations Record, op. cit., pp. 38–9.

14 Ibid., pp. 24–5.

15 The 36th Division, with a total strength of about 13,700, had been reorganized as a regimental combat team type division and equipped for amphibious operations. The artillery regiment was dropped, and a battalion of light artillery was made an organic part of each infantry regiment. Order of battle was as follows:

| | |
|---|---|
| Division Headquarters | 36th Division Tank Unit (four companies) |
| 222d Infantry | 36th Division Signal Unit (one company) |
| 223d Infantry | 36th Division Transport Unit |
| 224th Infantry | 36th Division Sea Transport Unit |

North of Australia Operations Record, op. cit. Annex I, Attached Table 1.

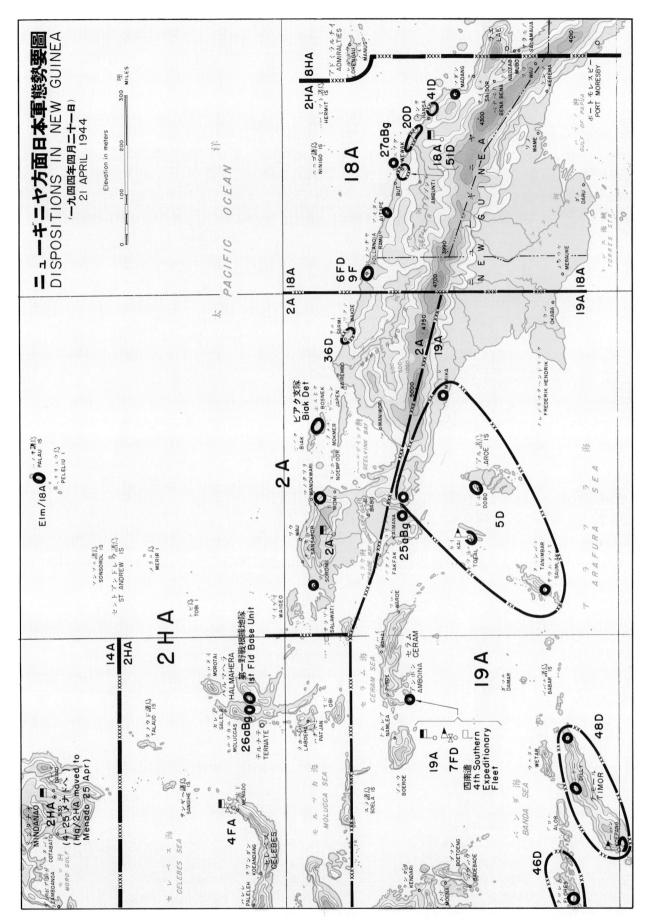

PLATE NO. 64
Dispositions in New Guinea, 21 April 1944

Sarmi. The 222d Infantry, reinf. was dispatched to Biak Island to begin organizing the defenses of that strategic position.

The Second Army's plan for the defense of Western New Guinea emphasized the importance of securing Geelvink Bay. This plan was based upon the availability of only two divisions, the troop strength originally allotted by the High Command. The three key positions in this defense scheme were the Sarmi-Wakde area, Biak Island, and Manokwari.[16] As finally decided on 8 January, the outline of planned strength dispositions was as follows:[17]

*Sarmi-Wakde*   One division (less one inf. regt.)
*East Japen*
*Koeroedoe I.*   One inf. regt. (reinf.)
*Noeboai*
*Biak*   One division (less one regiment)
*Manokwari*   One regt. (less one battalion)
*Wissel Lake*   One battalion

Under this plan, the 222d Infantry was to continue its interim mission of organizing the defenses of Biak Island until relieved by the 3d Division. The regiment would then proceed to garrison the east Geelvink Bay sector.

While initial attention was focussed on the Geelvink Bay area, the Second Area Army command was also concerned over the weak condition of the defenses of Hollandia, which lay just east of the 140th meridian in the Eighth Area Army zone of responsibility. An order to dispatch an element of the 36th Division to that sector was issued but was quickly revoked on the ground that it would weaken the defenses of Geelvink Bay without appreciably strengthening Hollandia.[18] A large section of the New Guinea coast between Wewak and Sarmi thus remained practically undefended. General Anami promptly dispatched a staff mission to Eighth Area Army headquarters at Rabaul to press for reinforcement of the Hollandia area, and a similar recommendation was communicated to Imperial General Headquarters during December. The 6th South Seas Detachment (two battalions), temporarily stationed on Palau, was dispatched by the High Command. No other action was taken, however, since both Eighth Area Army and Eighteenth Army, after the loss of Finschhafen, were more immediately concerned with checking further enemy penetration of the Dampier Strait region.

Though unsuccessful in obtaining action on Hollandia, General Anami continued to press the organization of defenses within the Second Army zone in Western New Guinea despite severe handicaps. Troop strength remained seriously short, and in addition the prospects of adequate air and naval support were discouraging. The 7th Air Division, with headquarters on Ambon, in the Moluccas, was the only air unit assigned to Second Area Army and was currently recuperating from heavy losses. Operations in eastern New Guinea between August and November, had cut down its strength to only about 50 operational aircraft.[19] This meager force was devoted almost exclusively to shipping escort missions in the rear areas.

The prospects for naval air support were no more encouraging. In case of an enemy attack directed at Western New Guinea, the Second Area Army could count upon the cooperation of the 23d Air Flotilla based at Kendari, in the Celebes, but the operational strength of this unit was likewise down to about 50 planes, and most of its experienced pilots had been transferred to the naval air forces at Rabaul during

---

16  Interrogation of Lt. Gen. Takazo Numata, Chief of Staff, Second Area Army.
17  Outline of Operational Preparations, Second Army, 8 Jan 44. ATIS Bulletin No. 1457, 20 Sept 44.
18  North of Australia Operations Record, op. cit., p. 48.
19  Statement by Lt. Col. Katogawa, previously cited.

the Solomons and Papuan campaigns.[20] Moreover, despite the withdrawal of the main defense line to Western New Guinea and the Carolines, the Navy continued to maintain its most efficient carrier flying units on land bases in the Rabaul area to serve as a forward strategic air barrier. This policy resulted not only in the steady depletion of the fleet air arm but in the immobilization of the carrier strength which otherwise might have been capable of providing air support at any threatened point of the main defense line.

The only naval surface combat forces in the immediate vicinity of the Second Area Army operational zone were the 16th Cruiser Division (*Ashigara, Kuma, Kitakami, Kinu*) and the 19th Destroyer Division (*Shikinami, Uranami, Shigure*), both under command of the Southwest Area Fleet with headquarters at Soerabaja. Charged with naval missions covering the area from the Indian Ocean to Western New Guinea, this fleet obviously had insufficient strength to provide support against an eventual enemy attack against the north coast of Dutch New Guinea. Meanwhile, the Second Fleet, containing the bulk of the Navy's battleships, was in the Truk area. Without attached carrier forces, however, its role was not offensive but merely to act as a fleet-in-being to deter enemy attack.

Naval ground forces in Western New Guinea under the command of the Fourth Expeditionary Fleet stationed at Ambon, were weak and uniformly small. During earlier operations in the southeast area, Army troops had received substantial support from naval base forces and special landing forces, but in Western New Guinea the naval base forces were too small. These were widely scattered at Hollandia, Wakde, Manokwari, Nabire, and Sorong.[21]

In the light of these unfavorable conditions, it was obvious that Second Area Army could not accomplish the organization of Western New Guinea defenses without substantial reinforcements of well-trained and well-equipped line units, as well as air strength. In mid-January, therefore, General Anami forwarded an urgent request to Tokyo for more troops. Imperial General Headquarters responded promptly with a plan to allot 15 infantry battalions, three heavy artillery regiments, and one tank regiment, in addition to the 14th Division, which the High Command now planned to assign to Second Area Army in place of the 3d Division.[22]

These reinforcements, together with the necessary service and supply elements, would boost the strength of the Area Army from approximately 170,000 to about 320,000 troops. The 14th Division was scheduled to arrive by the end of March, while the other combat units were to complete their movement to Western New Guinea by May.[23] The transport of service troops was to continue through July.

---

20 U. S. Strategic Bombing Survey (Pacific), Naval Analysis Division, *Interrogations of Japanese Officials*, 1946. Vol. II, pp. 287–8. (Interrogation of Capt. Hironaka Komoto, Staff Officer (Operations), 23d Air Flotilla.)

21 Statement by Comdr. Masataka Chihaya, Staff Officer (Operations), Fourth Expeditionary Fleet.

22 The previous plan to transfer the 3d Division was cancelled because the division could not be released from its commitments in Central China. The 14th Division, currently stationed in Manchuria, was formally reassigned to Second Area Army on 10 February. (1) North of Australia Operations Record, op. cit., p. 50. (2) Imperial General Headquarters Army High Command Record, op. cit., p. 188.

23 The 14th Division was to be deployed as previously planned for the 3d Division, i. e., the division main strength in the Biak area, and one regiment in the Manokwari area. The other combat reinforcements were to be deployed as follows: Sorong, three infantry battalions; Halmahera, nine infantry battalions; Area Army reserve, three infantry battalions (each of these forces to have appropriate supporting artillery and tank units). (Statement by Lt. Col. Katogawa, previously cited.)

As a further step to bolster the southern sector of the national defense zone, Imperial General Headquarters in December 1943 began contemplating an important modification of the command dispositions then in force. Principally to assure the mobility and economical use of air power and shipping resources, it was proposed to combine the Fourteenth Army in the Philippines and the Second Area Army in the Western New Guinea—eastern Dutch East Indies area under the higher command of Southern Army, at the same time restricting them to ground forces only and placing the Third and Fourth Air Armies, as well as shipping groups, directly under Southern Army command. Such a step was also deemed necessary to assure that Southern Army would transfer primary emphasis from the Asiatic mainland to the Pacific front, now clearly the decisive battlefront of the war.[24]

Before this proposal had a chance to reach concrete form, developments on the Central Pacific front temporarily usurped the attention of Imperial General Headquarters, with the result that the final orders directing the modification of the command set-up were not issued until 27 March 1944. The effective date of the new dispositions was fixed at 15 April.

## Setbacks to Defense Preparations

The suddenly increased tempo of the enemy advance in the Central Pacific during February gave rise to strong belief that an amphibious assault might develop against the Marianas or Carolines sector of the main defense line at any time.[25] This impending danger led the Army and Navy High Commands to press successfully for the transfer to military use of an additional 300,000 gross tons of non-military shipping during the months of February, March and April.[26]

First priority was assigned by Imperial General Headquarters to the movement of troops, munitions, and supplies to the Marianas and Carolines. Since military tonnage, despite the scheduled 300,000-ton increase, still fell below requirements, this decision necessitated the deferment of scheduled troop and supply shipments to other areas. In the latter part of February Imperial General Headquarters notified General Anami that the shipping allocation to Second Area Army was being temporarily suspended due to the urgency of the Central Pacific situation. This of course meant a critical delay in the program to reinforce Western New Guinea.[27]

---

24 (1) Imperial General Headquarters Army High Command Record, op. cit., pp. 255, 229. (2) Statement by Col. Takushiro Hattori, Chief, Operations Section, Imperial General Headquarters, Army Section.

25 Following the enemy invasion of Kwajalein on 1 February, a powerful American naval task force carried out a destructive two-day attack on the key Japanese fleet base of Truk in the Carolines on 17–18 February, while an enemy amphibious force simultaneously landed on Eniwetok in the western Marshalls. These startling developments had sharp repercussions in the Army and Navy High Commands. On 21 February General Tojo, already serving concurrently as Premier and War Minister, took over the post of Chief of Army General Staff from Field Marshal Sugiyama, and Navy Minister Admiral Shigetaro Shimada concurrently assumed the post of Chief of Navy General Staff, replacing Fleet Admiral Osami Nagano.

26 Principal Matters Pertaining to Changes in Military Shipping, op. cit.

27 On 10 March the chiefs of staff of all major subordinate commands under Second Area Army met at Davao for a conference on operational matters. In view of the suspension of the Area Army's shipping allocation, a major problem considered was an emergency plan for Western New Guinea to meet a possible enemy attack before the deployment of reinforcements to the theater could be carried out. Under this plan, the 36th Division in the Sarmi area was to prepare to move rapidly against an enemy force which might land to the east of Sarmi, but at the same time Second Army was to spread out its available forces to secure as many key points as possible. An implementing Second Army order issued 29 March called for the stationing of small units on Waigeo and Mapia Islands and at various points along the north coast of the Vogelkop Peninsula, while two companies were detached from the 222d Infantry on Biak to garrison Noemfoor and Sorong. The 2d Field Base Unit commander at Manokwari was placed in command of the Geelvink Bay defenses to the west of Biak. (1) North of Australia Operations Record, op. cit., pp. 49, 51, 81–3. (2) Interrogation of Lt. Gen. Takazo Numata, previously cited. (3) Second Army Operations Order No. 53, 29 Mar 44. ATIS Bulletin No. 1457, 20 Sep 44.

The suspension also led to changes in troop allocation plans. The 14th Division, previously allotted to Second Area Army, was reassigned on 20 March to the newly-activated Thirty-first Army for the defense of the Marianas and Carolines.[28] In its place Imperial General Headquarters early in April assigned the 35th Division to Second Area Army, directing employment of the division main strength on Western New Guinea.[29] However, the actual movement of the division main elements from China still had to await restoration of Second Area Army's shipping allocation.

In addition to, and partially as the result of, the shortage of shipping, slow progress in both the aircraft production and airfield construction programs seriously undermined the entire plan for the new Pacific defense line. Average monthly production for the period January-April 1944 was about 2,200.[30] This did not augur well for the attainment of the production goal of 40,000 planes for the fiscal year 1944. The prospects were even darker due to the fast dwindling cargo-carrying bottoms resulting from the transfer to the military of 30,000 tons and the tremendous losses from enemy action in recent months.

The ambitious air base construction program for Western New Guinea and the eastern Netherlands East Indies had meanwhile bogged down seriously. In these areas even combat units had been put to work as labor troops in an effort to carry out the plans formulated by Tokyo, but shortages of materials, transportation capacity, available field labor, and mechanized equipment, together with deficiencies in engineering technique, slowed down progress to a minimum. Less than one-third of the projected bases was completed by the time they were critically needed, and the funneling of effort into their construction materially delayed other operational preparations by the field forces. Of the 35 new airfields planned for the Western New Guinea area, only nine were available for use by the end of April 1944. All other installations used by the air forces during the Western New Guinea campaign had already been in existence prior to the start of the construction program.[31]

Despite the lack of progress in aircraft production and the building of new bases, the Army and Navy made serious efforts to replenish their first-line air strength, both in planes and pilots, in preparation for decisive

---

28 North of Australia Operations Record, op. cit., p. 50

29 In a directive dated 4 April supplementing the assignment order, Imperial General Headquarters specified that the 219th Infantry Regiment, currently in Japan, was to be detached from the 35th Division to garrison the St. Andrew Islands, lying between Palau and Western New Guinea. The regiment embarked from Yokohama on 6 April for Palau together with the 35th Division headquarters, which was to trans-ship at Palau for Western New Guinea. Since detachment of an entire regiment would seriously upset existing plans for the defense of the Geelvink Bay area, General Anami instituted negotiations with Imperial General Headquarters while the convoy was en route to Palau and succeeded in obtaining a modification of the 4 April directive. Imperial General Headquarters now agreed to the trans-shipment of the main strength of the 219th Infantry from Palau to Western New Guinea, leaving only one battalion to garrison the St. Andrew Islands. (1) *Dairikushi Dai Senkyuhyakuyonjuni-go* 大陸指第一九四二號 (Imperial General Headquarters Army Directive No. 1942) 4 Apr 44. (2) Statement by Lt. Col. Katogawa, previously cited. (3) Personal diary belonging to a member of 35th Division covering the period 1 Apr-16 Jul 44. ATIS Bulletin No. 1500, 12 Oct 44.

30 Monthly production figures during this period were: January, 1,815; February, 2,060; March, 2,711; April, 2,296. Data Bearing on the Termination of the Greater East Asia War, op. cit., p. 22.

31 The nine new airfields built in Western New Guinea under the September 1943 program were at Hollandia (Sentani and Cyclops), Sarmi (Sawar), Biak (Mokmer and Sorido), Noemfoor (Kamiri), Moemi, Manokwari and Sorong. In addition, six existing airstrips were improved. (1) 6th Air Division Operations Order (undated), ATIS Bulletin No. 1177, 22 Jun 44. (2) Statement by Comdr. Chihaya, previously cited.

Ever since the invasion of the Admiralties, the Eighteenth Army command had anticipated a new Allied amphibious operation against the Northeast New Guinea coast by March or April of 1944, but it had estimated that the target area would be somewhere to the east of Wewak.[41] With the carrier raid on Palau and the extension of Allied air attacks to Hollandia, a section of the Eighteenth Army staff saw an increasing possibility that the objective would lie farther west, not excluding even the distant Hollandia area. However, the estimate finally accepted still placed the most likely area of attack between Madang and Hansa Bay, including Karkar Island. Wewak was rated the next most probable target, with Hollandia least likely but not entirely excluded.[42]

One reason for Eighteenth Army's minimization of the immediate danger to Hollandia was the belief, based on past observation of General MacArthur's tactics, that landing operations in that area would not be attempted until advance bases had been taken, from which land-based Allied air forces (including fighters) could neutralize Japanese air bases to the west of Sarmi and also provide direct support to the landing forces.[43] With the most advanced Allied bases located at Saidor and in the Admiralties, almost 500 miles from Hollandia and over 600 miles from Sarmi, it was considered almost certain that General MacArthur's next move would be aimed at seizing a forward fighter base somewhere between Madang and Aitape, in preparation for a later invasion of Hollandia.

The fighter-escorted bomber raids on Hollandia in early April forced an upward revision of the calculated capabilities of enemy fighters from existing bases.[44] They did not modify Eighteenth Army's estimate of enemy offensive plans, however, since effective fighter range for the continuous type of support required in amphibious landing operations was still believed to be only about 300 miles. The possibility that carrier forces might be borrowed from the Central Pacific to provide tactical support was gravely underestimated since none of General MacArthur's previous invasion operations had been furnished such support.

Other enemy actions also were instrumental in strengthening Eighteenth Army's belief that the next blow would fall in the Madang—Wewak area. One was the unleashing in March of a heavy air offensive directed at the coastal area from Wewak eastward, with Wewak itself and Hansa Bay as the main targets; another was a marked augmentation of enemy motor torpedo boat activity from Dampier Strait west to Hansa Bay.[45]

Although the next Allied effort was thus expected to fall short of Hollandia, both Imperial General Headquarters and Second Area Army were strongly convinced that this valuable base would subsequently be attacked, possibly

---

41 Southeast Area Operations Record, Part III, op. cit. Vol. III, pp. 19–21.

42 Ibid., pp. 72-5.

43 Ibid., pp. 73–5.

44 "It was the opinion of our leaders.... that Allied fighter planes, which I presume were based at Nadzab, would not be able to accompany the long-range bombers due to their limited range.... However, we were completely fooled when these fighters were equipped with auxiliary tanks, enabling them to cover the rather long distance to Hollandia with ease." (Interrogation of Lt. Col. Nobuo Kitamori, Staff Officer (Communications), Second Area Army.)

45 Access to American operational documents during the preparation of this volume indicates that these enemy actions were part of a deliberate deception program instituted by General MacArthur's headquarters to cover the planned invasion of Hollandia and Aitape.

as early as June.[46] Not only were the major base facilities of Fourth Air Army located in the area, but Hollandia had become an important staging point on air transport routes to Japanese-held areas farther east,[47] as well as the chief port for logistic support of the Eighteenth Army. Huge amounts of military supplies were in open storage along the shore of Humboldt Bay. All these factors made it appear highly probable that the enemy eventually would seek to wrest Hollandia from Japanese control, especially since it would give the Allies a well-developed air and sea base, valuable as a staging-point for large-scale amphibious operations.

Despite growing awareness of the need to bolster Hollandia's defenses, Eighteenth Army was in no position to take immediate steps to that end. Although the Army Commander had issued orders on 10 March for a strengthening of Aitape, Wewak, and Hansa Bay, the Army was experiencing great difficulty in moving troops westward from Hansa Bay because of the shortage of sea transportation and heavy enemy air interference from forward bases at Nadzab and Saidor.[48]

Nevertheless, when Second Area Army assumed operational control of Eighteenth Army and Fourth Air Army on 25 March, General Anami promptly ordered Eighteenth Army to move as soon as possible to the west of Wewak and consolidate the defense of air bases, with particular emphasis on the installations at Aitape and Hollandia. Pursuant to this order, Lt. Gen. Adachi revised the existing plan for redeployment of Eighteenth Army forces along the following lines:[49]

*1. 51st Division to move to Hollandia instead of to Wewak.*

*2. 41st Division to assume the mission of garrisoning Wewak instead of Hansa Bay.*

*3. 20th Division to garrison Aitape, as previously planned.*

Eighteenth Army immediately threw its full effort into the execution of the revised plan, but from the outset it faced severe difficulties. Use of sea routes, normally traversable in a few days, was interdicted by Allied air and sea superiority, leaving no alternative but time-consuming movement overland. Roads were non-existent, and the native tracks leading west from Hansa Bay crossed two large rivers, the Ramu and Sepik, which were completely unfordable near the coast, and the mouths of which were flanked by broad stretches of almost impassable mangrove swampland.[50] Troop movements were further hampered by the necessity of keeping

---

46 (1) "The High Command believed that Wewak would be attacked before Hollandia.... Although we were convinced that the Allies would eventually attack Hollandia, we rather believed that they would attempt to acquire an important position somewhere east of Aitape (first)" (Interrogation of Lt. Gen. Jo Iimura, Chief of Staff, Southern Army.) (2) "Hollandia was expected to be attacked soon after a preliminary attack on Wewak. However, the attack on Hollandia was not expected until June." (Interrogation of Col. Arata Yamamoto, Senior Staff Officer Second Army.) (3) "A study made by Col. Kadomatsu, senior intelligence officer of Second Area Army, estimated that the Americans would land first at Hansa Bay and then at Hollandia. This estimate was based on a graph of all enemy landing operations." (Interrogation of Lt. Gen. Numata, previously cited.)

47 During January 1944, 94 transport missions were logged through Hollandia en route to Wewak, Madang, Hansa Bay, Rabaul, and other important bases to the east. Critical cargo, mail, and passengers were thus moved despite the Allied sea blockade. Transport Journal, Fourth Air Army Liaison Station, Hollandia, Jan 44. ATIS Enemy Publications No. 170, 14 Aug 44.

48 Statement by Lt. Col. Kengoro Tanaka, Staff Officer (Operations), Eighteenth Army.

49 Southeast Area Operations Record, Part III, op. cit. Vol. III, pp. 12-13, 32.

50 At the end of March, about 50 landing barges and 30 fishing and powered sailing vessels were available in this area. Most of these boats had to be used for ferrying munitions, ordnance and supplies. At the Sepik River, it was at first impossible to put across more than 50 troops per day on this account, although a maximum of 770 per day was later reached. It was estimated that it would take until early June to move across all Eighteenth Army forces. (1) Ibid., pp. 44, 49-50. (2) Statement by Lt. Col. Tanaka, previously cited.

constantly on the alert for an enemy surprise landing.

As later events proved, even had Eighteenth Army been able to adhere to its own timetable for these movements, they would not have been completed in time to meet the Allied attack at Hollandia. Given the most favorable conditions, the first echelon of the 51st Division, consisting of three infantry battalions, was not expected to reach Hollandia until late in May. The 20th Division meanwhile was held up on the east bank of the Ramu by the shortage of boats, and its first elements were unable to leave Hansa Bay for Aitape until early April.

As a stop-gap measure pending the arrival of the 51st Division, Eighteenth Army in early April dispatched Maj. Gen. Toyozo Kitazono, 3d Field Transport Unit commander at Hansa Bay, to Hollandia in order to assume direction of ground defense preparations by the miscellaneous army units already in that area. Just prior to Maj. Gen. Kitazono's arrival on April 10, Vice Adm. Yoshikazu Endo had temporarily transferred Ninth Fleet[15] headquarters from Wewak to Hollandia. Fourth Air Army headquarters also was still at Hollandia at this time but withdrew to Menado immediately after the Air Army's transfer to direct Southern Army command became effective on 15 April. This left Maj. Gen. Masazumi Inada, who had arrived on 11 April to take command of the 6th Air Division, the highest Army air commander.[52]

Due to the brief lapse of time between the arrival of the new commanders and the Allied assault on Hollandia, no local agreement for the coordinated use of all forces had yet been reached when the attack came.[53] These forces aggregated about 15,000, including all ground, air and naval personnel, of which about 1,000 were hospitalized ineffectives. Approximately 80 per cent of the total strength consisted of service units.[54] Combat air strength was

---

51 The Ninth Fleet had no ships of any importance and consisted only of the 2d and 7th Naval Base Forces, currently at Wewak. The 7th Naval Base Force had just completed a long and costly retreat from Lae-Salamaua via Madang and was shortly merged with the 2d Naval Base Force to form the 27th Special Naval Base Force. *Teikoku Kaigun Senji Hensei* 帝國海軍戰時編制 (Wartime Organization of the Imperial Navy) Navy General Staff, 1944.

52 Maj. Gen. Masazumi Inada had been relieved as 2d Field Base Unit commander at Manokwari to assume command of the 6th Air Division. Maj. Gen. Shikao Fujitsuka, Chief of Staff, Second Army, took over the 2d Field Base Unit.

53 Southeast Area Operations Record, Part III, op. cit. Vol. III, pp. 87–8.

54 Order of battle of Japanese forces in Hollandia at this time was as follows:

Army Ground and Service Units
  Hq., 3d Field Transport Command
  Hq., and 1st Bn., 6th South Seas Detachment
  68th Field AAA Bn.
  42d Independent Motor Transport Bn.
  49th Anchorage
  Elms, 31st Anchorage
  27th Field Depot (Ordnance, Mtr. Trans, Freight)
  54th Line of Communications Sector Unit
  4th Sea Transport Battalion
  79th and 113th Line of Communications Hospital
  Misc. signal, medical, ordnance, motor transport, field post office, and construction units.
Army Air and Air Service Units (no operational aircraft)
  Hq., 6th Air Division
  Hq., 14th Air Brigade
  Hq., Training Brigade
  68th, 78th, 63d, 248th, 33d, and 77th Fighter Regts.
  208th, 34th, and 75th Light Bomber Regts.
  7th Air Transport Unit
  Hq., 18th Air Sector Unit
  66th Field AAA Bn.
  66th and 39th Field Machine Cannon Cos.
  39th Field AAA Co.
  3d Searchlight Co.
  22d, 38th, and 209th Airfield Bns.
  14th Field Air Repair Depot
  14th Field Air Supply Depot
  Misc. navigation, intelligence, signal, repair, survey, and construction units.
Naval Units
  Hq., Ninth Fleet
  90th Naval Garrison Unit
  8th Naval Construction Unit

(1) North of Australia Operations Record, op. cit., pp. 92–3. (2) Chart of Forces Landed at Hollandia, issued by 54th Line of Communications Unit, Mar 44. ATIS Bulletin No. 1055, 20 May 44. (3) Misc. Order Files and Strength Charts of Units at Hollandia. ATIS Bulletins No. 1051, 19 May 44; No. 1054, 20 May 44; No. 1139, 8 Jun 44; No. 1177, 22 Jun 44; No 1187, 25 Jun 44; and No. 1284, 24 Jul 44.

also pitifully weak. The 6th Air Division had only a handful of aircraft still operational, and chief reliance was placed on the Navy's 23d Air Flotilla, which transferred its headquarters on 20 April to Sorong, on the Vogelkop Peninsula. The greater part of its strength began operating from a newly completed base on Biak.[55]

This was the situation of Hollandia's defenses when, on 17 April, the naval communications center at Rabaul radioed a warning that Allied landing operations might be expected imminently at some point on the New Guinea coast. Radio intercepts by Japanese signal intelligence revealed that Allied air units from Lae, Nadzab and Finschhafen were concentrating in the Admiralties,[56] and that a large number of enemy ships was moving in the Bismarck Sea, maintaining a high level of tactical radio traffic.

Two days later, on 19 April, a patrol plane of the Carolines-based First Air Fleet sighted a large enemy naval force, including aircraft carriers, moving north of the Admiralties. The same day, an army reconnaissance flight from Rabaul spotted a second convoy of about 30 transports, escorted by an aircraft carrier, two cruisers and ten destroyers, passing through the Vitiaz Strait. On the 20th, two large enemy groups—one a task force with four carriers and the other an amphibious convoy—were reported standing westward just north of the Ninigo Islands, about 200 miles due north of Wewak.

From the course which these forces were taking, no accurate prediction was yet possible as to where the enemy would land. However, the invasion force now turned suddenly southward and, on 21 April, launched simultaneous air strikes at three different places—Hollandia, Aitape, and the Wakde-Sarmi area. So violent were these attacks that the local forces in each area believed that their own sector would be the main target of invasion.

At Hollandia the enemy air preparation began at dawn on 21 April and continued without interruption until late afternoon. Wave after wave of both carrier and land-based planes, numbering approximately 600, pounded the area, inflicting severe damage, particularly on the three airfields located in the vicinity of Sentani Lake. The simultaneous attacks on Wakde and Sarmi, though less protracted, were equally devastating. Virtually all base installations in the three places were completely wrecked, and the last few operational aircraft of the 6th Air Division were destroyed. Combat air strength to the east of Sarmi was now reduced to nothing.

Beginning at 0530 on 22 April, carrier planes again struck at the Hollandia airfields and also at the beaches along Tanahmerah and Humboldt Bays. Combat ships entered both bays and laid down a heavy barrage of naval gunfire, while three carriers approached within approximately nine miles of the shore. Under cover of this close support, the enemy rapidly put ashore the largest landing force thus far thrown against any Japanese-held point in New Guinea.[57] Part of the force landed at Humboldt Bay, while a second contingent went ashore at Tanahmerah. (Plate No. 67) The Japanese forces defending both areas,[58] stunned by the

---

55 *Seibu Niyuginia oyobi Gohoku Homen no Kaigun Sakusen* 西部ニューギニヤ及濠北方面の海軍作戦 (Western New Guinea Area and North of Australia Area Naval Operations), op. cit., pp 6–7.

56 Southeast Area Operations Record, Part III, op. cit., Vol. III, p. 72.

57 Eighteenth Army estimated the strength of the Allied landing force at Hollandia at about two and a half divisions. (Statement by Lt. Col. Tanaka, previously cited.)

58 The Humboldt Bay sector was defended mainly by Eighteenth Army troops under Maj. Gen. Toyozo Kitazono, while the Tanahmerah Bay sector was defended by airfield troops under Maj. Gen. Inada. Southeast Area Operations Record, Part III, op. cit. Vol. III, pp. 92–3.

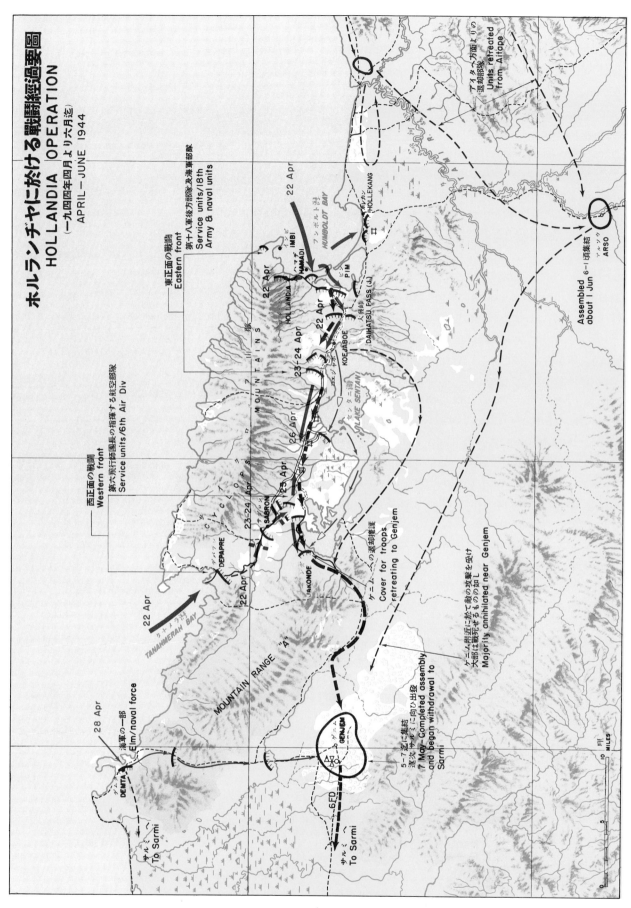

PLATE NO. 67
Hollandia Operation, April—June 1944

weight of the air and artillery preparation and without adequate prepared positions on which to make a stand, were forced to withdraw. By noon of 22 April, the entire beach and port areas around both Humboldt and Tanahmerah Bays were completely occupied by the enemy.

Simultaneously with the Hollandia invasion, Allied amphibious forces had also effected a landing in the strategic Aitape sector, about 125 miles east of Hollandia. At 0500 on 22 April, enemy warships began a two-hour bombardment of Aitape itself, of the Tadji airfield sector eight miles southeast of Aitape, and of Seleo Island, lying several miles offshore from Tadji. Under cover of this preparation, enemy troops landed near the Tadji airfield, where the Japanese garrison force of about 2,000, incapable of serious resistance, withdrew after a few skirmishes.[59] The airfield was immediately seized by the enemy, who had fighters based there by 24 April.

With the Japanese ground forces in both attack areas unprepared to offer a real defense, the initial reaction to the Allied landings necessarily was limited to air counterattacks. These were handicapped by the small number of planes available, but were prompt and partially effective. The 23d Air Flotilla, operating at extreme range from Sorong,[60] carried out night attacks with medium torpedo bombers against Allied surface craft on 22, 23, and 24 April. Elements of the 7th Air Division, which had advanced to Sorong from the Nineteenth Army area, joined in the attacks on the night of the 24th.

Still subjected to heavy enemy air and naval bombardment and lacking unified command, the defense forces on the Hollandia front had meanwhile fallen back on the Sentani Lake airfield sector. Here they were cut off from their ration and ammunition supplies, which were stored near the coast, and faced the hopeless prospect of conducting the defense of the airfields with less than a week's rations, very little small arms and machine gun ammunition, and no artillery. When enemy forces, advancing simultaneously from the Humboldt Bay and Tanahmerah beachheads, converged on the airfield sector on 26 April, the defenders were obliged to withdraw toward Genjem to escape being trapped, and all three airfields were occupied by the enemy.

On the night of 27 April, 23d Air Flotilla planes again attacked enemy shipping off Hollandia, claiming one light cruiser sunk and another large vessel damaged.[61] It was too late,

---

59 Order of battle of the Japanese forces in the Aitape area at the time of the Allied landing was as follows:

    20th Division Replacement Elms
    Elms 54th Line of Communications Sector Unit
    31st Anchorage Headquarters
    3d Debarkation Unit
    Elms 27th Field Ordnance and Freight Depots
    26th and 86th Airfield Cos.
    4th Airfield Construction Unit
    Elm 90th Naval Garrison Unit

(1) Ibid., pp. 110–111. (2) Various Personal Notebooks, Diaries, Order Files, and Official Strength and Situation Reports. ATIS Bulletins No. 1040, 16 May 44; No. 1054, 20 May 44; No. 1095, 29 May 44; No. 1121, 3 Jun 44; and No. 1177, 22 Jun 44.

60 The Biak airfield was already usable by reconnaissance and fighter planes but lacked a store of torpedoes and hence could not be used by the 23d Air Flotilla's torpedo bombers. These units were forced to operate from Sorong, 600 miles from Hollandia. (Statement by Comdr. Chihaya, previously cited.)

61 Western New Guinea Area and North of Australia Area Naval Operations, op. cit., p. 7.

however, for such attacks to have any appreciable effect on the ground situation, and due to the prohibitive losses inflicted by enemy night fighters and antiaircraft fire, the air offensive was discontinued.

By 7 May approximately 10,000 army and navy personnel had concentrated in the Genjem area, 20 miles west of Hollandia, where a large truck farming operation afforded limited food supplies.[62] Maj. Gen. Inada, 6th Air Division commander, now assumed command of all troops, organized them into several echelons, and initiated a general withdrawal toward Sarmi.[63]

While these developments were under way, Second Area Army headquarters at Davao had been giving urgent study to the situation created by the unexpectedly early Allied invasion of Hollandia. Immediately upon learning of the enemy landings on 22 April, General Anami took the optimistic view that the enemy had overreached himself by launching an amphibious assault at such great distance from his bases. He calculated that the local forces at Hollandia, despite deficiencies in training and equipment, would be able to offer at least partially effective resistance until adequate countermeasures could be taken by Eighteenth Army. In the meantime, he estimated that the morale of the garrison could be bolstered and its resistance stiffened by the dispatch of a small token force to Hollandia by Second Army. On the same day, therefore, he sent the following order to Lt. Gen. Teshima:[64]

*The Second Army Commander will immediately dispatch two battalions of infantry and one battalion of artillery (from the 36th Division) to the Hollandia area, where they will come under the command of the Eighteenth Army Commander.*

General Anami's optimistic estimate of the situation changed, however, as ensuing reports indicated that the enemy, using forces of considerable size, had easily established beachheads not only at Hollandia but at Aitape. The latter move obviously would render extremely difficult any attempt to move the main body of the Eighteenth Army westward to bolster Hollandia. At the same time, it was apparent that the loss of Hollandia, giving the enemy an advance base of operations against Western New Guinea while the Geelvink Bay defenses were still incomplete, would gravely imperil the southern sector of the absolute defense line.

In General Anami's judgment, these considerations dictated a more aggressive employment of Second Army forces. He estimated that the enemy's plans envisaged following up the Hollandia invasion by an assault on Biak or possibly Manokwari, by-passing Sarmi entirely or attacking it only as a secondary effort. Reinforcement of both Biak and the Manokwari area thus appeared vitally necessary. However, rather than pull back the forward strength of the 36th Division for this purpose, General Anami decided to risk waiting for the arrival of the 35th Division from China and Palau.[65]

---

62 These food supplies, however, were sufficient to last only for a few days. Southeast Area Operations Record, Part III, op. cit. Vol. III, p. 96.

63 Vice Adm. Yoshikazu Endo, Ninth Fleet Commander, had not been heard from since 22 April and was presumed killed in action. Naval personnel came under Maj. Gen. Inada's Command. (Statement by Rear Adm. Kawai, previously cited.)

64 (1) North of Australia Operations Record, op. cit., pp. 96-7. (2) Interrogation of Lt. Gen. Numata, previously cited.

65 General Anami planned to reinforce Biak, currently garrisoned only by the main strength of the 222d Infantry /36th Division, with the main strength of the 219th Infantry/35th Division, coming from Palau. (Cf. n. 27, p. 239). The 35th Division main elements, en route from China, were to take over from the 2d Field Base Unit the task of organizing the defenses of Noemfoor, Manokwari and Sorong, with division headquarters at Manokwari. The new plans further called for the immediate reorganization of all service units in the Western New Guinea area into provisional combat battalions. (1) North of Australia Operations Record, op. cit., pp. 89, 103-6. (2) Second Army Operations Order No. 68, 25 Apr 44. ATIS Bulletin No. 1457, 20 Sep 44.

Meanwhile, in view of the lesser danger to Sarmi, he felt that it was pointless to keep the 36th Division idle in that area when it might be used in an effort to smash the enemy at Hollandia, in conjunction with an Eighteenth Army attack from the east. He therefore decided to modify the original plan for dispatch of a token force in favor of a major counteroffensive by the main strength of the 36th Division.

Pursuant to this decision, Second Area Army ordered the Second Army on 24 April to prepare to send the main strength of the 36th Division from Sarmi to Hollandia.[66] On the same date, General Anami radioed an urgent recommendation to Imperial General Headquarters, Southern Army, the Combined Fleet, and the Southwest Area Fleet that all forces at Sarmi be committed in an attempt to retake Hollandia, and that strong naval forces be rushed to Western New Guinea to block any Allied leap-frog operation toward the Geelvink Bay area.

Neither Southern Army nor Imperial General Headquarters reacted favorably to the recommendation despite the dispatch of Lt. Gen. Takazo Numata, Second Area Army chief of staff, to Singapore in an effort to press the plan upon Southern Army headquarters.[67] At the same time, both Lt. Gen. Teshima, Second Army commander, and Lt. Gen. Hachiro Tanoue, 36th Division commander, indicated that they likewise questioned the advisability of the plan. Finally, on 27 April, the Combined Fleet replied that naval strength adequate to support the plan would not be available until about the middle of May.[68]

Although unable to gain support for his plan to commit the 36th Division in an immediate counterattack against the enemy at Hollandia, General Anami allowed his order of 24 April to remain in effect so that the division would continue preparations facilitating its eventual use as a mobile force to be moved quickly to any threatened sector. Nor did he rescind the earlier order for dispatch of a token force to the Hollandia area. The 36th Division assigned this mission to two infantry battalions of the 224th Infantry Regiment, reinforced by half the regimental artillery battalion. After completing its preparations, this force, under Col. Soemon Matsuyama, 224th Infantry commander, started out from the Sarmi area on 8 May via overland routes. By this date, the enemy was in firm possession of Hollandia and was already using the airfields for operational purposes.

Meanwhile, far to the east, the Eighteenth Army command independently revised its own operational plans in the light of the radically altered situation created by the Hollandia and Aitape landings. The main body of the Army, numbering close to 55,000 troops, including air force ground personnel and naval units, now found itself cut off from all outside sources of supply and deprived, by a combination of geography, the enemy, and insurmountable difficulties of movement, of every possibility of rejoining the Second Area Army forces west of Hollandia for the crucial defense of Western New Guinea.

Faced by the certainty that starvation and disease would gradually destroy his forces even if they remained passively in their present positions, Lt. Gen. Adachi decided that it was preferable to undertake active operations before the fighting strength of the troops was entirely dissipated. More important, he saw the possi-

---

66 (1) North of Australia Operations Record, op. cit., pp. 98–9. (2) Interrogation of Lt. Gen. Numata, previously cited.

67 Interrogation of Lt. Gen. Iimura, previously cited.

68 North of Australia Operations Record, op. cit., p. 100.

密林の死闘（ニューギニヤ戰線）　佐藤　敬　昭和一八・一一

Original Painting by Kei Sato

PLATE NO. 68
Deadly Jungle Fighting : New Guinea Front

bility that bold counterattacks by Eighteenth Army against the enemy's rear might force the diversion of Allied forces eastward, thus hampering the massing of enemy strength against the dangerously weak defenses of Western New Guinea.[69]

Acting swiftly to implement his decision, Lt. Gen. Adachi issued orders to the 20th, 41st and 51st Divisions on 26 April to prepare to move forward for a counterattack against the enemy beachhead at Aitape. On 7 May, advance elements of the 20th Division, then the farthest west of the Eighteenth Army's forces, began advancing toward Aitape from the Wewak area and, by early June, had driven in enemy outposts to reach the Driniumor River, about 12 miles from the main objective at Tadji airfield.[70]

Hollandia nevertheless was irrevocably lost, depriving the Japanese forces of their most valuable remaining air base and port on the northern coast of New Guinea. On the other hand, General MacArthur's forces had won an important forward base of operations seriously jeopardizing Japanese hopes of holding the absolute defense zone and the approaches to the Philippines.[71]

## Failure of the Reinforcement Plan

Almost on the eve of the Allied invasion of Hollandia, a temporary easing of the shipping situation finally made it possible for the Japanese High Command to act on its long-delayed plan to move substantial troop reinforcements to Western New Guinea. Early in April, Imperial General Headquarters restored the shipping allocation of the Second Area Army and on 9 April directed the immediate movement from China of the main elements of the 35th Division.[72]

Pursuant to this directive, the Navy's General Escort Command organized a special convoy, designated Take (竹) No. 1 and consisting of nine large transports.[73] The convoy was to carry, in addition to the 35th Division main elements, the 32d Division which had been

---

69 No formal orders were received by Eighteenth Army either from Second Area Army or from Imperial General Headquarters directing Lt. Gen. Adachi to take any specified course of action as a result of the Hollandia-Aitape landings. He was left full discretion to shape Eighteenth Army's future operational plans according to local circumstances. His decision to counterattack Aitape was also dictated by the Army's desperate supply situation. In late April, the Army had only two months' rations on hand and, even counting upon additional food supplies obtained locally, would face wholesale starvation by October at the latest. (1) Statement by Lt. Col. Tanaka, previously cited. (2) Southeast Area Operations Record, Part III, op. cit. Vol. III, p. 112.

70 These advance elements fought successful actions against enemy outpost positions at Ulau, 9–16 May, east of Yakamul, 16–24 May, and west of Yakamul, 2–5 June, thence pushing on to the Driniumor River. These operations covered the assembly of the main Army strength west of Wewak and reconnoitered a line of departure for the projected counterattack. Southeast Area Operations Record, Part III, op. cit. Vol. III, pp. 157–64.

71 "By advancing to Hollandia (direct)... the Allies cut the length of time required by one-third. Had they advanced to Wewak, then to Aitape, and then to Hollandia, we would have had time to prepare the defenses of Sarmi-Wakde, Biak, and Manokwari.... As it was, there was very little time to prepare for the defense of Sarmi. Biak and Manokwari were also placed well within bomber range." (Interrogation of Maj. Gen. Akinosuke Shigeyasu, Staff Officer (Operations), Second Area Army.)

72 The 35th Division elements awaiting shipment at Shanghai were the 220th and 221st Infantry Regiments and the 4th Independent Mountain Artillery Regiment. (Statement by Lt. Col. Katogawa, previously cited.)

73 Until February 1944 Japan's surface escort system was weak, partially due to the lack of escort forces and partially to the failure to develop an effective command system for individual convoys. In March 1944 the Navy adopted the policy of using large convoys, at the same time concentrating scattered escort forces into strong units. Under the new system, convoy formations were to be commanded by officers of rear admiral's rank with good sea records. Convoy headquarters, however, were usually undermanned because of the shortage of young staff officers.

assigned to Fourteenth Army for the reinforcement of Mindanao. Protected by an unusually large naval escort, the convoy sailed from Shanghai on 17 April en route to Manila.

On the night of 26 April, four days after the start of the Hollandia invasion, the *Take* convoy encountered its first disaster in the waters northwest of Luzon. In a sudden attack by enemy submarines, one of the transports carrying one regiment of the 32d Division was sunk with the loss of virtually the entire regiment.[74] The rest of the convoy continued on to Manila, where it arrived 29 April.

In the interim between the convoy's departure from Shanghai and its arrival at Manila, Imperial General Headquarters had suddenly altered the assignment of the 32d Division, transferring it to direct command of Second Area Army. This was due to realization that unless swift action was taken under the still unimplemented plan to strengthen Second Area Army by 15 battalions, the mounting danger to shipping movements into forward areas might completely bar execution of the reinforcement plan. Hence, when the *Take* convoy resumed its voyage from Manila on 1 May, it still carried the 32d Division.

To lessen the danger of enemy submarine attack, the convoy took a special route laid out by the Third Southern Expeditionary Fleet. In broad daylight on 6 May, however, the convoy was struck again as it neared the northeastern tip of the Celebes. Enemy torpedoes hit and sank three transports in rapid succession.

Although rescue operations were relatively successful, the 32d Division was reduced to only five infantry battalions and one and a half artillery battalions, while the two infantry regiments of the 35th Division were down to four battalions, with only a battery of artillery.[75] The surviving ships of the convoy, carrying these troops, put in at Kaoe Bay, Halmahera, on 9 May.

Meanwhile, the definitive loss of Hollandia had seriously compromised Second Area Army's hopes of safely moving reinforcements into Western New Guinea, even from the nearby Halmahera area. The enemy now had an operating base within easy fighter range of Sarmi, Biak, and the Geelvink Bay area, while Allied bombers could strike at Sorong and Japanese ports and bases in the Moluccas. Japanese air strength was totally inadequate to meet this challenge. In view of the insufficient progress of the aircraft production program, no large air reinforcements could be allocated to Western New Guinea, and material defects, lack of proper maintenance, and other causes rendered unserviceable a large proportion of those few aircraft which were sent out from the Homeland.

The extension of the radius of Allied air control, coupled with the increasingly bold incursions by enemy submarines into heretofore Japanese-controlled waters, so augmented the menace to Japanese sea transportation that it appeared seriously questionable whether any fresh troops could be moved into the threatened sectors of Western New Guinea. General Anami faced the discouraging prospect, therefore, of defending that portion of the national defense zone with little more than his current strength.

### Revision of Defense Plans

Imperial General Headquarters was now called upon to make a difficult decision of

---

74 (1) Western New Guinea Area and North of Australia Area Naval Operations, op. cit., p. 4. (2) North of Australia Operations Record, op. cit., p. 102.

75 Ibid., p. 103.

strategy. In view of the loss of Hollandia and the obvious difficulty of moving adequate reinforcements into Western New Guinea, a minority in the Army General Staff began broaching the idea of pulling back the perimeter of the absolute defense zone in the southern area from Western New Guinea to the Philippines.[76] On the other hand, the Army High Command was aware that General Anami, despite the rejection of his proposal for an all-out counterattack to retake Hollandia, remained inclined toward a decisive defense of the forward positions in the Geelvink Bay area.

Imperial General Headquarters was in no way disposed to consider an outright revision of the national defense zone at this stage, but at the same time it decided that General Anami must be restrained from pouring the bulk of the reinforcement divisions into the Geelvink Bay sector instead of using them in the weakly defended Vogelkop—Halmahera zone. With the Navy section's reassurance that the Bay sector would have vital value in future naval operations, the Army Section dispatched a directive to Southern Army headquarters on 2 May, the main points of which were as follows:[77]

*1. The line to be secured in the Western New Guinea area is designated as a line connecting the southern part of Geelvink Bay, Manokwari, Sorong, and Halmahera.*
*2. Strategic points on and in the vicinity of Biak Island will be held as long as possible.*
*3. Necessary troops will be withdrawn to Biak from the Sarmi district as quickly as possible.*

Before Southern Army dispatched implementing orders to Second Area Army, General Anami had learned of the serious losses suffered by the *Take* convoy in the 6 May attack. He nevertheless wired both Southern Army and Imperial General Headquarters urging that some of the remaining ships of the convoy, despite the risk, be sent on at least to Sorong, and preferably as far as Manokwari, to complete the movement of the 35th Division. The 32d Division was to be retained for the defense of Halmahera.

Because of the grave risk entailed in sending major fleet units into waters dominated by Allied air power, General Anami's request was flatly rejected by Imperial General Headquarters. Moreover, the serious reduction of 32d and 35th Division strength led the High Command to modify the order of 2 May in favor of a further contraction of the projected main line of resistance. An Imperial General Headquarters directive to Southern Army on 9 May stated as follows:[78]

*1. The line to be secured in the Western New Guinea area will be a line extending from Sorong to Halmahera.*
*2. The area covering the lower part of Geelvink Bay, Biak, and Manokwari will be held as long as possible.*

Southern Army on 11 May dispatched implementing orders to the Second Area Army, directing that the 35th Division be stationed in the Sorong area.

There was now a serious and fundamental conflict of opinion between Imperial General Headquarters and General Anami with respect to defense strategy for Western New Guinea. On the one hand, the 9 May directive discarded the idea of a decisive defense of the Geelvink Bay positions, envisaging their use only to delay the enemy advance as long as possible. On the other hand, General Anami continued to hold that the forward line, including Biak, must be aggressively and determinedly defended even if adequate reinforcements were unavailable.[79]

---

76 Statement by Col. Hattori, previously cited.
77 (1) Imperial General Headquarters Army High Command Record, op. cit., pp. 215–17. (2) Statement by Col. Hattori, previously cited.
78 Imperial General Headquarters Army High Command Record, op. cit., p. 218.
79 North of Australia Operations Record, op. cit., pp. 109–12.

Apart from the strategic consideration that Sorong would be difficult to defend if attacked, and might even be by-passed entirely once the Geelvink Bay area was in enemy hands, General Anami was influenced by other factors. First, the transfer of 36th Division troops from the Sarmi sector to Biak would be difficult in view of the lack of shipping and enemy air control over the area. Second, orders to fight delaying actions on the forward positions rather than defend them to the last would be meaningless and detrimental to morale, since the possibilities of safe evacuation would be slight. Third, General Anami had discovered that in the Combined Fleet's plan, the waters between Palau and Western New Guinea was considered to be a probable theater for a decisive naval battle, and thus he felt that a premature relinquishment of the Geelvink Bay area, giving the enemy valuable land air bases close to the theater of action, would seriously harm the Navy's chances.[80]

General Anami consequently decided to take advantage of what leeway was left him by Imperial General Headquarters and Southern Army directives to continue to throw the bulk of his strength into the defense of the key forward positions. Since Imperial General Headquarters had strongly vetoed the dispatch of merchant shipping into the Western New Guinea area, General Anami now opened negotiations with the Fourth Southern Expeditionary Fleet at Ambon to effect the transport to Western New Guinea by warships of the 35th Division troops stranded on Halmahera and Palau. The plan for the deployment of these units upon their arrival was now modified as follows:[81]

1. *35th Division Hq. and Special Troops—Manokwari*
2. *one regiment (less one bn)—Biak*
3. *one regiment (less one bn)—Sorong*[82]
4. *one regiment (less one bn)—Manokwari area*

On 14 May General Anami proceeded to Second Army headquarters at Manokwari and personnally informed the Army commander, Lt. Gen. Teshima, of these developments, instructing him to hold the Geelvink Bay area at all costs and to continue to secure the Sarmi area as a lifeline held out toward the Eighteenth Army forces cut off to the east of Hollandia. At the same time, General Anami dispatched messages to both the Combined Fleet and the Southwest Area Fleet outlining his intentions, substantially as follows:[83]

*The fact that the Navy is preparing to wage a decisive battle in the waters near Geelvink Bay in the near future is a source of gratification to the Second Area Army. Although at this time changes in the main line of resistance have been ordered by higher authority, the Area Army is resolved to hold the Geelvink region at all costs. It is thus ready to give all possible assistance to the naval air forces in this area and to cooperate fully in the decisive naval battle. Moreover, the Area Army has expressed its opinion to higher authorities that the Army air forces should assemble as much strength as possible to cooperate with the Navy at the time of the decisive battle.*

By this time the Navy's preparations for a showdown battle were well under way. On 3 May, Admiral Soemu Toyoda formally assumed command of the Combined Fleet in succession to Admiral Koga, and on the same day an Imperial General Headquarters Navy Directive to the Combined Fleet ordered plans to be laid for the so-called *A-Go* Operation. The

---

80 (1) North of Australia Operations Record, op. cit., pp. 109–10. (2) Statement by Lt. Col. Katogawa, previously cited.

81 North of Australia Operations Record, op. cit., pp. 103–6.

82 General Anami's decision to station the one infantry regiment at Sorong was a mere token compliance with Imperial General Headquarters and Southern Army directives. (Statement by Lt. Col. Katogawa, previously cited.)

83 North of Australia Operations Record, op. cit., pp. 109–10.

essentials of this directive were:[84]

*1. The Commander-in-Chief, Combined Fleet, will swiftly prepare the naval strength required for decisive battle and, during or after the latter part of May, will apprehend and crush the main strength of the enemy fleet in the waters extending from the Central Pacific to the Philippines and Western New Guinea.*

*2. Decisive battle will be avoided, except under specified circumstances, until the required strength has been prepared.*

*3. This battle shall be designated A-Go Operation.*

Under the plans elaborated by the Combined Fleet, the First Mobile Fleet,[85] and the First Air Fleet were assigned the principal roles in the projected battle. The former assembled its surface strength at Tawitawi in the Sulu Archipelago on 16 May, while the land-based units of the First Air Fleet continued to be widely deployed in the Marianas and Carolines to take advantage of any tactical opportunity that might arise. Tawitawi was chosen as the main staging point for the First Mobile Fleet because of its proximity to both the refueling facilities of Balikpapan and the sea area which the Navy High Command expected to be the scene of the decisive battle. It was also safely beyond the range of enemy land-based air power and afforded greater security against Allied intelligence than other anchorages in the Philippines.

In the midst of these preparations, however, the Western New Guinea front flared into action again as General MacArthur's forces, less than a month after the invasion of Hollandia, launched a new amphibious assault against the Wakde-Sarmi area guarding the coastal approach to the vital Geelvink Bay region.

## Wakde—Sarmi

With the enemy in firm possession of Hollandia, it was fully apparent that the Allied assault on the heart of Japan's Western New Guinea defenses in the Geelvink Bay area would not long be delayed. Second Area Army estimated that Biak would be the next major objective of General MacArthur's forces, but defense preparations were also hastened in the Wakde-Sarmi coastal sector to meet the possibility that the enemy might first attempt to seize Japanese air bases there to facilitate fighter support of subsequent operations against Biak or Manokwari.[86]

Under Second Area Army's original plans formulated late in 1943, the Wakde-Sarmi sector, roughly 145 miles west of Hollandia, was to be the forward bastion of the defenses of Geelvink Bay. Engineer and construction units, assisted by combat troops of the 36th Division, had been intensively engaged in building airfields, roads, bridges, and various base installations in the area since January 1944.[87] Highest priority was given to airfield construction. By the time of the Hollandia invasion, one of four projected airfields had been completed near Sawar, seven miles below Sarmi, and another was under construction at nearby Maffin Bay.[88]

The Wakde Islands, lying two and a half miles off the coast about ten miles east of Maffin Bay, were at the eastern extremity of the sector.

---

84 *Daikaishi Dai Sambyakushichijusan-go* 大海指第三百七十三號 (Imperial General Headquarters Navy Directive No. 373) 3 May 44.

85 Administratively the First Mobile Fleet consisted of the Second Fleet (battleships, cruisers and destroyers) and the Third Fleet (carriers, cruisers and destroyers).

86 Interrogations of Maj. Gen. Shigeyasu and Col. Yamamoto, previously cited.

87 Engineer and construction units were grouped under the command of Maj. Gen. Shigeru Yamada, 4th Engineer Group commander.

88 36th Division Airfield Construction Bulletin, 1 Mar 44. ATIS Bulletin No. 1206, 28 Jun 44.

Small, flat and coral-fringed, the islands were not suited for defense against amphibious assault, but on the main island of Insoemoear was an airstrip just under 5,000 feet in length, maintained and used by the Navy but which the Fourth Air Army used both as an operational base and as a dispersal and relay field.[89]

Japanese troop strength in the Wakde-Sarmi area at the end of April totalled approximately 14,000.[90] The 36th Division under Lt. Gen. Hachiro Tanoue, less the 222d Infantry Regiment already stationed on Biak, was the principal combat force, supplemented by various construction units, airfield and antiaircraft personnel, and supply elements.[91] These forces were concentrated mainly between Maffin Bay and Sarmi, with only the 9th Company of the 224th Infantry Regiment, a mountain artillery platoon, some antiaircraft and airfield units, and the 91st Naval Garrison Unit stationed on Insoemoear Island guarding Wakde airfield.[92]

Prior to the Allied invasion of Hollandia, the development of adequate ground defenses in the Wakde-Sarmi area had been seriously hampered by the funnelling of the main effort of the local forces, including combat troops, into the long-range airfield construction program. The enemy's sudden penetration to within a little more than 100 miles of Wakde led Lt. Gen. Tanoue to order an immediate shift of emphasis to the organization of defenses against amphibious attack. Time was already short, however, and Second Area Army's order directing the dispatch of the 224th Infantry Regiment (Matsuyama Detachment) toward Hollandia seriously curtailed the number of combat troops immediately available.

With only the 223d Infantry, remaining elements of the 224th and miscellaneous supporting troops at his disposal, Lt. Gen. Tanoue decided to center his defensive dispositions in the Maffin Bay-Sawar sector, leaving

---

89 During March the 45th Fighter and 61st Bomber Regiments operated from Wakde. At the end of March the 45th Fighter Regiment was withdrawn to Moemi, on the east coast of the Vogelkop Peninsula, and the 61st Bomber Regiment to Galela, Halmahera. Field Diary of 20th Airfield Battalion, Wakde Expeditionary Unit, 1–31 Mar 44. ATIS Bulletin No. 1148, 11 Jun 44, p. 8.

90 In addition to these 14,000 troops, about 3,000 survivors of the Hollandia fighting, who succeeded in getting back to the Maffin Bay-Sarmi area, were integrated into the 36th Division combat forces during the latter phases of the fighting in that area. (Statement by Maj. Gen. Shintaro Imada, Chief of Staff, 36th Division.)

91 Order of battle of the forces in the Wakde-Sarmi area at the end of April was as follows:

36th Division (less 222d Infantry)
    Headquarters
    223d Infantry (reinf.)
    224th Infantry (reinf.)
    36th Division Tank Unit
    Division Special Troops
Headquarters, 4th Engineer Group
16th, 17th Mobile Lumber Squads
20th Airfield Bn.
16th, 103d Airfield Units
228th Independent Motor Transport Co.
53d Field Antiaircraft Bn.
42d Field Machine Cannon Co.
4th Field Searchlight Bn.
Elms 24th Signal Regt.
11th Debarkation Unit
54th, Special Water Duty Co.
91st Naval Garrison Unit

(1) North of Australia Operations Record, op. cit., Annex No. 1 and Attached Table 1. (2) Cohoku Sakusen Kiroku Furoku Dai Ichi: Dai Ni-Gun Sarumi Biaku Nunhoru oyobi Maru Sento Gaishi 濠北作戰記錄附錄第一第二軍サルミビアクヌンホル及びマル戰闘概史 (North of Australia Operations Record, Supplement 1: General Outline of Second Army Operations at Sarmi, Biak, Noemfoor and Maru,) 1st Demobilization Bureau Jul 46, p. 3. (3) Western New Guinea Area and North of Australia Area Naval Operations, op. cit., p. 8. (4) Intelligence Report No. 7, 36th Division, 25 Jan 44, Supplement II, Attached Chart No. 5. ATIS Bulletin No. 1277, 22 Jul 44.

92 ATIS Bulletin No. 1148, 11 Jun 44, p. 7.

the coastal stretch east of the Tor River and opposite Wakde unguarded. (Plate No. 69) A 36th Division order issued on 8 May set forth the essentials of the operational plan substantially as follows:[93]

*1. The division will assume new dispositions and prepare to destroy the enemy at close range.*

*2. The Right Sector Unit will consist of the 16th Airfield Survey and Construction Unit, with the remaining elements and part of the artillery of the 224th Infantry Regiment attached. Main strength of the unit will secure the Mt. Irier and Mt. Sakusin sectors, and an element will secure the Toem area.*

*3. The Central Sector Unit will consist of the 223d Infantry (less 2d Battalion), with the Division Tank Unit (less one platoon) and the 103d Airfield Survey and Construction Unit attached. It will secure the area from the Mt. Saksin—Mt. Irier line to the Sawar River.*

*4. The Left Sector Unit will be composed of the 2d Battalion, 223d Infantry, with one tank platoon attached. It will secure the area from the Sawar River to Sarmi, including the Mt. Hakko position.*

*5. Above units will hold as many mobile reserves as possible. Enemy landing forces will be smashed at the beach.*

*6. The 4th Engineer Group Commander will supervise the construction of fortifications in the various sectors.*

*7. The Division command post is at Mt. Hakko but will later move to Mt. Saksin.*

While these dispositions were hastily being put into effect, enemy air attacks on the Sawar and Wakde airfields, begun soon after the fall of Hollandia, increased in frequency and intensity. Since the meager remaining combat strength of the Japanese air forces in the Western New Guinea area had already retired to rear bases less vulnerable to attack, these raids appeared to be only a precautionary measure to ensure that the fields could not be used. The enemy bombing offensive reached maximum violence in the middle of May, with apparent emphasis on the coastline of Maffin Bay. At the same time, frequent appearances by Allied destroyers and motor torpedo boats in the coastal waters near Sarmi gave further indication that an early landing might be attempted.[94] On 16 May Lt. Gen. Tanoue communicated the following estimate of the situation to his subordinate commanders:[95]

*On the basis of the daily increasing severity of enemy air attacks, the constant activity of warships off the coast.... and the relative situation of our and the enemy's forces, it appears highly probable that landings are being planned near Wakde Island and Sarmi Point.*

This estimate proved true sooner than anticipated. At 0400 on 17 May, an enemy task force of heavy surface units began a fierce three-hour bombardment of Insoemoear Island, interspersed with heavy air strikes, and at 0700 Allied troops began landing operations on the mainland opposite Insoemoear, in the sector between Toem and Arara. The main Japanese forces, concentrated as they were to the west of the Tor River, were unable to offer any opposition to the landing.[96] By the evening of 17 May, the Allied forces had established a firm

---

93 (1) 36th Division Operations Order No. A–121, 8 May 44. ATIS Bulletin No. 1148, 11 Jun 44, pp. 1–2. (2) Statement by Maj. Naoshi Hanami, Staff Officer (Intelligence), 36th Division.

94 The activities of enemy torpedo boats and destroyers became so persistent that Lt. Gen. Tanoue issued an order on 12 May directing that each sector unit commander station a platoon of 75 mm howitzers in selected coastal positions to fire on enemy craft, and that armed patrols make a thorough search of the coast to mop up enemy agents and coast watchers. 36th Division Operations Order No. A 125, 12 May 44. ATIS Bulletin No. 1137, 7 Jun 44.

95 36th Division Staff Intelligence Report No. 17, 16 May 44. ATIS Bulletin No. 1137, 7 Jun 44.

96 The only Japanese forces located to the east of the Tor were a two-gun artillery platoon and a small infantry element of the Right Sector Unit disposed there in compliance with the 8 May operation plan of the 36th Division. These troops withdrew at the beginning of the violent enemy naval gunfire preparation, and the Allied landing in the Toem-Arara sector was thus completely unopposed. (Statement of Maj. Hanami, previously cited.)

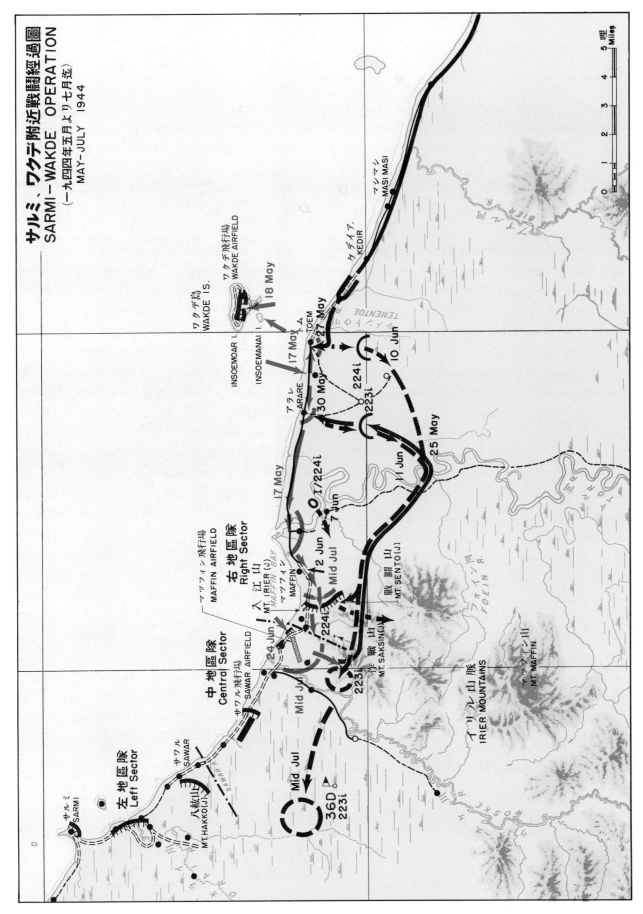

PLATE NO. 69
Sarmi—Wakde Operation, May—July 1944

beachhead between the Tementoe and Tor Rivers.

At 2200 the same day, Lt. Gen. Tanoue ordered his forces to prepare for an attack to wipe out the enemy beachhead, simultaneously ordering the Matsuyama Detachment, then to the east of Masi Masi on its way toward Hollandia, to return to the Toem area as quickly as possible and attack the enemy from the east.[97] As these preparations were getting under way, the enemy on 18 May moved strong amphibious forces across to Insoemoear Island and wiped out the small Wakde garrison in brief but sharp fighting.[98] With the capture of Wakde airfield, the enemy achieved what seemed to be his main strategic objective.

Although not certain as to the exact situation east of the Tor River, Lt. Gen. Tanoue rushed his attack preparations and, on the night of 18 May, ordered the Right Sector Unit to cross the river and begin a preliminary attack.[99] The following day an order was received from the Second Army Commander directing an all-out attack on the enemy in the Toem sector. Accordingly, Lt. Gen. Tanoue at 1000 on 19 May issued a new order, the essential points of which were as follows:[100]

> 1. *It is estimated that the strength of the enemy forces, which landed in the vicinity of Toem, is about two infantry regiments and about 100 tanks and armored cars.*
>
> 2. *The Matsuyama Detachment is returning and will attack the enemy. The Right Sector Unit is also preparing to attack.*
>
> 3. *The division main strength will annihilate the enemy in the vicinity of the Tor River.*
>
> 4. *The Central Sector Unit will cross the Tor River by dawn of 22 May and prepare to attack Toem. Special assault and incendiary units will be organized. The main strength of the artillery will remain in their present positions on the coast.*
>
> 5. *The division command post will move to Mt. Saksin before dawn on 22 May.*

Execution of the attack plan, however, necessitated moving the main strength of Lt. Gen. Tanoue's forces across the wide and unbridged Tor River. Since movement along the coastal road and a crossing near the river mouth would be exposed to enemy naval gunfire, the 36th Division commander decided to move his units inland through jungle and mountainous terrain to cross the Tor at its confluence with the Foein River, and then swing back toward the enemy beachhead. In preparation for this operation, Lt. Gen. Tanoue on 20 May ordered the 36th Division bridging unit to move landing craft and collapsible boats via the Foein River to the proposed crossing-point on the Tor.[101]

While the Central Sector Unit (223d Infantry) was beginning its difficult march to attack positions east of the Tor, the Right Sector Unit commenced operations in the Maffin sector on 19 May. A reconnaissance to the east of Maffin village on that date revealed that the enemy had already crossed the Tor River near its mouth and established a small bridgehead. By 21 May, this Allied element

---

97 36th Division Operations Order No. A-134, 17 May 44. ATIS Bulletin No. 1167, 19 Jun 44.
98 North of Australia Operations Record, Supplement I, op. cit., p. 4.
99 36th Division Operations Order No. A-138, 18 May 44. ATIS Bulletin No. 1167, 19 Jun 44.
100 36th Division Operations Order No. A-140, 19 May 44. ATIS Bulletin No. 1181, 15 Jun 44.
101 (1) 36th Division Operations Order No. A-147, 20 May 44. ATIS Bulletin No. 1179, 22 Jun 44. (2) Field Message, 36th Division Bridging Unit Commander, 21 May 44. ATIS Bulletin No. 1167, 19 Jun 44.

had pushed to a point about one mile east of Maffin, stoutly resisted by rear echelon troops of the 224th Infantry, which had been organized into a provisional battalion. By 24 May, the enemy occupied Maffin village and continued to advance westward, threatening the Mt. Irier-Mt. Saksin positions occupied by the main strength of the Right Sector Unit. The Allied force, however, halted at the east bank of the Maffin River and did not launch an immediate assault on the hill positions. Lt. Gen. Tanoue took advantage of this momentary lull to order the Left Sector Unit (2d Battalion, 223d Infantry) to move immediately to the Mt. Irier line and bolster the rapidly weakening Right Sector Unit. Pending arrival of these reinforcements on 29 May, the Right Sector Unit hastily organized defenses in depth on the high ground west of Maffin.[102]

While the Right Sector Unit was heavily engaged in the Maffin area, the 223d Infantry, delayed in its overland march to the assigned crossing point on the Tor and handicapped by the slowness with which the assault boats were brought forward, finally crossed the Tor on 25 May, three days behind schedule. Meanwhile, the Matsuyama Detachment completed its forced march back from the east, closed up to the right bank of the Tementoe River, and prepared to attack the enemy's left flank at Toem.

On the night of 27 May, the Matsuyama Detachment crossed the Tementoe River and launched a surprise attack on Toem. This operation met with limited success. A part of the detachment penetrated as far as the beach, forcing a number of the enemy to flee offshore in landing boats. The enemy's artillery and naval gunfire reaction to this attack was extremely violent. Heavy casualties were sustained, and before dawn on 28 May, Col. Matsuyama, fearing that the narrow salient might be pinched off, withdrew the advance elements and allowed his exhausted troops a breathing spell.[103]

Meanwhile, the 223d Infantry, having completed its arduous trek from the Sawar area, began assembling in the jungle about two miles south of Arara on the night of 27 May and prepared to attack. Completion of the assembly and attack preparations consumed three days, and it was the night of 30 May before the regiment was ready to strike in force. The attack penetrated the outer part of the enemy perimeter, but again the Japanese lacked the means to exploit the initial success. The regiment withdrew before dawn on 1 May but kept up nightly raiding attacks thereafter.[104]

While the 223d and 224th Infantry continued their pressure on the enemy's Toem-Arara beachhead, the situation in the Maffin area improved.[105] The Right Sector Unit commander, having been reinforced by the 2d Battalion, 223d Infantry, on 29 May, decided to recapture Maffin village. This attack was carried out on the night of 31 May but did not succeed. On the night of 2 June, however, the effort was renewed, and Maffin was reoccupied. The Japanese continued the attack eastward against increasing enemy resistance but did not succeed in wiping out the Allied bridgehead across the mouth of the Tor.

---

102 Statement by Maj. Hanami, previously cited.
103 Statement by Col. Soemon Matsuyama, Regimental Commander, 224th Infantry Regiment.
104 Statement by Col. Naoyasu Yoshino, Regimental Commander, 223d Infantry Regiment.
105 The Japanese pressure on the Toem-Arara beachhead caused a slowing down of the enemy's operations in the Maffin area and facilitated the subsequent seizure of the initiative by the Japanese. (Statement by Maj. Hanami, previously cited.)

Division, under command of Col. Naoyuki Kuzume, continued to constitute the combat nucleus of the garrison, the remainder of which consisted of rear echelon, service, and construction units. In addition to the Army troops, 2,000 naval personnel were on the island, bringing the aggregate strength of the forces on Biak to approximately 12,000.[114]

Five days after the enemy landings at Hollandia, Col. Kuzume took initial action to organize and dispose his forces to meet a possible amphibious attack. These dispositions were laid down in an operations order issued on 27 April, the essentials of which were as follows:[115]

*1. The Biak Detachment will destroy at the water's edge any enemy force attacking this island. The detachment main strength will be disposed along the south coast immediately.*

*2. Rear area forces will be converted into combat units. The detachment will assume command of Navy ground troops.*

*3. Coastal sectors of responsibility are designated as follows:*

   *19th Naval Garrison Unit—Bosnek to Opiaref*
   *1st Bn, 222d Infantry—sector east of Opiaref*
   *2d Bn, 222d Infantry—Sorido to Bosnek*
   *3d Bn, 222d Infantry—detachment reserve*
   *Tank Company—take position at Arfak Saba and prepare to move against enemy landing.*

*4. Detachment headquarters will be $2\frac{1}{2}$ miles north of Jadiboeri.*

On the heels of this order, the Allied air forces on 28 April carried out a heavy raid on the Sorido airfield sector, marking the beginning of a month-long air assault which constantly hampered the progress of defensive preparations.[116] From 17 May, when the Allied landing in the Wakde area took place, the bombings increased sharply in violence and assumed the characteristics of pre-invasion softening-up operations.

In view of the intense enemy concentration on the Sorido-Mokmer airfield sector, Col. Kuzume decided on 22 May to shift the operational center of gravity of the detachment to the west. The 1st Battalion, 222d Infantry, was relieved of its mission in the sector east of Opiaref and sent to replace the naval garrison unit in the Bosnek sector. The naval troops were, in turn, shifted westward into the Sorido

---

114 Order of battle of the forces on Biak at the time of the Allied landing on 27 May was as follows:

Army Forces
  222d Infantry (less 2 cos.)
  Elms 36th Div. Sea Transport Unit
  Elms 14th Div. Sea Transport Unit
  17th, 107th, 108th, Airfield Constr. Units
  Elm 109th Airfield Constr. Unit
  Elms 248th Independent Motor Trans. Co.
  15th Formosan Special Labor Group
  41st Special Land Duty
  50th and 69th Construction Cos.
  3d Btry, 49th Field Antiaircraft Bn.
  1st Branch, 36th Division Field Hospital
  30th Field Ordnance Depot Branch
  Elms 24th Signal Regt.
  5th, 12th Mobile Lumber Squads
  Elm 47th Anchorage Hq.

Navy Forces
  Elms 28th Naval Base Force
  33d and 105th Antiaircraft Units
  19th Naval Garrison Unit
  202d Civil Engineer Unit

(1) Mimeographed Organization Tables of Biak Garrison, 29 Apr 44. ATIS Bulletin No. 1274, 19 Jul 44. (2) Miscellaneous documents published in following ATIS Bulletins: No. 1176, 21 Jun 44; No. 1231, 6 Jul 44; No. 1249, 11 Jul 44; No. 1283, 24 Jul 44.

115 Biak Detachment Operation Plan, 27 Apr 44. ATIS Bulletin No. 1266, 16 Jul 44.

116 The narrow Sorido—Mokmer airfield sector was attacked frequently by as many as 150 Allied planes at one time. Diary of Petty Officer Seishichi Kumada, 202d Pioneer Unit. ATIS Bulletin No. 1265, 16 Jul 44.

airfield sector, while the tank company was brought over from Arfak Saba and assembled in the area northwest of Mokmer airfield.[117]

On 25 May, Lt. Gen. Takazo Numata, Chief of Staff of Second Area Army, flew in to Biak from General Anami's headquarters at Menado in order to inspect the condition of the Biak defenses.[118] Also on the island at this time was Rear Adm. Sadatoshi Senda, commander of the 28th Naval Base Force, who had come from his headquarters at Manokwari to inspect the local naval forces. All indications pointed to the conclusion that the enemy's assault was not far off.

The attack, however, came even sooner than anticipated. At 0500 on 27 May, a powerful enemy naval force estimated at three battleships, two cruisers, and ten destroyers, accompanied by a number of troop transports, appeared off the south coast of Biak and began a fierce artillery bombardment of shore installations. After two hours of preparatory shelling, enemy troops estimated at about one division began landing operations in the vicinity of Bosnek at 0700.[119] (Plate No. 70)

Although the landing could not be prevented, the topographical features of the Bosnek sector were highly favorable to the defense once the enemy was ashore. The coastal plain was a narrow strip extending only 400–800 yards inland, where it was hemmed in by a steep, 250-foot coral ridge, in which cave positions and artillery emplacements had been built. Because of the extreme narrowness of the beach and the few entrances inland, deployment of the large enemy landing force was bound to be difficult. Col. Kuzume decided to seize this momentary advantage and swiftly ordered the 1st and 3d Battalions, 222d Infantry, to carry out an attack all along the Bosnek beachhead during the night of 27 May.[120]

Heavy casualties were inflicted on the enemy in this attack, but the Japanese force was too small to achieve any marked reduction of the beachhead. On 28 May the Allied troops drove vigorously to the west, and an infantry battalion, supported by amphibious tanks and heavy naval gunfire, succeeded in pushing into Mokmer against weak resistance by elements of the 3d Battalion and the 14th Division Shipping Unit.[121] A further enemy advance was successfully blocked by the main strength of the 2d Battalion, 222d Infantry disposed west of Mokmer.

Col. Kuzume now decided to commit his tank force in an effort to roll up the extended enemy flank. During the night of 28–29 May, nine

---

[117] (1) 1st Battalion, 222d Infantry Operations Order, 22 May 44. ATIS Bulletin 1228, 6 Jul 44. (2) Interrogation of Lt. Gen. Numata, previously cited.

[118] After a two-day stay on Biak, Lt. Gen. Numata was about to take off from Mokmer Airdrome on the morning of 27 May on his return flight to Menado when the Allied attack began. Enemy shelling of the airfield prevented the take-off, and Lt. Gen. Numata remained on the island until 10 June. Although not the ranking officer during this period, Col. Kuzume remained in operational command.

[119] (1) North of Australia Operations Record, op. cit. Supplement I, p. 15. (2) The south coast of Biak, where the airfields were concentrated, was regarded as the most probable enemy landing point, and the Japanese defenses were strongest in that sector. Some possibility was also seen of a landing in the vicinity of Sawabas on the opposite side of the island, north of Bosnek, but troop strength was inadequate to organize that area. (Statement by Lt. Col. Katogawa, previously cited.) (3) Interrogation of Lt. Gen. Numata, previously cited.

[120] (1) 1st Battalion, 222d Infantry Operations Order, 27 May 44. ATIS Bulletin No. 1182, 24 Jun 44. (2) *Biaku Sento Gaiyo* ビアク戰闘概要 (Summary of Biak Battle), Second Army Headquarters, Nov 45, pp. 2–3.

[121] The sudden appearance of the enemy in the Mokmer sector led to the mistaken belief that a second amphibious landing had been made.

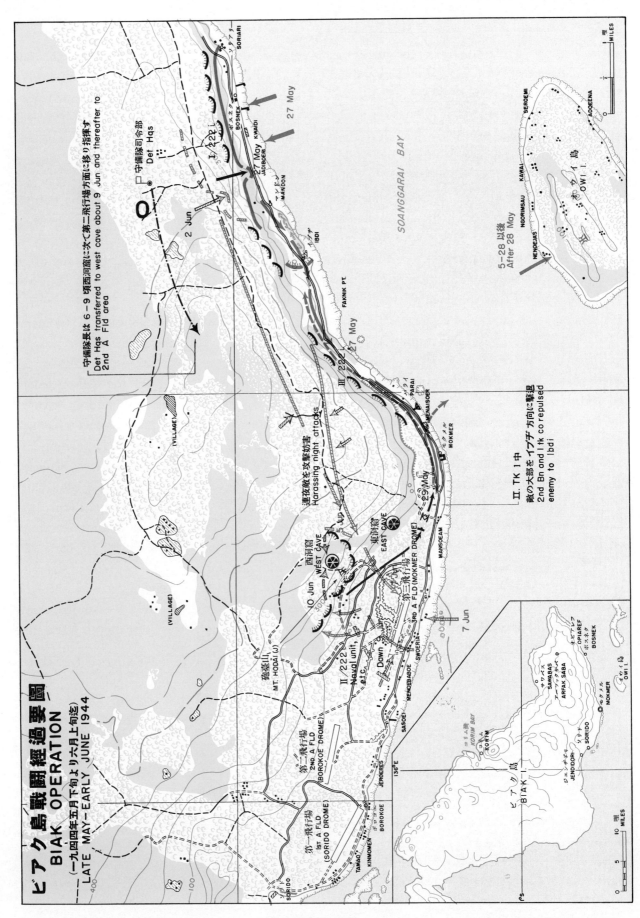

PLATE NO. 70
Biak Operations, May–June 1944

tanks of the 36th Division Tank Unit assembled in defilade north of Mokmer airfield, and at 0610 on the 29th attacked toward Mokmer, in support of the 2d Battalion, 222d Infantry. Although severe reaction by enemy aircraft, artillery, and armor resulted in the destruction of seven of the tanks, the attack succeeded in pushing the enemy completely out of the Mokmer sector. At the end of the action, Col. Kuzume's troops firmly held both Mokmer and Parai and had restored the line as far east as Ibdi.[122]

The 23d Air Flotilla and 7th Air Division meanwhile threw what strength they could muster into attacks on the enemy landing force. On 27 May four Army heavy bombers and nine Navy fighters carried out a daylight attack against fierce air opposition, all but four fighters failing to return.[123] The Combined Fleet on the same day ordered the First Air Fleet to dispatch strong reinforcements to the 23d Air Flotilla,[124] and on 29–30 May the flotilla carried out fresh attacks on the Biak landing force.[125] Though losses were again great, the air offensive continued to harass the enemy until 8 June.

On the ground, the situation appeared highly favorable.[126] The enemy was now crowded into a small beachhead on the narrow coastal shelf in the vicinity of Bosnek, and Col. Kuzume, with considerable uncommitted strength at his disposal, was confident that a determined attack would succeed in pushing the enemy back into the sea. Local enemy air and naval control, however, presented a serious obstacle to complete success. To overcome this, Lt. Gen. Numata and Rear Adm. Senda, on May 29, dispatched a joint message to higher Army and Navy headquarters urging the immediate commitment of fleet and air strength in the Biak battle. The message stated:[127]

*The officers and men on Biak Island are firm in their resolution to crush the enemy. However, our operations are severely restricted by the uncontested superiority of the enemy's fleet and air units. We believe that the immediate commitment of our air forces and, if possible, some fleet units would give us a splendid opportunity to turn the tide of battle in the whole Pacific area in our favor.*

Meanwhile, independent pressure by higher echelons of the Army and Navy operational command had already brought the issue of throwing additional strength into the defense of Biak squarely before Imperial General Headquarters.

### The *Kon* Operation

The operational policy laid down by the Army Section of Imperial General Headquarters on 9 May had clearly stipulated that the main line of resistance in the Western New Guinea area would henceforth be the line

---

122 (1) Situation of the Tank Battle, 29 May 44, Biak Detachment. ATIS Bulletin No. 1270, 19 Jul 44. (2) Summary of Biak Battle, op. cit., pp. 2–3.

123 Three enemy planes were reported shot down, and several landing craft set afire. Western New Guinea Area and North of Australia Area Naval Operations, op. cit., p. 11.

124 These reinforcements consisted of 70 fighters, 16 carrier-type bombers, and four reconnaissance planes. *A-Go* Operation Record, op. cit., p. 8.

125 Western New Guinea Area and North of Australia Area Naval Operations, op. cit., p. 11.

126 The only unfavorable development at this time was a shortage of rations and, to a lesser extent, of ammunition. The rapid enemy landing in the Bosnek sector had overrun vast stocks of supplies piled near the beach preparatory to dispersal to inland dumps. The naval shelling also destroyed considerable quantities of stores. (Interrogation of Lt. Gen. Numata, previously cited.)

127 This message was addressed to Southern Army, Second Area Army, Fourth Air Army, Fourth Southern Expeditionary Fleet, Southwest Area Fleet and Combined Fleet. North of Australia Operations Record, op. cit., p. 120.

Sorong-Halmahera, with the Geelvink Bay positions, including Biak, becoming a forward barrier to be held as long as possible by the forces already present. Though the Imperial General Headquarters Navy Section first gave unreserved consent to the Army Section's policy, it was soon realized that the Navy's own planning for the *A-Go* Operations placed new emphasis on the strategic importance of Biak.

The over-all plan of the *A-Go* Operation formulated by the Combined Fleet in early May envisaged challenging the enemy's main fleet strength in decisive battle in the general area of the Palau Islands, then estimated to be the most probable target of the enemy's next move in the Central Pacific.[128] This rendered it vitally important to retain possession of the Biak airfields, which were strategically located to provide land-based air support of the fleet operations. Conversely, enemy seizure and use of the airfields would seriously impair the Navy's ability to retain air control over the battle zone and thus diminish the chances of victory.[129]

These considerations led the Southwest Area Fleet, charged with implemental planning for support of the *A-Go* Operation, to initiate discussions with Southern Army in mid-May concerning a proposal for the dispatch of troop reinforcements to Biak aboard naval combat ships. The Allied invasion of the Wakde-Sarmi area, which followed immediately, spurred the negotiations, and by about 20 May a tentative joint plan had been drawn up between the two headquarters envisaging the transport to Biak by Southwest Area Fleet of the 2d Amphibious Brigade, then at Zamboanga, Mindanao.[130]

The tentative plan had not yet been referred to the central Army and Navy commands for approval, however, when the unexpectedly early invasion of Biak occurred on 27 May. Spurred into action by this menacing development, Southern Army and Southwest Area Fleet, on the 28th, dispatched an urgent joint recommendation to Imperial General Headquarters and the Combined Fleet, in substance as follows:[131]

*1. It is recommended that the 2d Amphibious Brigade be transported to Biak immediately aboard two battleships.*

*2. If this is not feasible, it is recommended that the brigade be transported to Manokwari via Sorong or direct to Manokwari, for trans-shipment to Biak, by the 16th Cruiser Division and the 19th and 27th Destroyer Divisions.*

---

128 The operational planning staff of the Combined Fleet estimated that there was a slightly smaller probability of an enemy invasion of the Marianas area. In the outline plan of the *A-Go* Operation, the decisive fleet battle areas were designated as (a) the Palau area and (b) the western Carolines. The plan provided that, should the enemy move toward the Marianas or into both the Marianas and one of the above areas simultaneously, that portion of the enemy in the Marianas area would be attacked only by the base air forces in the Marianas. The main factor in this concept of operations was the acute shortage of fleet tankers which made it impossible to give logistical support to any large-scale operation in the Philippine Sea at this time. (1) Combined Fleet Top Secret Operation Order No. 76, 3 May 44. ATIS Limited Distribution Translation No. 39, Part VIII, p. 170. (2) Statement by Capt. Toshikazu Ohmae, Staff Officer (Operations), First Mobile Fleet.

129 Ibid.

130 The 2d Amphibious Brigade was one of several special units of this type organized and stationed at strategic points in readiness to move, by naval ships, to any sector invaded by the enemy. These units were developed to offset Japan's inability to garrison all sectors of its overextended area of operations with adequate troop strength. Authorized wartime strength of an amphibious brigade was 5,400. It was made up of three infantry battalions, a machine cannon unit, a tank unit, and appropriate service elements.

131 Statement by Col. Horiba, previously cited.

The joint recommendation gained the immediate concurrence of Combined Fleet and the Navy Section of Imperial General Headquarters, but the final consent of the Army High Command was obtained only after a joint staff conference on 29 May, attended by the chiefs of both the Army and Navy General Staffs.[132] On the same day, instructions were radioed to Combined Fleet and Southern Army directing them to execute the proposed reinforcement plan, designated as the Kon (渾) Operation.

As finally agreed upon, the Kon plan called for the transport of the 2d Amphibious Brigade, to be released by Southern Army to temporary command of the Second Area Army, from Mindanao to Korim Bay, on the north central coast of Biak. Embarkation preparations were swiftly completed, and the main convoy, consisting of a transport group carrying the main strength of the brigade and a screening group of cruisers, destroyers and one old battleship,[133] sortied from Davao on 2 June under the command of Rear Adm. Naomasa Sakonju. (Plate No. 71) Remaining elements of the brigade embarked simultaneously at Zamboanga aboard other naval ships. The scheduled date of debarkation on Biak was set at 4 June.

On 3 June, when the force was still 600 miles northwest of Biak, a scout plane reported a strong American carrier group approaching the waters east of Biak, and at the same time Rear Adm. Sakonju radioed to Combined Fleet that the convoy had been detected by an enemy submarine and was being tracked by two B-24 bombers. Unwilling to risk so many ships under these circumstances, Combined Fleet ordered the Kon Operation suspended and directed most of the screening group to return to Davao. The transport groups and the 27th Destroyer Division were ordered to proceed to Sorong.

On 4 June it was discovered that the reported approach of an enemy carrier force was erroneous, and the Combined Fleet ordered resumption of the operation, this time employing only six destroyers with Rear Admiral Sakonju flying his flag aboard the *Shikinami*, one of the destroyers. In order to make the run from Sorong to Biak, however, three warships first had to be refueled, necessitating a 700-mile round-trip to Ambon since no fuel was available at Sorong. This delayed the final departure for Biak until midnight of 7 June.[134]

---

132 The Army General Staff adhered to the line of the 9 May directive, taking the stand that it was tactically and strategically unfeasible to commit additional troops to the defense of Biak in view of the enemy's possession of air bases at Hollandia and Wakde. The Navy's strong insistence on the necessity of holding Biak, however, finally won the consent of the Chief of Army General Staff. (Statement by Col. Hattori, previously cited.)

133 Composition of the Kon Force was as follows:
From Davao:     Transport Group
                   16th Cruiser Division: *Aoba* and *Kinu*
                   19th Destroyer Division: *Shikinami, Uranami, Shigure*
                Screening Group
                   5th Cruiser Division: *Myoko, Haguro*
                   27th Destroyer Division: *Harusame, Shiratsuyu, Samidare*
                   10th Destroyer Division: *Asagumo, Kazagumo*
                   Independent Unit: *Fuso* (BB)
From Zamboanga: Independent Group
                   (two minelayers, two submarine chasers, one mine-sweeper, one armed transport)

(1) Western New Guinea Area and North of Australia Area Naval Operations, op. cit., p. 12. (2) USSBS, *Interrogations of Japanese Officials*, op. cit. Vol. II, p. 450. (Interrogation of Capt. Momochiyo Shimanouchi, Staff Officer (Operations), 16th Cruiser Division.)

134 The transport group consisted of the 19th Destroyer Division and carried only a portion of the 2d Amphibious Brigade, numbering about 600. The 27th Destroyer Division was its screening group. Western New Guinea Area and North of Australia Area Operations, op. cit., p. 15.

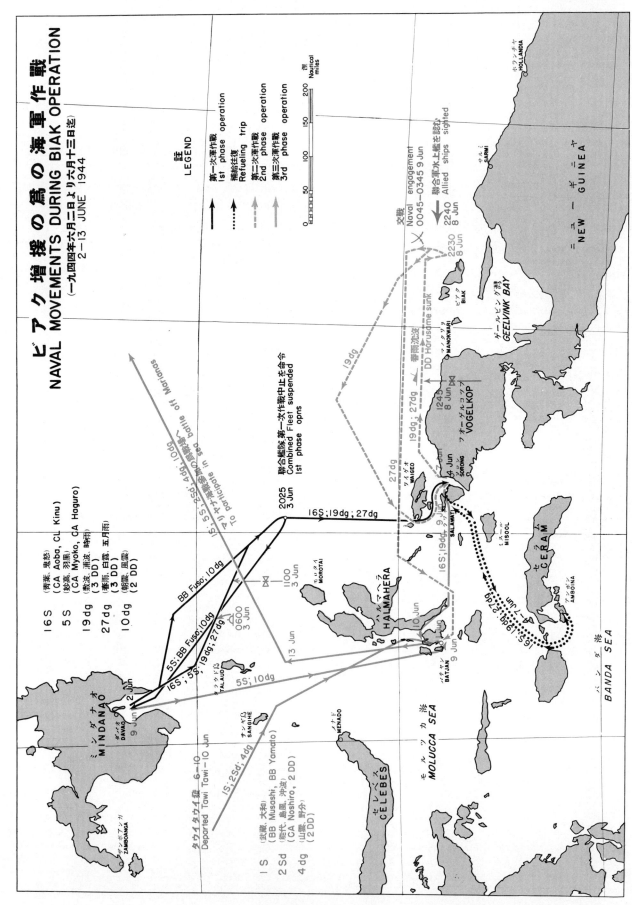

PLATE NO. 71
Naval Movements During Biak Operation, 2-13 June 1944

At 1245 on 8 June, the *Kon* formation, then 200 miles northwest of Biak, was suddenly attacked by an Allied fighter-bomber force of about 50 planes. The destroyer *Harusame* was heavily hit and sank in five minutes, while minor damage was sustained by other ships of the convoy. The formation pressed on, however, also refusing to be deterred by a report at 1800 the same day that an enemy task force of one battleship, four cruisers and eight destroyers was approachin geast of Biak at high speed. At 2230 the destroyer-transports arrived off Korim Bay and prepared to dash in to debark the troops.

Barely ten minutes later, a destroyer of the screening force signalled the approach of the Allied task force. Seriously outweighed, the *Kon* formation, without effecting the debarkation, retired westward at full speed with the enemy force in pursuit. Two hours later, a brisk three-hour gunfire and torpedo action ensued, but no serious damage was received, and at 0345 on 9 June, the *Kon* force disengaged and returned to Sorong and Halmahera.[135]

Despite this second failure, the Navy was still determined to carry out the *Kon* plan. On 9 June, Vice Adm. Jisaburo Ozawa, Commander-in-Chief of the First Mobile Fleet, who had been holding the bulk of the Navy's surface and carrier strength at Tawitawi in readiness for the *A-Go* Operation, sent a message to the Combined Fleet stating in substance:[136]

*The battle of Biak has taken an unfavorable turn. If we should lose the island, it would greatly hinder our subsequent operations. I am therefore in favor of sending reinforcements, especially since this might draw the American fleet into the anticipated zone of decisive battle and enable us to launch the "A-Go" Operation. I am prepared to dispatch the 2d Carrier Division to the Biak area to support the reinforcement plan, though this is the maximum strength which I can divert at this time.*

Upon receipt of this message, Admiral Toyoda, Commander-in-Chief of the Combined Fleet, swiftly decided to reinforce the *Kon* force for a new attempt to move troops to Biak. Although rejecting Vice Adm. Ozawa's recommendation to use the 2d Carrier Division, he issued an order on 10 June directing the addition to the *Kon* force of the 1st Battleship Division, comprising the 64,000-ton *Yamato* and *Musashi* and the 2d Destroyer Squadron. Vice Adm. Matome Ugaki, 1st Battleship Division commander, was placed in command of the augmented task force and was directed to carry out the operation as follows:[137]

*1. Crush enemy reinforcement convoys in the Biak area, and destroy by bombardment the enemy forces on Owi Island.*

*2. Move the 2d Amphibious Brigade to Biak.*

*3. Operate so as to draw the enemy's fast carrier forces to the decisive battle area, if the situation permits.*

Vice Adm. Ugaki designated Batjan anchorage, in the Moluccas, as the rendezvous point, and by the morning of 12 June the *Yamato*, *Musashi*, and other ships of the force had assembled in readiness for the start of the operation.[138] On 11 June, however, carrier-planes from a powerful enemy task force had begun a sustained attack on Japanese bases in the Marianas. On 13 June a surface force stood off Tinian and Saipan Islands, subjecting the coast defenses to heavy preparatory shelling. With the enemy's intentions now clear, Admiral Toyoda, at 1730 the same day, suspended all fleet commitments, including the *Kon* opera-

---

135 Western New Guinea Area and North of Australia Area Naval Operations, op. cit., pp, 15–6.
136 *A-Go* Operation Record, op. cit., p. 51.
137 The 4th Destroyer Division (*Yamagumo* and *Nowaki*), already accompanying the 1st Battleship Division, was formally enrolled in the *Kon* force on 11 June. Ibid., pp. 33–4, 104.
138 Ibid., p. 34.

サイパン島大津部隊の奮戦　橋本八百二　昭和十九

Original Painting by Yaoji Hashimoto

PLATE NO. 72
Fierce Fighting of Otsu Unit in Saipan

In coordination with this advance, enemy troops on the Ibdi front also pushed westward, ignoring the coast road to take to the high ground northwest of Ibdi.[148] This latter maneuver threatened to outflank the 3d Battalion, 222d Infantry and the 14th Division Transport Unit holding positions along the coast road west of Ibdi.

The situation of the Biak Detachment was now desperate. Heavily outnumbered, the detachment was being split in half by the enemy's two-pronged westward advance. In order to relieve this critical situation, the naval ground units garrisoning the coastal sector from Borokoe to Sorido and the airfield construction units from the three airfields were ordered to move immediately to the West Cave sector and prepare to meet the enemy advance.[149]

Between West Cave and the advancing enemy lay dense jungle terrain believed to be impassable to any large force. Enemy air power, however, paved the way for the ground advance by flattening the forest cover with bombing attacks. Even though Japanese combat patrols continually harrassed the enemy flanks, the advance was so rapid that by 5 June it had reached the summit of the coastal terrace overlooking Mokmer airfield, between East and West Caves. The enemy was now operating in the detachment's rear area, causing supply and communications to break down completely. East and West Caves were cut off from each other, while the 1st Battalion, 222d Infantry, and detachment headquarters were still far to the east in the area north of Bosnek, endeavoring to harass the enemy rear.

On 7 June, enemy troops in the heights above Mokmer airfield launched a vigorous attack toward the field, supported by heavy concentrations of artillery and strong tank elements. At the height of the battle, an enemy landing force was put ashore on the south side of Mokmer drome, and by nightfall the two arms of this pincers movement had overrun the field. Col. Kuzume's forces were now split into three widely separated segments: one in the area north of Bosnek (Biak Detachment headquarters and 1st Battalion, 222d Infantry); another in the East Cave sector (elements of the 2d and 3d Battalions, 222d Infantry, shipping units and other miscellaneous service elements); and a third in the West Cave and Mokmer airfield sector (main strength of the 2d Battalion, 222d Infantry, airfield construction units, naval garrison unit and miscellaneous service troops).

Lt. Gen. Takazo Numata now assumed personal command of the troops in the West Cave sector and issued orders for a final attack by all available units to retake Mokmer airfield. Just before dawn on 9 June, the attack was launched. The 2d Battalion, 222d Infantry, advanced about halfway down the airstrip before being halted by the firmly entrenched enemy. A company of naval garrison troops, attacking down the center, infiltrated completely across the airfield to reach the sea in the enemy rear, but inadequate strength made it impossible to press this advantage, and the company was forced to withdraw to the sector west of the airfield.[150]

In view of the deteriorating situation in the western sector, Col. Kuzume on 8 June directed the 1st Battalion, 222d Infantry, to cease operations north of Bosnek and hurry west to assist in the efforts to regain Mokmer airfield.

---

148 Interrogation of Lt. Gen. Numata, previously cited.

149 The naval units mustered a force of about company strength, while the airfield construction units were able to put about 300 men into the line. These units participated in the defense of Mokmer airfield, operating out of West Cave. Ibid.

150 (1) Ibid. (2) Summary of Biak Battle, op. cit. p. 10.

Detachment headquarters itself left the sector north of Bosnek and set out for West Cave on the 9th. Meanwhile, in the airfield sector, the enemy quickly followed up his advantage and by 10 June had compressed the defenders into a semi-circular area in the immediate vicinity of West Cave.

With the loss of Mokmer airfield, effective Japanese resistance on Biak came to an end.[151] Small reinforcements, which moved forward from Manokwari by small craft, arrived too late to exert any effect on the tide of battle.[152] The cancellation of the *Kon* Operation on 13 June ended any remaining hope that Biak could be held. On the 14th, the enemy began operational flights from Mokmer airfield, and the next day strong Japanese positions on Hodai (砲臺) Mt. had to be evacuated under heavy enemy pressure.

By 16 June, Japanese resistance was reduced to the last-ditch defense of East and West Caves and of numerous smaller cave positions scattered along the face of the coral terrace from a point north of Parai west to Hodai Mt. Exhaustion of rations and water finally forced the defenders of West Cave to abandon this position on 22 June and scatter into the hills of central Biak. By the end of the month, the East Cave position had also fallen to the enemy, and the Biak garrison was reduced to small groups of starving and exhausted men hiding in the interior, obtaining food at night from native gardens, and suffering from exposure, malnutrition, and disease.[153]

The loss of Biak finally sealed the fate of both the Eighteenth Army forces cut off in Northeast New Guinea and the 36th Division remnants still putting up sporadic resistance in the vicinity of Sarmi.[154] Meanwhile, the enemy, firmly in possession of the Biak airfields, moved to consolidate his control of the Geelvink Bay area by attacking Noemfoor Island, halfway between Biak and the Second Army strongpoint at Manokwari.

Noemfoor's strategic value lay in the existence on the island's northwest coast of two airfields: Kamiri field, a first-class completed air base; and Kornasoren field, which was only partially completed. The island was garrisoned by a small force organized around a nucleus of six infantry companies, supplemented by a number of miscellaneous supporting and service

---

151 Prior to the unsuccessful attempt to retake Mokmer airfield on 9 June, the morale of the Biak defenders had been very high. Under the combined impact of shortage of rations and water, disease, and tactical failure, the detachment first showed signs of defeat on 9 June, and its disintegration was very rapid thereafter. (1) Interrogation of Lt. Gen. Numata, previously cited. (2) Summary of Biak Battle, op. cit., p. 10.

152 The first reinforcements to arrive were the headquarters and two companies of the 2d Battalion, 221st Infantry, which landed at Korim Bay, on the north coast of Biak, on 4 June. These troops had reached the West Cave area by 8 June but were kept in reserve during the abortive 9 June attack to retake Mokmer airfield. On 16 June about 700 additional reinforcements of the 2d Battalion, 219th Infantry, landed at Korim Bay. These troops did not reach the vicinity of West Cave until 23 June. Summary of the Biak Battle, op. cit., pp. 5, 10, 13.

153 Under orders to return to Second Area Army headquarters, Lt. Gen. Numata left West Cave on 10 June and departed Korim Bay on the 14th by landing craft. He arrived at Manokwari on 19 June. After leaving West Cave, Col. Kuzume was killed in action on 2 July north of Borokoe airfield. Rear Adm. Senda died the following December after spending seven months hiding in the jungle.

154 The Japanese forces in the Sarmi area were obliged to become totally self-sufficient. While they still had military supplies, however, they conducted sporadic defensive operations against the enemy and held out until the end of the war, although they were powerless to prevent Allied development and use of the Maffin airfield.

units.[155] The force, designated the Noemfoor Defense Detachment, was commanded by Col. Suesada Shimizu, regimental commander of the 219th Infantry, 35th Division, who arrived on Noemfoor on 8 June. Col. Shimizu disposed his forces in fourteen strongpoints scattered around the perimeter of the island.[156]

At 0540 on 2 July, an enemy naval force estimated to consist of four cruisers and four destroyers approached Noemfoor and began shelling the south coast. This proved to be a diversionary operation, and the actual landing was accomplished on the northwest coast, where approximately 2,000 enemy troops, accompanied by tanks and with strong air and naval gunfire support, went ashore squarely in the Kamiri airfield sector.

Although the terrain of the island was characterized by the same coral terraces which had been used to great advantage by the defenders of Biak, the Noemfoor garrison, scattered around the entire perimeter of the island, was too small and dispersed to organize an effective defense. Col. Shimizu's forces retired inland before the enemy attack, which swiftly carried beyond Kamiri airfield. On 3 July the enemy was suddenly and heavily reinforced,[157] and two days later Kornasoren field was also overrun. After 5 July, the garrison forces were cut off from all outside contact, and their activity was limited to harassing night raids from the interior.

## Aitape Counterattack

While the enemy drove an ever-deepening wedge into the main defenses of Western New Guinea, Lt. Gen. Adachi's isolated Eighteenth Army forces far to the east doggedly continued to mass their strength in the area west of Wewak in preparation for the planned counterattack on Aitape.

The Eighteenth Army commander planned to use approximately 20,000 of his total 55,000 troops as attack forces, employing 15,000 for logistical support and holding there maining 20,000 in the Wewak area for its defense.[158] By the beginning of June, 20th Division elements had carried out a reconnaissance in force and had established positions on the east bank of the Driniumor River, while the 41st and 51st Divisions labored painfully to move up men and supplies to the forward assembly area, located to the west of But.

---

155 Order of battle of the major units on Noemfoor at the time of the Allied landing was as follows:

| | |
|---|---|
| Headquarters, 219th Infantry | 8th Independent Bn. (Provisional) |
| 3d Bn., 219th Infantry | 102d, 117th and 119th Airfield Constr. Units |
| 7th Co., 219th Infantry | 248th Independent Motor Transport Co. |
| One Infantry Gun Co. | Elm 47th Airfield Bn. |
| Elms 9th Co., 222d Infantry | 36th Airfield Co. |
| Elms 6th Co., 222d Infantry | 41st Antiaircraft Machine Cannon Unit |

Noemfoor Detachment Operation Orders No. A-2, 27 May 1944; No. A-22, 6 Jun 44; No. A-31, 16 Jun 44; No. A-34, 17 Jun 44; No. A-37, 20 Jun 44; No. A-39, 25 Jun 44; and No. A-40, 28 Jun 44. ATIS Bulletins No. 1360, 18 Aug 44; No. 1326, 6 Aug. 44, and No. 1289, 26 Jul 44.

156 Disposition of Units and Expected Landing Areas, Noemfoor Island, 25 Jun 44. ATIS Bulletin No. 1287, 25 Jul 44.

157 The Japanese force first learned from Allied radio broadcasts on 13 July that this reinforcement had been accomplished by dropping parachute troops on Kamiri field. North of Australia Operations Record, op. cit. Supplement I, pp. 26–7.

158 The assault forces were to be composed of 6,600 men of the 20th Division, 10,700 of the 41st Division, and 2,860 in Army reserve (of which 2,000 were from the 51st Division). The main body of the 51st Division was included in the Wewak Defense Force. (1) Southeast Area Operations Record, Part III, op. cit. Vol. III, pp. 165. (2) Statement by Lt. Col. Tanaka, previously cited.

Original Painting by Ryohei Koiso

PLATE NO. 73
Japanese Staff Conference: West Cave, Biak

Throughout the first part of June these preparations continued. Weakened by lack of adequate rations, soaked by interminable rains, and forced to operate at night to escape air attack, officers and men alike toiled at transportation duties. Equipment and supplies had to be carried by hand over the last portion of the trek to the assembly area, and the heavy labor exacted of the men resulted in a high death toll from sickness, malnutrition, and exhaustion.[159] Despite these efforts, the volume of ammunition and provisions collected in the forward area by mid-June, when Lt. Gen. Adachi had hoped to complete all preparations, was still below requirements for the start of the offensive.

On 20 June, in view of the steadily deteriorating situation on the Western New Guinea front and to allow General Anami to concern himself only with that critical area, Imperial General Headquarters abruptly transferred Eighteenth Army from Second Area Army command to the direct command of Southern Army.[160] The following day General Terauchi, Southern Army Commander-in-Chief, dispatched an order to Lt. Gen. Adachi stating that the mission of the Eighteenth Army would henceforth be limited to a "delaying action at strategic positions in Eastern New Guinea."

The Southern Army order, which fully conformed to the views of Imperial General Headquarters, confronted Lt. Gen. Adachi with a difficult decision. The terms of the order clearly released him from any obligation to carry out the Aitape attack The Eighteenth Army Commander, however, remained determined to make the most effective use of his forces while they still retained fighting power, with the objective of diverting as much enemy strength as possible away from the Western New Guinea battlefront. This, he decided, could not be accomplished by anything short of a large-scale counterattack on Aitape. He therefore allowed the attack plan to stand unchanged.

On 23 June Lt. Gen. Adachi's headquarters, which had been located at Boikin, about 20 miles west of Wewak, began to displace forward to the assembly area behind the Driniumor River line. By 30 June the concentration of the attack forces, now reduced by malnutrition and disease to about 17,000 troops, was almost completed. The Eighteenth Army Commander, by 3 July, had formulated his attack plans, deciding that the assault on the enemy's forward positions on the west bank of the Driniumor River would be made on 10 July by the 20th Division and the 237th Infantry Regiment of the 41st Division. The assault troops, after over-running the forward positions, were to push westward until they contacted the main enemy positions east of Aitape and act as a covering force for the build-up of Army forces in preparation for the final assault on Aitape. On 5 July, he issued a message to the troops, in which he clearly set forth the purpose of the operation as follows:[161]

> ....The presence of the enemy in Aitape affords us a last favorable chance to display effectively the fighting power which this Army still possesses, and to contribute toward the destruction of the enemy's strength. It is obvious that, if we resort from the first to mere delaying tactics, the result will be that we shall never be able to make effective use of our full strength. The forces which survive this operation will be adequate to carry on delaying tactics.
>
> We are resolved to annihilate the enemy in the Aitape area by an all-out effort. And by our success, we shall help to lift the morale of the Japanese forces on every front and make a valuable contribution to the over-all campaign at this critical juncture when our comrades are courageously fighting in Western New Guinea. Thus, we shall display the true merit of the Imperial Army.

---

159 Approximately 5,000, or one-third of the total troops assigned to logistic support duties, died of these causes. (Statement by Lt. Col. Tanaka, previously cited.)
160 Imperial General Headquarters Army High Command Record, op. cit., p. 219.
161 Southeast Area Operations Record, Part III, op. cit. Vol. III, pp. 180–4.

According to plan, at 2200 on the night of 10 July, the attack forces swung into action.[162] (Plate No. 74) After a ten-minute artillery preparation, the 20th Division and the 237th Infantry Regiment crossed the Driniumor River about two miles upstream from its mouth and launched a fierce assault against strong Allied positions on the opposite bank of the river.[163] During the crossing, enemy artillery set up such an intense and accurate barrage that the forward units, particularly the 20th Division, suffered heavy losses.[164] The assault units nevertheless continued to attack until the positions were reduced and a sizeable bridgehead established. The 237th Infantry, attacking in a column of battalions, enveloped the whole group of enemy positions downstream from the crossing and swung north toward the Paup coast. At the conclusion of the action on the morning of 11 July, the Japanese forces began preparing for the next phase of the attack, utilizing forest cover to regroup the forward units and assemble supplies and reinforcements. The 237th Infantry threw out strong patrols toward the coastal sector in preparation for a renewal of the offensive.[165]

On 14 July, Lt. Gen. Adachi decided to exploit the initial success of his forces by committing all reserve units as soon as possible, even though most of these units had not yet arrived in the forward area.[166] He planned to send the main body of the 41st Division, reinforced by the 66th Infantry released from Army reserve, across the Driniumor River in the zone of the 237th Infantry and increase the pressure on the Paup coast. The 20th Division was ordered to swing south and

---

162  Order of battle of forces participating in the initial assault was as follows:

| 20th Division | 41st Division |
|---|---|
| 78th Infantry Regt. | 237th Infantry Regt. |
| 80th Infantry Regt. | 1st Battalion, 41st Mountain Artillery Regt. |
| 79th Infantry Regt. | 3d Company, 8th Independent Engineer Regt. |
| 26th Field Artillery Regt. | |
| 20th Engineer Regt. | |
| 2d Company, 33d Independent Engineer Regt. | |

A small coastal detachment consisting of one company from the 237th Infantry supported by a battery from the 41st Mountain Artillery Regt. and some infantry cannons from the 237th Infantry, attacked across the mouth of the Driniumor and penetrated as far as Chakila before being annihilated by an enemy counterattack on 15 July. (1) Southeast Area Operations Record, Part III, op. cit. Vol. III, pp. 197–201, 282. (2) Operations Order No. 67, 1st Battalion, 41st Mountain Artillery Regiment, 9 July 44. ATIS Bulletin No. 1392, 27 Aug. 44. (3) Statement by Lt. Col. Tanaka, previously cited.

163  These positions were believed to constitute the enemy outpost line guarding the main defenses near Aitape. Enemy strength holding this line was estimated at about three infantry battalions, with supporting artillery. Ibid.

164  The 1st Battalion, 78th Infantry, of the 20th Division, alone lost 300 men in effecting this river crossing.

165  Southeast Area Operations Record, Part III, op. cit. Vol. III, pp. 205–9.

166  Units held in reserve were as follows:

| 20th Division | 41st Engineer Regt. |
|---|---|
| 79th Infantry | 8th Independent Engineer Regt. (less 3d Co) |
| 41st Division | Eighteenth Army Reserve |
| 238th Infantry (less one bn.) | 66th Infantry Regt. |
| 239th Infantry | 37th Independent Engineer Regt. |
| 41st Mountain Artillery Regt. (less 1st Bn.) | 12th AAA Headquarters and 62d AAA Bn. |

(1) Southeast Area Operations Record, Part III, op. cit. Vol. III, pp. 209–15, 223. (2) Statement by Lt. Col. Tanaka, previously cited.

and occupy the enemy positions in the vicinity of Afua and Kwamrgnirk, in the foothills of the Toricelli Mountains, where it would be joined by its reserve regiment (the 79th). The division was then to attack northward, sweeping the left bank of the Driniumor.[167]

While these plans and preparations were under way, small enemy units began infiltrating the Japanese lines from 12 July and regained some riverside positions. An enemy tank unit also counterattacked the 237th Infantry in the coastal sector on 13 July, causing heavy casulties. The small, isolated skirmishes which accompanied these maneuvers went largely unnoticed until it suddenly became apparent that the entire bridgehead was endangered. Since the main body of the 41st Division and the 66th Infantry had still not arrived in the forward area, Lt. Gen. Adachi had no mobile reserve to counter the enemy threat. By 15 July, all the original crossing points on the Driniumor had fallen to the Allied forces, leaving the 20th Division and the 237th Infantry marooned several miles beyond the west bank.

Although Lt. Gen. Adachi remained determined to commit his remaining combat forces across the Driniumor, the situation continued to develop unfavorably. On 17 July an attempt by the 1st Battalion, 239th Infantry, to retake one of the crossing points ended in failure. On 22 July the 237th Infantry was driven out of the Paup coastal sector by a crushing enemy counterattack, and the enemy continued to bolster his positions on the west bank of the Driniumor.

Meanwhile, in the south, the 20th Division, in compliance with earlier orders, had already gathered in the Afua area, where it was joined on 18 July by the 79th Infantry. The division began attacking toward Kwamagnirk on 20 July, meeting with initial success.

By this time, however, the deterioration of the fighting strength of both the 20th Division and the 237th Infantry, as well as the serious food situation, made the Army's original plan of conserving the main effort for the final assault on Aitape an impossibility. In addition, the rapid concentration of enemy forces around the Driniumor River convinced Lt. Gen. Adachi that the enemy was going to put up its main resistance in that area. Therefore, Lt. Gen. Adachi decided to commit his entire forces, including the Army reserves and the 41st Division, against the Allied right flank in the vicinity of Afua, thereby exploiting the initial success of the 20th Division. On 21 July, the 66th Infantry and on 26 July, the 41st Division were ordered to assemble immediately in the Afua area and join the 20th Division in its northward attack.[168]

The units involved in this move had almost reached the Driniumor line in preparation for the frontal assault. Due to the change in plan, they were now forced to make their way through five more miles of dense jungle along the east bank of the river to Afua. It was 1 August before the 41st Division was able to join the 20th in the Afua offensive.[169]

The attack of these two divisions finally succeeded in enfilading the entire enemy line and carried the attack forces downstream about two miles. Japanese losses from severe enemy artillery fire had been so heavy, however, that the attack forces were not strong enough to exploit their success. By 4 August the strength

---

167 Southeast Area Operations Record, Part III, op. cit. Vol. III, pp. 210-2.

168 (1) Ibid., pp. 220-8. (2) 41st Division Operations Order No. 224, 26 July 44. ATIS Bulletin No. 1435, 7 Sep 44.

169 Southeast Area Operations Record, Part III, op. cit. Vol. III, pp. 235-8, 249.

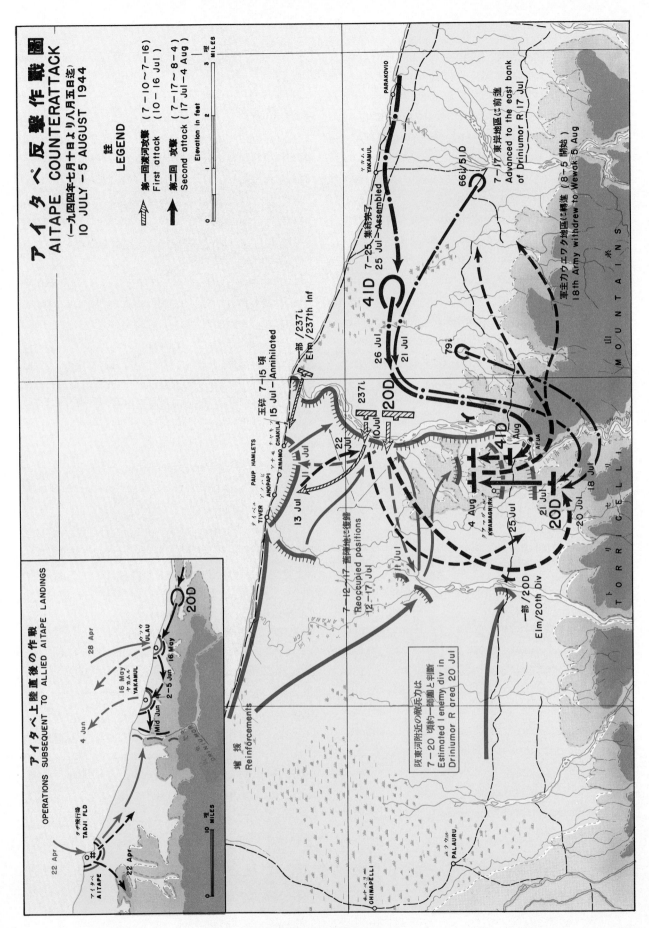

PLATE NO. 74
Aitape Counterattack, 10 July—5 August 1944

of each infantry regiment did not exceed 100 men, and some were down to as few as 30.[170] In addition, all field rations were exhausted, and the only food available was foraged plants and vegetables and small amounts of enemy provisions found in evacuated positions.

With his troops obviously unable to continue the operation, Lt. Gen. Adachi on 4 August ordered cessation of the attack. The remnants of the attack forces were directed to return to the But-Wewak area and consolidate the defenses there in the hope of denying at least that area to the enemy. This marked the end of the last major effort of the Japanese forces in New Guinea. This effort had cost 13,000 lives, but exerted little effect in checking General MacArthur's westward advance.

## End of the New Guinea Campaign—Sansapor

The curtain had not yet rung down on the last phase of Eighteenth Army's counterattack against Aitape when General MacArthur's forces, on the heels of Admiral Nimitz' moves into Tinian and Guam in the Marianas, launched a new advance to complete the conquest of western New Guinea and obtain forward air bases for the final drive toward the Philippines.

This new move, directed at Sansapor, 200 miles west of Noemfoor on the northwest coast of Vogelkop Peninsula, was preceded by large-scale strategic air operations. On 27 July American carrier planes again attacked the Palau Islands, while a formidable force of land-based aircraft operating from Western New Guinea bases simultaneously struck at Japanese airfields on Halmahera, destroying about 100 7th Air Division planes newly assembled in that area.[171] Enemy planes also ranged over the Banda and Flores Seas as part of the same strategic preparation.

On 30 July, an Allied invasion force of approximately one division followed up these preliminary operations with a landing at Opmarai Point, about ten miles northeast of Sansapor, simultaneously seizing the small islands of Amsterdam and Middleburg, about three miles offshore. These landings were completely unopposed since the only Japanese unit in the area was a line of communications guard platoon at Sansapor. Units of the 35th Division and the 2d Amphibious Brigade, though only 65 miles away at Sorong, were unable to move against the enemy force because of a serious shortage of landing craft and the fact that effective combat strength was down to only a few hundred troops.[172]

This complete lack of opposition enabled the enemy to consolidate his position without interference and to carry out the speedy construction of airfields bringing virtually all of the Moluccas within the range of Allied fighter planes. This meant that, even without the support of carrier air strength, General MacArthur's forces were now in a position to undertake an invasion of Halmahera, the last barrier in the way of a direct thrust into the southern Philippines.

---

170 Total casualties in the combat forces numbered about 5,000 killed in action and about 3,000 who died of disease and malnutrition. (Statement by Lt. Col. Tanaka, previously cited.)

171 Western New Guinea Area and North of Australia Area Naval Operations, op. cit., p. 19.

172 North of Australia Operations Record, op. cit., Supplement 1, p. 30.

# CHAPTER XI
# PHILIPPINE DEFENSE PLANS

## Strategic Situation, July 1944

So ominous for Japan's future war prospects were the defeats inflicted by Allied arms in the Marianas and Western New Guinea in the early summer of 1944 that, in mid-July, they culminated in the second shake-up of the Army and Navy High Commands in five months, and the first major political crisis since the Tojo Government had taken the nation into war.[1]

On 17 July Admiral Shigetaro Shimada resigned as Navy Minister in the Tojo Cabinet, and a day later, simultaneously with the public announcement of the fall of Saipan, General Tojo tendered the resignation of the entire Cabinet, yielding not only his political offices as Premier and War Minister, but also stepping down as Chief of Army General Staff. General Kuniaki Koiso (ret.), then Governor-General of Korea, formed a new Cabinet on 22 July with Field Marshal Sugiyama as War Minister and Admiral Mitsumasa Yonai as Navy Minister. General Yoshijiro Umezu was named Chief of Army General Staff.[2] The shake-up was not finally completed until 2 August when Admiral Shimada also yielded his post as Chief of Navy General Staff to be replaced by Admiral Koshiro Oikawa.

While these changes in Japan's top-level war command were still hanging fire, the planning staffs of the Army and Navy Sections of Imperial General Headquarters were already giving urgent attention to the revision of future war strategy in the light of the new situation created by the parallel Allied thrusts into Western New Guinea and the Marianas. Strategically, as well as in virtually every other respect, the situation was darker than at any time since the outbreak of the Pacific War.

The "absolute" defense zone defined by Imperial General Headquarters in September 1943 had in fact been penetrated at two vital points. In the south, General MacArthur's forces, within six months of their break-through via the Vitiaz and Dampier Straits, had pushed one arm of the Allied offensive more than one thousand miles along the north coast of New Guinea, coldly by-passing and isolating huge numbers of Japanese troops along the axis of advance. The capture of Hollandia gave the enemy a major staging area for further offensive moves, and his land-based air forces, with forward bases on Biak and Noemfoor, were in a position to extend their domination over the Moluccas, Palau, and the sea approaches to the southern Philippines.

In the Central Pacific, the northern prong of the Allied offensive had by-passed the Japanese naval bastion at Truk to penetrate the planned defense perimeter at a second vital

---

[1] This chapter was originally prepared in Japanese by Maj. Toshiro Magari, Imperial Japanese Army. Duty assignments of this officer were as follows: Faculty, Japanese Military Academy, 7 Jul 41—10 Dec 42; Army Staff College, 10 Dec 42—31 Jul 44; Staff Officer (Operations), Thirteenth Army, 31 Jul 44—15 Aug 45. All source materials cited in this chapter are located in G–2 Historical Section Files, GHQ FEC.

[2] *Asahi Nenkan* 朝日年鑑 (Asahi Yearbook) Asahi Newspaper Co., Tokyo, Jun 46, p. 136.

point in the Marianas, only 1,500 miles from the home islands themselves. The seizure of Saipan not only placed the Volcano and Bonin Islands within easy range of Allied tactical bomber aircraft, but threatened Japan itself with intensified raids by the new B-29, already operating from western China. More serious still, the crippling losses suffered by the Navy in the Philippine Sea Battle, especially in air strength, gave the enemy unquestioned fleet and air supremacy in the Western Pacific.

Because of the vast expansion of the areas menaced by Allied sea and air activity, it was necessary to abandon all projects for the dispatch of major reinforcements to segments of the outer defense line which still remained intact, notably Palau and Halmahera. Japanese lines of communication with the southern area were pushed back into the inner waters of the South and East China Seas, and even these relatively protected, interior routes were now subject to increased danger since the acquisition of new bases in the Marianas enabled enemy submarines to step up and prolong their operations against convoys moving along the inner shipping lanes. (Plate No. 75)

Transport losses due to enemy undersea attacks, particularly in the waters adjacent to the Philippines, had already assumed grave proportions before the loss of the Marianas.[3] Vital military and raw materials traffic between Japan and the southern area was seriously affected, and by the summer of 1944 fuel reserves in the homeland had dwindled to a critically low point, while southbound troops and materiel began to pile up at Manila, the central distribution point for the entire southern area, for lack of transport. Personnel replacement depots in the Manila area were so overcrowded that local food supplies ran short and the troops had to be placed on reduced rations.[4]

The shortage of fuel reserves in Japan Proper had a hampering effect on the operational mobility of Japan's remaining fleet strength. Soon after returning to home bases from the Philippine Sea Battle, the First Mobile Fleet was obliged to split its forces, dispatching most of its surface strength to Lingga anchorage, in the Dutch East Indies, where sufficient fuel was available, while the carrier forces remained in home waters to await aircraft and pilot replacements.[5]

## Importance of the Philippines

With the main Pacific defense line breached at two points and its remaining segments incapable of being reinforced adequately to ensure

---

3 Initial steps to combat the growing submarine menace to Japanese shipping were taken in the latter part of 1943. On 1 November, Fourteenth Army was ordered by Imperial General Headquarters to cooperate with the Navy in providing security for convoys in the waters adjacent to the Philippines by the assignment of Army aircraft to escort and patrol duty. On 15 November the Navy established the General Escort Command and launched serious study of measures to strengthen the convoy system and improve submarine detection devices. (1) U. S. Strategic Bombing Survey (Pacific), Naval Analysis Division, *Interrogations of Japanese Officials*, 1946. Vol. II, pp. 440–1. (Interrogation of Capt. Atsushi Oi, Staff Officer (Operations), General Escort Command; and Comdr. Kiyoshi Sogawa, Imperial General Headquarters, Navy Section.) (2) *Hito Sakusen Kiroku Dai Niki* 比島作戦記録第二期 (Philippine Operations Record, Phase Two) 1st Demobilization Bureau, Oct 46, p. 27. American Editor's Note: The success of American submarine "wolf-packs" in these waters was made possible largely by radio intercepts and prompt intelligence transmittal by coast-watcher teleradio stations established in the islands in increasing numbers despite severe Japanese counter-intelligence measures.

4 *Hito Sakusen Kiroku Dai, Sanki Dai Ikkan Hito ni okeru Dai Juyon Homengun no Sakusen Jumbi* 比島作戦記録第三期第一巻比島に於ける第十四方面軍の作戦準備 (Philippine Operations Record, Phase Three, Vol. I: Operational Preparations of the Fourteenth Area Army in the Philippines) 1st Demobilization Bureau, Oct 46, p. 30.

5 (1) *Hito Homen Kaigun Sakusen Sono Ni* 比島方面海軍作戦其二 (Philippine Area Naval Operations, Part II) 2d Demobilization Bureau, Oct 47, p. 37. (2) Statement by Capt. Toshikazu Ohmae, Staff Officer (operations), First Mobile Fleet.

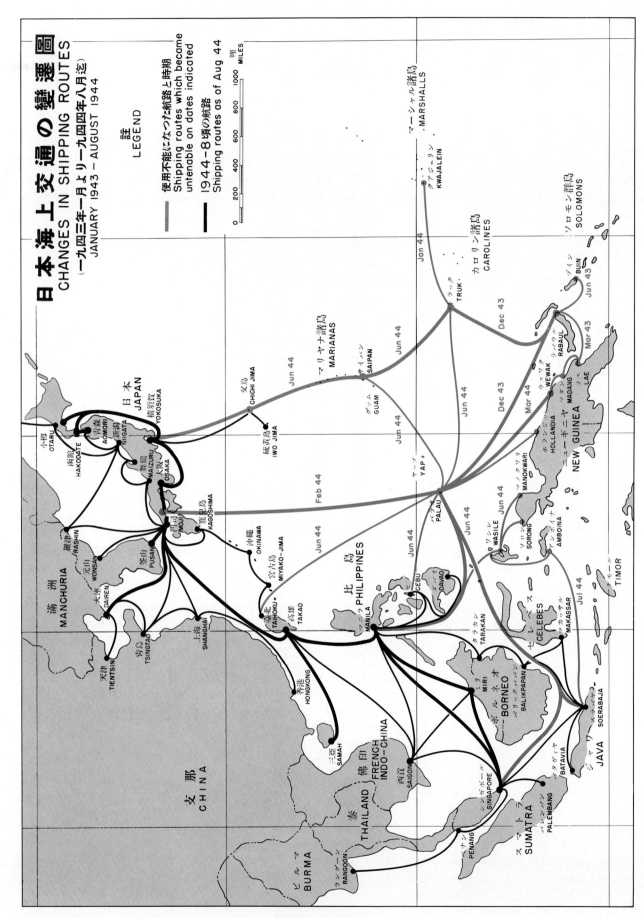

PLATE NO. 75
Changes in Shipping Routes, January 1943—August 1944

successful resistance to Allied assault, it was obvious that Japan must now fall back upon its inner defenses extending from the Kuriles and Japan Proper through the Ryukyu Islands, Formosa, and the Philippines to the Dutch East Indies. Plans and preparations must swiftly be completed with a view to the eventual commitment of the maximum ground, sea and air strength which Japan could muster in a decisive battle to halt the enemy advance when it reached this inner line.

Because of their key strategic position linking Japan with the southern area of natural resources, the Philippines naturally assumed a position of primary importance in the formulation of these plans. Just as their initial conquest had been necessary to guard the Japanese line of advance to the south in 1941–2,[6] so was their retention of paramount importance to the defense of the Empire in 1944.

Were the Philippines lost, the already contracted supply lines over which flowed the fuel and other resources essential to continued prosecution of the war would be completely severed, and all of Japan's southern armies from Burma to the islands north of Australia would be cut off from the homeland. At the same time, the enemy would gain possession of a vital stepping-stone toward the heart of the Empire and a staging area adequate to accommodate the vast build-up of forces and materiel required for mounting the final assault upon the Japanese home islands.[7] (Plate No. 76)

Imperial General Headquarters estimates of Allied offensive plans also underlined the importance of the Philippines. As early as March 1944, a careful study of the enemy's

---

[6] Cf. Chapter IV, pp. 48–9.

[7] The various considerations which made the Philippines of central and primary importance in the formulation of Japan's defensive war plans in the summer of 1944 are set forth in the following interrogations made subsequent to the surrender by key staff officers of the Army and Navy Sections of Imperial General Headquarters:

(1) Viewed from the standpoint of political and operational strategy, holding the Philippines was the one essential for the execution of the war against America and Britain. With the loss of these islands, not only would Japanese communications with the southern regions be severely threatened, but the prosecution of strategic policies within the southern regions as far as supply and reinforcements were concerned would be of paramount difficulty....The islands were also essential and appropriate strategic bases for the enemy advance on Japan. After their capture, the advantage would be two to one in favor of the enemy.... (Interrogation of Lt. Gen. Shuichi Miyazaki, Chief, 1st Bureau (Operations), Imperial General Headquarters, Army Section, 1944–5.)

(2) To shatter American war plans, the Army held it necessary to maintain the Philippines to the end and to fight a decisive battle with the Americans, who planned to recapture the islands. Furthermore, the Philippines were absolutely necessary to the security of traffic between Japan Proper and the southern area. (Interrogation of Lt. Gen. Seizo Arisue, Chief, 2d Bureau (Intelligence), Imperial General Headquarters, Army Section, 1942–5.)

(3) Japan recognized that the Philippines were important as a line of communications center on the route to the South Pacific and that they must be held at all costs. (Interrogation of Lt. Gen. Hiroshi Nukada, Chief, Transport and Communications Bureau, Imperial General Headquarters, Army Section, 1943–5.)

(4) The Philippines were regarded as the supply distribution point for the occupied areas in Java, Sumatra, Borneo, Burma, Malaya, and New Guinea. They were also a key point in the chain which linked these areas with the homeland. In July 1944 these islands became the key defensive position. Retaining control of the Philippines was necessary to link the southern areas with Japan. (Interrogation of Col. Sei Matsutani, Chief, War Policies Board, Imperial General Headquarters, Army Section, 1943–4.)

(5) After the fall of Saipan, the Philippines became the last line of national defense. The major portion of the Fleet was committed at Leyte. This was considered the final stand, and the loss of the Philippines left no hope for the successful continuation of the war. (Interrogation of Rear Adm. Tasuku Nakazawa, Chief, 1st Bureau (Operations,) Imperial General Headquarters, Navy Section, 1943–4.)

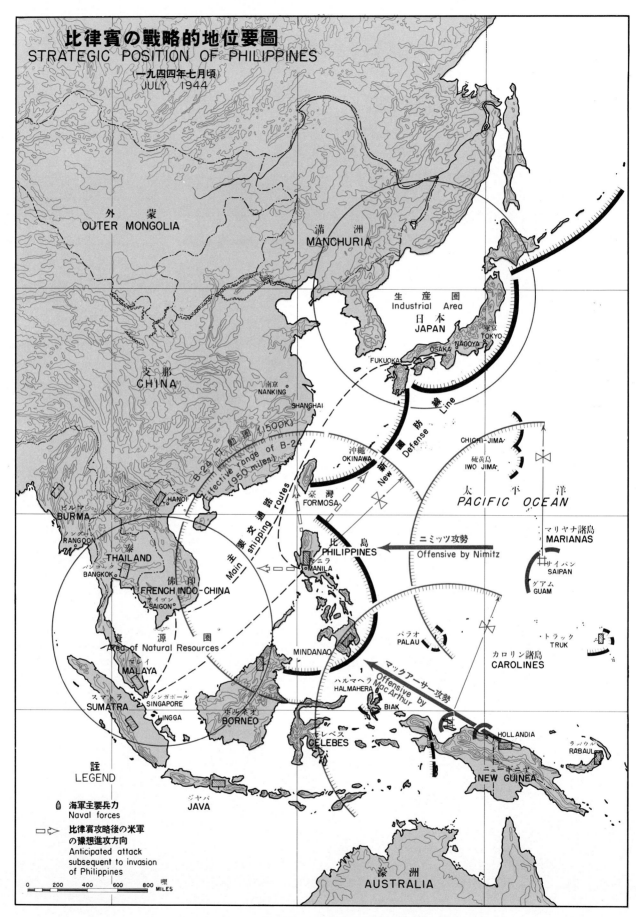

PLATE NO. 76
Strategic Position of Philippines, July 1944

probable future strategy arrived at the conclusion that there was only slight possibility of a direct advance upon Japan from the Central Pacific, primarily because the absence of land bases within fighter range of the main islands would make effective air support of an invasion via that route extremely difficult. Instead, it was considered most probable that the enemy offensives from New Guinea and the Central Pacific would first converge upon the Philippines in order to sever Japan's southern line of communications, and that, with these islands as a major base, the advance would then be pushed northward toward Japan via the Ryukyu Island chain.[8] Land-based air power would be able to support each successive stage of this advance.

The enemy thrust into the Marianas in June caused Imperial General Headquarters to re-examine the possibilities of a direct advance upon the homeland from that direction, but while a capability was accepted, the High Command did not diverge from its previous estimate that the enemy's most probable course would be to undertake reconquest of the Philippines as a prior requisite to the invasion of Japan Proper.[9]

On the basis of these estimates, Imperial General Headquarters decided that top priority in preparations for decisive battle along the inner defense line must be assigned to the Philippines. It was anticipated that the enemy would launch preliminary moves against Palau and Halmahera about the middle of September in order to secure advance supporting air bases, and that the major assault on the Philippines would come sometime after the middle of November.[10]

With the short space of only four months remaining before the anticipated invasion deadline, all Japan's energies now had to be concentrated on the task of transforming the Philippines into a powerful defense bastion capable of turning back and destroying the Allied forces.

## Local Situation

For almost a year and a half following the completion of the Japanese occupation of the Philippines in June 1942, little attention had been given either by Imperial General Headquarters or by the occupying forces to preparations against an ultimate Allied reinvasion. Japan's full war energies were thrown into the outer perimeter of advance to meet steadily intensifying Allied counterpressure, and the development of a strategic inner defense system went neglected until the establishment in September 1943 of the "absolute defense zone" embracing areas west of Marianas—Carolines—Western New Guinea line.

Under the plans worked out for this zone, as outlined earlier,[11] the Philippines were to play the role of a rear base of operations, i. e., an assembly and staging area for troops and supplies and a concentration area for air reserves, to support operations at any threatened

---

8 (1) *Daihonyei Rikugun Tosui Kiroku* 大本營陸軍統帥記錄 (Imperial General Headquarters Army High Command Record) 1st Demobilization Bureau, Nov 46, p. 210. (2) Statement by Col. Takushiro Hattori, Chief, Operations Section, Imperial General Headquarters, Army Section.

9 Slightly less probability was seen of a direct enemy invasion of Formosa or of the Ryukyu Islands, by-passing the Philippines. The homeland was rated third in order of probability, and the Kuriles last. (Ibid.)

10 (1) *Hito Homen Kaigun Sakusen Sono Ichi* 比島方面海軍作戰其一 (Philippine Area Naval Operations, Part I) 2d Demobilization Bureau, Aug 47, p. 7. (2) Statement by Col. Ichiji Sugita, Staff Officer (Operations), Imperial General Headquarters, Army Section.

11 Cf. Chapter X, pp. 232-3.

point on the main defense perimeter from the Marianas south to Western New Guinea and the Banda Sea area. To implement these plans, Imperial General Headquarters in October directed the Fourteenth Army[12] to complete the establishment of the necessary base facilities by the spring of 1944.

Major emphasis in this program was laid upon the construction of air bases. The Army alone planned to build or improve 30 fields in addition to 13 already in operational use or partially completed.[13] The Navy projected 21 fields and seaplane bases to be ready for operational use by the end of 1944, expanding its total number of Philippine bases to 33.[14] Line of communications and other rear-area base installations were also to be expanded and improved.

To speed up the execution of the program, Imperial General Headquarters dispatched additional personnel to the Philippines in November, and ordered the reorganization and expansion of the 10th, 11th, and 17th Independent Garrison Units, currently stationed on Mindanao, the Visayas, and northern Luzon respectively, into the 30th, 31st, and 32d Independent Mixed Brigades, with a strength of six infantry battalions each, plus normal supporting elements. In addition, the 33d Independent Mixed Brigade was newly activated at Fort Stotsenberg, Luzon, by combining various garrison units.[15]

In the political sphere, Japan sought to win increased Filipino cooperation in October by setting up an independent government under the presidency of Jose P. Laurel. A treaty of alliance concluded simultaneously with the inauguration of the new government provided for close political, economic and military cooperation " for the successful prosecution of the Greater East Asia War " and was supplemented by attached "Terms of Understanding" which stipulated:[16]

*The principal modality of the close military cooperation for the successful prosecution of the Greater East Asia War shall be that the Philippines will afford all kinds of facilities for the military actions to be undertaken by Japan, and that both Japan and the Philippines will closely cooperate with each other in order to safeguard the territorial integrity and independence of the Philippines.*

In accordance with these provisions, the Laurel Government promulgated orders to ensure cooperation with local Japanese military commanders, and steps were taken through local administrative agencies to recruit Filipino labor for use in carrying out the airfield construction program and the improvement of defense installations. As General MacArthur's forces steadily forged ahead toward the Philippines in the spring of 1944, however, cooperation with the Japanese armed forces gradually broke down, giving way to sabotage and active hostility.[17]

Anti-Japanese feeling and discontent were heightened by food shortages. Prior to the

---

12  On 29 June 1942, following the completion of the campaign to occupy the Philippines, the Fourteenth Army was removed from the command of Southern Army and placed directly under Imperial General Headquarters. Cf. Chapter VI, p. 113.

13  Philippine Operations Record, Phase Two, op. cit., p. 41.

14  (1) Ibid., p. 42.  (2) Philippine Naval Operations Part I, op. cit., pp. 2, 6.

15  (1) Philippine Operations Record, Phase Two, op. cit., pp. 53–4, 75.  (2) File on reorganization of forces under 16th Group (Philippines). ATIS Bulletin No. 1631, 21 Dec 44, pp. 1–3.

16  Foreign Affairs Association of Japan, *Japan Yearbook 1943–44*. Tokyo, Dec 44, pp. 1031–2.

17  The labor recruiting program lagged so badly that, in the summer of 1944, President Laurel issued a proclamation reminding the Filipinos that they were obligated by the treaty of alliance with Japan to cooperate in the execution of defense measures. Despite this reminder, results remained unsatisfactory. (Statement by Maj. Mikio Matsunobe, Staff Officer (Intelligence), Fourteenth Area Army.)

war a substantial volume of food products had been imported, but as the intensification of enemy submarine warfare cut down shipping traffic, these imports almost ceased. Commodity prices soared to inflation levels, and Filipino farmers refused to deliver their prescribed food quotas to government purchasing agencies. Allied short-wave propaganda broadcasts effectively played upon this unrest by emphasizing Allied economic and military superiority and the certainty of Filipino liberation.

The local situation was deteriorating so rapidly that a report drawn up by Imperial General Headquarters at the end of March 1944 summarized conditions in the following pessimistic terms:[18]

*Even after their independence, there remains among all classses in the Philippines a strong undercurrent of pro-American sentiment. It is something steadfast, which cannot be destroyed. In addition, the lack of commodities, particularly foodstuffs, and rising prices are gradually increasing the uneasiness of the general public. The increased and elaborate propaganda disseminated by the enemy is causing a yearning for the old life of freedom. Cooperation with and confidence in Japan are becoming extremely passive, and guerrilla activities are gradually increasing.*

It was these guerrilla activities, which the small Japanese occupation forces had never been able to stamp out, that posed the most serious potential threat to military operations. In the spring of 1944 the total strength of the organized guerrillas was estimated at about 30,000, operating in ten "battle sectors".[19] Allied submarines and aircraft operating from Australia and New Guinea brought in signal equipment, weapons, explosives, propaganda leaflets and counterfeit currency for the use of the guerrilla forces,[20] and liaison and intelligence agents arrived and departed by the same means.

Particularly dangerous to the Japanese forces was the gathering and transmission by the guerrillas of intelligence data to the Allies. A network of more than 50 radio stations, at least five of which were powerful enough to transmit to Australia and the United States, kept feeding out a constant flow of valuable military information: identifications and locations of Japanese units, troop movements, locations and condition of airfields, status of new defense construction, arrival and departure of

---

18 Imperial General Headquarters, Army Section Report, *Saikin ni okeru Hito Jijo* 最近に於ける比島事情 (Recent Situation in the Philippines) 31 Mar 44, p. 1.

19 The geographical locations of these sectors, and the respective leaders of the guerrilla forces in each, were as follows:

| | | |
|---|---|---|
| 1st Battle Sector | North Luzon | Maj. Russell Volkmann |
| 2d Battle Sector | Pangasinan, Tarlac, Nueva Vizcaya | Maj. Robert B. Laphan |
| 3d Battle Sector | Bulacan, Pampanga, Zambales | Maj. Edwin Ramsey |
| 4th Battle Sector | Tayabas, Laguna, Batangas, Cavite | Capt. Bernard Anderson |
| 5th Battle Sector | Sorgosen, Camarines | Lt. Col. Salvador Escudero |
| 6th Battle Sector | Panay | Lt. Col. Macario Peralta |
| 7th Battle Sector | Negros | Lt. Col. Salvador Abcede |
| 8th Battle Sector | Cebu, Bohol | Col. James Cushing |
| 9th Battle Sector | Leyte, Samar | Maj. Gen. Ruperto Kangleon |
| 10th Battle Sector | Mindanao | Brig. Gen. Wendell Fertig |

Ibid., Table No. 7.

20 (1) Ibid., pp. 4–5. (2) Field Diary and Intelligence Reports, Iloilo Military Police Unit, 1–31 Aug 44. ADVATIS Translation Nos. 24, 25 Dec 44.

aircraft, ship movements and defense plans.[21]

Intelligence agents operated boldly in virtually every part of the Philippines, but the greatest activity appeared to be concentrated in the area around the Visayan Sea and on Mindanao, a fact which suggested the probability of an eventual Allied landing in that area.[22] Japanese local units repeatedly undertook campaigns to eliminate the guerrillas and silence their radio stations but as soon as the troops withdrew after a clean-up expedition, guerrilla activity would spring up anew.

Until the summer of 1944, direct military action by the guerrillas was generally limited to sporadic hit-and-run attacks on small Japanese units in out-of-the-way areas and on supply columns.[23] Such harassing tactics did not affect the overall dispositions of the Japanese forces, but they required that small garrisons keep constantly on the alert. Moreover, as enemy invasion became a more and more imminent probability, the activities of the guerrillas grew bolder and more flagrant, aided by the fact that the Japanese forces were increasingly preoccupied with defensive preparations and were obliged to concentrate troop strength in anticipated areas of attack.

### Southern Army Defense Plans

Between October 1943 and March 1944, military preparations in the Philippines remained confined to the development of the islands as a rear operational base for support of decisive battle operations along the Marianas—Carolines—Western New Guinea line. No plans were yet considered for fortifying the Philippines themselves against enemy invasion, partially because Japan's resources were already heavily taxed in order to complete preparations along the main forward line, and partially because of belief that the Allied advance could be stopped at this forward barrier.

In March, however, the first indications of a change in strategic thinking with regard to the Philippines appeared. On 27 March Imperial General Headquarters, Army Section ordered a revision of the command set-up for the southern area, expanding Southern Army's operational control to take in the Fourteenth Army in the Philippines, the Second Area Army in Western New Guinea and the eastern Dutch East Indies, and Fourth Air Army. These new dispositions were to become effective 51 April.[24] The same order directed Fourteenth Army to institute defense preparations, particulaly on Mindanao, and on 4 April Imperial General Headquarters transferred the 32d Division to Fourteenth Army for the purpose of reinforcing the southern Philippines.[25]

Consequent upon the revision of command, Southern Army drew up new operational plans

---

21 Recent Situation in the Philippines, op. cit., pp. 7–8. American Editor's Note: These operations were carried out by clandestine sections of General MacArthur's intelligence system, i.e., the A. I. B. (Allied Intelligence Bureau) and P. R. S. (Philippines Regional Section). See G-2 Historical Section, GHQ FEC, General Intelligence Series: Vol. I, "The Guerrilla Resistance Movement in the Philippines" and Vol. II, "Intelligence Activities in the Philippines During the Japanese Occupation."

22 Ibid., p. 9.

23 Interrogation of Col. Shujiro Kobayashi, Chief, Operations Section, Fourteenth Area Army.

24 Cf. Chapter X, p. 239.

25 The 32d Division sailed from Shanghai in the *Take* convoy on 17 April together with the 35th Division destined for Western New Guinea. While the convoy was en route to Manila, Imperial General Headquarters suddenly decided to reassign the 32d Division to Second Area Army for the purpose of reinforcing Halmahera, and the division therefore did not disembark in the Philippines. On 28 April, an Imperial General Headquarters order formally transferred the 32d Division to Second Area Army and, in its place, assigned the 30th Division, then in Korea, to Fourteenth Army. (1) Imperial General Headquarters Army High Command Record, op. cit., pp. 230-1. (2) *Dai Ni Homengun Dai Juyon Homengun Ido Hyo* 第二方面軍第十四方面軍移動表 (Table of Movements of the Second and Fourteenth Area Armies).

荷揚作業（フィリッピン戰區）　寺内萬次郎　昭和二十三年三月十七日

Original Painting by Manjiro Terauchi

PLATE NO. 77
Unloading Operations, Philippine Area

covering its expanded zone of responsibility and specifying the missions of subordinate forces. An essential feature of these plans was the emphasis placed upon strengthening the Philippines, not merely as a rear supporting base, but as a bastion against eventual direct invasion by the enemy. The main points were as follows:[26]

*1. Southern Army's main line of defense will be the line connecting Burma, the Andaman and Nicobar Islands, Sumatra, Java, the Sunda Islands, the north coast of New Guinea west of Sarmi, Halmahera and the Philippines. The Philippines, Halmahera, Western New Guinea, Bengal Bay and the Burma sectors of this line are designated as "principal areas of decisive battle." The Philippines shall be the "area of general decisive battle."[27]*

*2. The forces defending the sectors designated as "principal areas of decisive battle" (Fourteenth Army, Second Area Army, Seventh Area Army, Burma Area Army) will, in cooperation with the Navy, strengthen combat preparations and annihilate the enemy if and when he attacks on those fronts. The forces holding other sectors of the Army's main defense line will secure key points and repulse enemy attacks.*

*3. Ground and air forces in the Philippines will be reinforced, and in the event that the enemy offensive reaches this area, Southern Army will mass all its available ground and air strength there for the general decisive battle.*

*4. The Fourth Air Army will be responsible for operations in the Philippines and eastern Dutch East Indies (including Western New Guinea); and the Third Air Army will be responsible for operations to the west of and including Borneo. In the event of enemy attack on the Pacific sector of decisive battle, however, the full strength of both Air Armies will be concentrated on that front and annihilate the enemy.[28]*

Thus, even while preparations were under way for a decisive defense of the Marianas—Western New Guinea line, Imperial General Headquarters, Army Section and Southern Army already had begun to envisage an even greater decisive battle in the Philippines, which would spell the fate of Japan's entire conquered empire in the south. The enemy's startling advance to Hollandia, which occurred while the Southern Army's plans were in the final stage of preparation, served to underline the new emphasis given to the Philippines.

The substance of these plans was communicated to the commanders of the various armies under Southern Army control at a conference specially summoned for that purpose at Singapore on 5 May. In mid-May Field Marshal Hisaichi Terauchi, Southern Army Commander-in-Chief, transferred his headquarters to Manila in order to exercise closer control over operations on the Army's eastern decisive battlefront, and at the same time the 3d Shipping Transport headquarters, which controlled all ocean transportation within the Southern Army area, displaced from Singapore to Manila.[29]

---

26 (1) *Nampo Gun Sakusen Kiroku* 南方軍作戰記錄 (Southern Army Operations Record) 1st Demobilization Bureau, Jul 46, pp. 136-8. (2) Statement by Col. Kazuo Horiba, Chief, Operations Section, Southern Army. (3) *Nampo Gun Sakusen Keikaku Taiko* 南方軍作戰計畫大綱 (Outline Policy of Southern Army Operation Plan) 1 May 44.

27 The term "area of general decisive battle" was used in a dual sense. First, it denoted Southern Army's intention to commit virtually all its strength in the Philippines in the event of enemy invasion, even at the cost of abandoning its commitments on other fronts within the Army's zone of responsibility. For the Southern Army this was considered the final decisive battle. Second, it was intended to convey the strategic concept of the Philippines as an area in which the Army and Navy would completely coordinate their forces in a general decisive battle. This latter concept became the central principle of the *Sho-Go* Operation plans elaborated by Imperial General Headquarters in July. (Statement by Col. Horiba, previously cited.)

28 If, prior to an attack on the Philippines, the enemy launched offensive operations on the Burma front or against the Palembang area, the main strength of the Fourth Air Army, conversely, was to be shifted to the western front to reinforce the Third Air Army. Southern Army Operations Record, op. cit., p. 137.

29 Statement by Maj. Jiso Yamaguchi, Staff Officer (Operations), Southern Army.

Southern Army meanwhile began pressing for action to increase troop and air strength in the Philippines to more adequate levels. At the beginning of May, Fourteenth Army's combat ground forces consisted of only one division (16th) and four independent mixed brigades, with one additional division (30th) already allocated by Imperial General Headquarters late in April and scheduled for early transfer form Korea. As against this meager strength, the operations staff of Southern Army estimated that fifteen field divisions would be required for decisive battle operations in the Philippines, in addition to eight independent mixed brigades for security control and garrison duty.[30] A large-scale reinforcement of the Fourth Air Army was also considered vitally necessary.[31]

Southern Army recognized, however, that the prior demands of reinforcing the Marianas—Western New Guinea line left no immediate possibility of boosting troop strength in the Philippines up to the level of its estimated requirements. No formal representations were therefore made to Tokyo, although the Army's views were informally communicated to Imperial General Headquarters staff officers who visited Singapore and, subsequently, Manila for liaison purposes. The need of allocating sufficient shipping to Southern Army to permit moving troops from other sectors of its own responsible area to the Philippines was also stressed in these conversations.[32]

The Fourteenth Army Commander, Lt. Gen. Shigenori Kuroda,[33] had meanwhile taken initial steps in April to regroup the forces already at his disposal with a view to ultimate defense against invasion. The 16th Division (less the 33d Infantry and other minor elements designated Army reserve) was transferred from Luzon to Leyte and, with the 31st Independent Mixed Brigade attached, was made responsible for the defense of Visayas. The 32d and 33d Independent Mixed Brigades were directed to undertake defense preparations in northern and southern Luzon, respectively. The 30th Division, upon arrival from Korea, was to be assigned to the defense of Mindanao, reinforced by the 30th Independent Mixed Brigade, already in the Mindanao area.[34]

While the Army was carrying out these preliminary moves to revitalize the defenses of the Philippines, the major elements of the Navy were fully occupied in preparations for the planned decisive battle operations in the Western Pacific.[35] The 3d Southern Expeditionary Fleet, which had been responsible since January 1942 for local naval security in Philippine waters, had only small forces and was unable to take more than limited measures to

---

30 (1) Statement by Col. Horiba, previously cited. (2) Fourteenth Army, in an earlier estimate submitted to Imperial General Headquarters in March, had placed troop requirements for securing the Philippines at a minimum of seven field divisions, with 24 additional infantry battalions to maintain public order and combat guerrilla forces. Philippine Operations Record, Phase Two, op. cit., p. 60.

31 For this purpose, Fourth Air Army headquarters was to be moved back from Menado to the Philippines to effect a reorganization of the Army's component air groups, using reinforcements to be sent out gradually from the Homeland. Southern Army Operations Record, op. cit., p. 138.

32 Statement by Col. Horiba, previously cited.

33 Lt. Gen. Kuroda assumed command of Fourteenth Army on 19 May 1943.

34 The east coast of Mindanao was regarded, at this time, as a probable landing point in case of enemy invasion of the Philippines. The 30th Division was therefore ordered to deploy its troops in the Surigao area when the division arrived in the latter part of May. (Statement by Lt. Gen. Gyosaku Morozumi, Commanding General, 30th Division.)

35 Cf. discussion of Combined Fleet preparations for the "A-Go" Operation, Chapter X.

strengthen the islands' sea defenses.[36]

The situation in regard to air strength also remained unsatisfactory pending the execution of plans to reinforce the Fourth Air Army. Of the Air Army's existing components, the 6th Air Division had lost virtually all of its remaining strength at Hollandia, while the 7th Air Division was fully committed in the Ceram area.[37] Active operations on the Burma front meanwhile barred any early transfer of Third Air Army strength to the Philippines area.[38] Naval air strength was chiefly limited to the 26th Air Flotilla, which had moved back from the Rabaul area to Davao in February for reorganization and training.[39]

## Battle Preparations No. 11

The enemy's unexpectedly early penetration to Hollandia in April brought wider recognition that no time must be lost in strengthening the defenses of the Philippines. The main Western New Guinea defense line under preparation in the Geelvink Bay area was still incomplete and inadequately manned, and serious doubt began to be felt that it would succeed in stopping General MacArthur's accelerated drive toward the Philippines.

To meet this danger, Imperial General Headquarters, Army Section in the middle of May ordered Southern Army to carry out a program of operational preparations in the Philippines, designated as *Battle Preparations No. 11*.[40] The Army High Command recognized that air power would be of key importance in defending so large an island area and therefore assigned first priority in this program to preparations for large-scale air operations. Army ground forces were charged with full responsibility for carrying out the airfield construction program, which was expanded to provide for 30 new fields in addition to those projected in October 1943.[41]

A sufficient number of additional fields were to be ready for use by the end of July to permit the deployment of four air divisions, and subsequent construction was to proceed rapidly enough to enable two more air divisions to be deployed in the Philippines by the end of 1944. Already established fields, such as those at Manila, Clark, Lipa, Bacolod, Burauen, Del Monte and Davao, were to be maintained as air bases.[42]

Shortly prior to the issuance of *Battle*

---

36 Since early in 1944, the Navy had been preparing Tawitawi and Guimaras anchorages, in the Philippines, to accommodate major elements of the Combined Fleet in support of planned operations in the Marianas and Carolines areas. Preparations had also been started in March to establish facilities for accommodating command posts of the Combined Fleet and First Air Fleet at Davao. Philippine Naval Operations, Part I, op. cit., pp. 2–3.

37 *Nanto Homen Sakusen Kiroku Sono San: Dai Juhachi Gun no Sakusen* 南東方面作戰記錄其三第十八軍の作戰 (Southeast Area Operations Record, Part III: Eighteenth Army Operations) 1st Demobilization Bureau, Sep 46. Vol. III, pp. 78–9.

38 Southern Army Operations Record, op. cit., pp. 161–2.

39 The 26th Air Flotilla was assigned to the First Air Fleet on 5 May but did not participate in the Philippine Sea Battle of 19–20 June. *A-Go Sakusen* あ號作戰 ("A-Go" Operations) 2d Demobilization Bureau, Aug 47, pp. 4, 20, 86–7.

40 Imperial General Headquarters Army High Command Record, op. cit., p. 232.

41 Of the 30 Army airfields projected in October 1943, six had been generally completed by May 1944, and 24 were still under construction. Of the 21 projected Navy fields, 15 were still incomplete.

42 Each air base consisted of several airfields, each of which was an integral part of the base. The advantages of this arrangement were; (a) closer and more effective coordination of defense measures; (b) more concentrated and efficient use of air strength; (c) better command and maintenance facilities. Bases varied in size from those capable of accommodating a full air division down to bases which could accommodate half a division. Imperial General Headquarters Army High Command Record, op. cit., p. 206.

*Preparations No. 11*, Imperial General Headquarters had taken initial steps to reinforce air strength in the Philippines, ordering the transfer of the 2d and 4th Air Divisions from the Second Air Army in Manchuria. The 4th Air Division was directly assigned to the Fourth Air Army, while the 2d was assigned to Southern Army, which subsequently placed it under Fourth Air Army command. Late in May the first increment of these reinforcements arrived in the Philippines, and on 1 June Fourth Air Army headquarters effected its planned transfer from Menado to Manila.[43] Before leaving Manchuria, the 2d and 4th Air Divisions were reorganized, most of the flying units being assigned to the 2d[44] and base maintenance units to the 4th.[45]

Under *Battle Preparations No. 11*, steps were also taken to bolster Fourteenth Army troop strength. During June 15,000/20,000 filler replacements were transported to the Philippines, and Fourteenth Army was ordered to reorganize and increase its four independent mixed brigades to divisions, using these replacements to fill them up to division strength.[46] The new divisions and their locations were as follows:

| Ind. Mixed Brig. | Division Designation | Headquarters |
|---|---|---|
| 30th | 100th Division | Davao, Mindanao |
| 31st | 102d Division | Cebu |
| 32d | 103d Division | Baguio, Luzon |
| 33d | 105th Division | Las Baños, Luzon |

In addition to these units, two new independent mixed brigades, the 54th and 55th, were activated on Luzon, the 55th remaining in Central Luzon and the 54th transferring to Zamboanga via Cebu shortly after organization was completed.[47] The 58th Independent Mixed Brigade, just organized in Japan Proper, was also assigned by Imperial General Headquarters order to Fourteenth Army as a further step to strengthen the ground forces in the Philippines.[48]

Despite the shift of a portion of the airfield construction program over to the 4th Air Division, a large part of the Army ground forces still had to be allocated for this purpose instead of to the immediate preparation of ground defenses against invasion. Fourteenth Army, however, was keenly aware of the detrimental consequence which the same course had produced with respect to the tenability of the Western New Guinea defense line, and decided that August must be fixed as the deadline for the switch-over of all ground forces to preparations for ground operations.[49]

---

43 *Hito Koku Sakusen Kiroku Dai Niki* 比島航空作戦記録第二期 (Philippine Air Operations Record, Phase Two) 1st Demobilization Bureau, Oct 46, pp. 4–5.

44 The 6th and 10th Air Brigades of the 2d Air Division moved forward to the Philippines in June, followed by the 7th and 13th during July and August. The 2d Air Division also assumed command of the 22d Air Brigade, already in the Philippines. Report on reinforcements sent to the Philippines, prepared by the 1st Demobilization Bureau in reply to memorandum of the United States Strategic Bombing Survey, 27 Oct 45. Submitted 14 Nov 45.

45 The 4th Air Division, upon its arrival, was assigned the primary mission of executing part of the airfield construction program. Two air reconnaissance companies, with about 20 aircraft, were assigned to the division, however, and allocated to anti-submarine patrol duty.

46 The organization of these new divisions differed from the standard Japanese infantry division in that each had two infantry brigades made up of four independent infantry battalions, with an approximate over-all strength of 10,000 troops. *Rikugun Butai Chosa Hyo* 陸軍部隊調査表 (Table of Army Units) War Ministry, 28 Oct 45, Part I, pp. 36–8.

47 (1) *Dairikumei Dai Sennijukyu-go* 大陸命第千二十九號 (Imperial General Headquarters Army Order No. 1029) 15 Jun 44. (2) Philippine Operations Record, Phase Three, op. cit. Vol. I, pp. 10–11.

48 (1) Imperial General Headquarters Army High Command Record, op. cit., pp. 259–60. (2) Philippine Operations Record, Phase Three, op. cit. Vol. I, p. 12.

49 *Hito Sakusen Kiroku Dai Sanki Dai Nikan Furoku Reite Sakusen Kiroku* 比島作戦記録第三期第二巻附錄レイテ作戦記錄 (Philippine Operations Record, Phase Three, Vol. II, Supplement: Leyte Operations Record) 1st Demobilization Bureau, Oct 46, pp. 9–10.

Efforts were also launched to increase the efficiency of the line of communication system and to accumulate reserves of military supplies. One of the first moves was the formation of the Southern Army Line of Communications Command on 10 June.[50] This headquarters took over command of all line of communication units in the Philippines and, in addition, was charged with responsibility for logistical support of the entire Southern Army.

Concurrently with the execution of the Army's *Battle Preparations No. 11*, the Navy also took steps to reinforce its Philippine defenses, especially in air strength. After suffering heavy losses in June and early July at the hands of enemy carrier forces in the Central Pacific, the 61st Air Flotilla of the land-based First Air Fleet[51] was ordered back to Philippine bases and immediately began reorganizing and replenishing its strength with replacements arriving from Japan.[52]

Meanwhile, on 12 July, Southwest Area Fleet transferred its headquarters from Surabaya to Manila in order to assume closer control of naval base and surface forces in the Philippines.[53]

The sharp acceleration of defense preparations in the Philippines made it necessary to allocate additional shipping for military use. In August 105,000 tons of general non-military ships were made available to the Army and earmarked for employment in reinforcing the Philippines. At the same time, in order to speed the importation to Japan of oil and critical raw materials from the southern area, the Government in July transferred 200,000 tons of shipping from general freight transport between Japan and China—Manchuria to the southern shipping route, and ordered the conversion of 232,000 gross tons of cargo ships into oil tankers.[54]

To achieve maximun utilization of shipping space, a central coordinating control body was established in July, composed of representatives of the War, Navy and Transportation Ministies. This body permitted a more flexible system whereby military shipping, which might return to Japan empty after discharging troops or supplies at Manila for example, could be diverted to Singapore or some other southern port to pick up critical cargo for the Homeland. Similarly, non-military freighters hitherto sent out empty to southern ports could be used to carry military traffic as far as Manila.[55]

Along with these measures, the Navy, after considerable experimentation, achieved a more effective system of convoy protection. The number of escort ships was increased to 80, almost three-fourths of which operated on the

---

50 (1) *Dairikumei Dai Sennijuichi-go* 大陸命第千二十一號 (Imperial General Headquarters Army Order No. 1021) 7 Jun 44. (2) Philippine Operations Record, Phase Three, op. cit. Vol. I, p. 27.

51 The other major elements of the First Air Fleet at the end of June were : 22d Air Flotilla, stationed on Truk ; 23d Air Flotilla, which had been transferred from the control of the Southwest Area Fleet on 5 May and was deployed in the Ceram-Halmahera area ; and the 26th Air Flotilla, which was still at Davao. The 62d Air Flotilla had been organized in Japan and assigned to the First Air Fleet early in 1944. Just before the Philippine Sea Battle in June, it was transferred temporarily to direct Combined Fleet command. "*A-Go*," Operation, op. cit., pp. 11–14.

52 Philippine Naval Operations, Part I, op. cit., p. 30.

53 Southwest Area Fleet had previously made preparations to transfer its headquarters to Kendari, in the Celebes, or to Halmahera, in order to facilitate support of operations in Western New Guinea. With the invasion of Hollandia, however, these plans were cancelled in favor of a transfer to Manila. Ibid., pp. 28–9.

54 *Gun Hoyu Sempaku Hendo ni kansuru Shuyo Jiko* 軍保有船舶變動に關する主要事項 (Principal Matters Pertaining to Changes in Military Shipping) Shipping Division, General Maritime Bureau, Ministry of Transportation, 20 Jan 46.

55 Statement by Capt. (Navy) Oi, previously cited.

southern route under the 1st Escort Force headquarters, located at Takao, Formosa.[56] Four escort carriers, converted from merchant ships, were also made available for escort duty, and new air groups, with radar-equipped planes, were organized exclusively for patrolling shipping lanes.[57] Navy seaplanes were also being equipped with a newly-perfected magnetic device for the detection of submerged submarines.[58]

### Central Planning for Decisive Battle

Initial steps to gird the Philippines against eventual enemy invasion were thus already under way when the penetration of the main Western New Guinea defense line and of the Marianas made it evident that Japan must now prepare to wage an all-out decisive battle in defense of the inner areas of her Empire.

Although the Philippines were expected to be the first of these inner areas to be attacked, Imperial General Headquarters also had to consider the possible contingency that the enemy might strike alternatively at Formosa or the Ryukyu Islands, or possibly even at the Japanese home islands themselves. Comprehensive operational plans therefore had to be worked out to cover all these possibilities.[59]

The basic strategic principle adopted as the foundation of these plans was that whichever of the inner areas first became the object of invasion operations by the main strength of the enemy would be designated as the "decisive battle theater," and that as soon as this theater was determined, all available sea, air and ground forces would be swiftly concentrated there to crush the enemy. Because of the necessity of central control and coordination, the decision as to when and where to activate decisive battle operations was reserved to the highest command level, Imperial General Headquarters.[60]

Detailed matters discussed in connection with the plans included problems relating to the preparation and concentration of all three arms, an undertaking which exceeded in scale anything attempted by the Japanese High Command since the initial war operations in December 1941. A further vital topic of discussion centered around the most effective employment of air, sea and ground forces during the various stages of an enemy invasion.

The first major problem concerned the employment of air forces, the most mobile arm and therefore the one which could be most rapidly concentrated at any point of attack. The High Command estimated that the Allied forces would employ the same tactical pattern of invasion which they had established in the Marshalls, at Hollandia and at Saipan, i. e., carrier-based planes would first endeavor to gain air superiority by neutralizing Japanese base air forces; second, while Allied aircraft maintained control of the air, naval surface units would seek to destroy ground defense positions near the beach by concentrated shelling; and third, troop transports would begin disembarking the

---

56 *Beikoku Senryaku Bakugeki Chosa Oboegaki Dai Nijuyon-go ni taisuru Kaito* 米國戰略爆擊調査覺書第二十四號に對する回答 (Reply to United States Strategic Bombing Survey Memorandum No. NAV-24) Navy Ministry, 26 Nov 45, Chart C.

57 USSBS, *Interrogations of Japanese Officials*, op. cit. Vol. II, pp. 440-1. (Interrogations of Capt. Oi, and Comdr. Sogawa, both previously cited.)

58 Ibid., pp. 309-10. (Interrogation of Capt. Shunji Kamiide, Commander, 901st Air Group, Imperial Japanese Navy.)

59 General data regarding Imperial General Headquarters planning of the decisive battle operations were furnished by Col. Hattori, Col. Sugita, and Capt. Ohmae, all previously cited.

60 Imperial General Headquarters Army High Command Record, op. cit., pp. 251-2.

assault troops.

Against these Allied tactics the Japanese heretofore had followed the practice of committing most of their available air strength in attacks directed at the enemy carriers during the first phase of invasion. The plane losses suffered in such attacks, however, generally were so high that insufficient strength remained to carry out effective attacks against the enemy's troop transport during the third and critical phase. The transports consequently were able not only to approach the landing points without having suffered any appreciable damage at sea, but to ride relatively unmolested in anchorage while the troops debarked.

In the light of this past experience, Imperial General Headquarters concluded that a change in air tactics was necessary. It was estimated that the most effective results would be obtained if the employment of the main strength of the Air forces were withheld until the third phase of the enemy attack and the maximum strength were then thrown simultaneously against both troop transports and carriers. In accordance with this plan, the High Command decided to employ the main strength of the Army Air forces in attacks against troop transports and the main portion of the Navy Air forces against enemy carriers.[61] These assignments were subsequently embodied in a Army-Navy Central Agreement covering air operations, issued on 24 July.

Withholding air attacks until the third phase necessitated the fortification of airfields to withstand enemy bombing and strafing attacks during the first two phases. Additional construction was therefore planned,[62] and as an added precautionary measure to reduce losses during the first and second phases, it was decided to disperse air units at fields staggered in depth.

The employment of naval surface forces, the next most mobile element, constituted the second major problem confronting Imperial General Headquarters in preparing the plans for decisive battle. The problem was rendered doubly difficult by the fact that the Combined Fleet, as a result of its heavy losses in the Philippine Sea Battle on 19–20 June, was so depleted in both aircraft carriers and, to an even greater degree, in carrier-borne air forces, that its ability to wage a modern-type sea battle was seriously impaired. No more than six carriers, of which only one was a regular, first-class carrier, remained in operation, plus two battleships fitted to launch aircraft by catapult.[63] The Fleet, however, still possessed considerable surface firepower, including the two 64,000-ton super-battleships *Musashi* and *Yamato*, five other battleships, 14 heavy cruisers, seven light cruisers and about 30 destroyers.[64]

---

61  Ibid., pp. 266–7.

62  Since 1943 some fortification work had been done on airfields within the national defense sphere. Concrete revetments had been constructed to protect fuel and ammunition and control installations. Efforts were now renewed to complete this work in the Philippines. (Statement by Col. Sugita, previously cited.)

63  The *Zuikaku*, of about 29,800 tons, was the only regular carrier remaining. The *Chiyoda, Chitose, Zuiho* and *Ryuho*, all with an approximate tonnage of 14,000, had been converted from seaplane and submarine tenders and were classed as light carriers. The *Junyo*, 27,500 tons, was a converted merchant ship. The *Ise* and *Hyuga*, converted from battleships, had no flight decks and carried only 22 aircraft which were launched by catapult. (1) *Kakukan Kozokuryoku To Ichiranhyo* 各艦航續力等一覽表 (Table Showing Radius of Action of Naval Ships) 2d Demobilization Bureau, 19 Jul 47, p. 3. (2) *Japanese Naval Vessels at the End of the War*, 2d Demobilization Bureau, Apr 47, p. 2.

64  The size, structure, and armament of the *Musashi* and *Yamato* were one of the Japanese Navy's most closely guarded secrets. The five other battleships of the fleet included two old, slow-speed ships, the *Fuso* and *Yamashiro*. The *Fuso* had participated in naval actions since early 1944 under direct Combined Fleet command. On 10 September, it was assigned together with the *Yamashiro* to the Second Battleship Division. Philippine Naval Operations, Part II, op. cit., p. 37.

驅潜艇の活躍　藤本東一郎　昭和一八

Original Painting by Toichiro Fujimoto

PLATE NO. 78
Subchaser in Action

In spite of the serious weakness in aircraft carriers, the Navy High Command opposed adoption of passive defense tactics on the ground that the Fleet would face annihilation at some later date under still more unfavorable conditions if the enemy succeeded in occupying any of the inner areas. The Third and Fifth Fleets, both of which were in the Inland Sea, would be cut off from indispensable fuel supplies of the southern area, while the Second Fleet, which had left home waters on 9 July for Lingga Anchorage, would be separated from its source of ammunition resupply in the homeland. Moreover, the sea areas in which the Navy must operate would thereafter be within range of superior enemy land-based air forces.

On the basis of this reasoning, the Navy Section of Imperial General Headquarters decided in favor of risking the full remaining strength of the Fleet in bold offensive action. Surface forces, supported by land-based air strength, would launch a concerted attack designed to catch and destroy the enemy fleet of invasion transports at the points of landing. The assault was to be facilitated by a diversionary move to draw off the enemy naval forces covering the landing operations.

Because of the absolute necessity of preventing enemy penetration of the inner defense line and the inadequate sea and air forces available to oppose such a penetration, Imperial General Headquarters also considered the possible initiation of *tokko*, or special-attack, tactics for the purpose of destroying the enemy at sea.[65]

In planning the most effective method of using the ground forces, the Army Section of Imperial General Headquarters gave particular attention to a revision of the hitherto accepted tactical concepts of defense against enemy landing operations. Almost complete reliance had hitherto been placed upon strong beach positions, with little or no emphasis on secondary defenses. The primary tactical principle had been to destroy the enemy troops from these beach positions as they attempted to come ashore. The successive defeats suffered since Tarawa, however, had demonstrated that such positions could not be decisively held under the type of devastating preparatory naval bombardment employed by the Allied forces.

As a result of the careful studies made of this problem over a period of some months, the Army Section decided that new tactics of defense should be employed in the ground phase of the projected operations. These tactics involved: (1) preparation of the main line of resistance at some distance from the beach to minimize the effectiveness of enemy naval shelling: (2) organization of defensive positions in depth to permit a successive wearing down of the strength of the attacking forces; and (3) holding substantial forces in reserve to mount counterattacks at the most favorable moment.[66]

Instructions based upon the conclusions reached by Imperial General Headquarters were subsequently communicated to all armies in the field.

### Army Orders for the *Sho-Go* Operations

By the latter part of July, the basic plans covering the decisive battle operations to be conducted along the inner defense line had been completed. Imperial General Headquarters designated these operations by the code name *Sho-Go* (捷號), meaning "victory," and proceeded to issue implementing orders to the various Army and Navy operational commands. The basic order governing Army operations was issued by the Army Section of Imperial General Headquarters on 24 July, stating in

---

65 Cf. Chapter XVII for detailed discussion of special attack tactics and organization.
66 Imperial General Headquarters Army High Command Record, op. cit., pp. 262–3.

part as follows:[67]

*1. Imperial General Headquarters is planning to initiate decisive action against anticipated attack by the enemy's main force during the latter part of the year.... The decisive battle area is expected to be Japan Proper, the Nansei (Ryukyu) Islands, Formosa or the Philippines. The zone of decisive action and the date of the initiation of operations will be designated by Imperial General Headquarters.*

*2. To accomplish their respective missions, the Commander-in-Chief, Southern Army, the Commander, Formosa Army, the Commander-in-Chief, General Defense Command, the Commander, Fifth Area Army, and the Commander-in-Chief, China Expeditionary Army will swiftly prepare for decisive action in cooperation with the Navy.*

A directive implementing the above order was issued the same day by the Army Section, Imperial General Headquarters, including the following instructions:[68]

*1. Army commanders will generally complete preparations for decisive action in their respective area by the following dates: Philippine area (Sho Operation No. 1); end of August: Formosa and Nansei Islands (Sho Operation No. 2); end of August: Japan Proper, excluding Hokkaido (Sho Operation No. 3); end of October; Northeastern area (Sho Operation No. 4); end of October.*

Air operations were specifically dealt with in a Army-Navy Central Agreement which was appended to this directive. The principal stipulations of this agreement were as follows:[69]

*1. Operational Objective:*

*The Army and Navy Air forces will complete preparations for decisive action by mid-August. In the event of an enemy invasion, the total Air forces of both the Army and Navy shall be concentrated in the area of decisive action and will engage and destroy the invading forces through coordinated action.*

*Imperial General Headquarters shall determine the zone wherein decisive action shall be executed.*

*2. Disposition and Employment of Air Forces:*

*a. The basic disposition of Army and Navy Air forces shall be as follows:*

*Northeast area—Twelfth Air Fleet; 1st Air Division.*

*Japan Proper (excluding Hokkaido)—Third Air Fleet; Air Groups attached to Third Fleet (if stationed in Japan); Training Air Army; 10th Air Division; 11th Air Division; and 12th Air Division.*

*Nansei Islands and Formosa Area—Second Air Fleet; 8th Air Division.*

*Philippines, Western New Guinea—Halmahera, and Central Pacific areas—First Air Fleet*[70]*; Fourth Air Army.*

*Present dispositions will be maintained on other fronts.*

*b. Plans for employment of Air forces:*

*For Sho Operation No. 1, the Navy shall concentrate the First and Second Air Fleets in the Philippines, hold the Third Air Fleet in reserve, and transfer the Twelfth Air Fleet to Japan Proper. Besides the total strength of the Fourth Air Army, the Army shall send as reinforcements to the Philippines two fighter regiments and one heavy bomber regiment from the Training Air Army, one fighter regiment, one light bomber regiment, and one heavy bomber regiment from the 8th Air Division, and two fighter regiments from the Fifth Air Army in China. The 1st Air Division shall be held as strategic reserve.*

*3. Allocation of Missions and Command Dispositions:*

---

67 *Dairikumei Dai Senhachijuichi-go* 大陸命第千八十一號 (Imperial General Headquarters Army Order No. 1081) 24 Jul 44.
68 Imperial General Headquarters Army High Command Record, op. cit., pp. 252–3.
69 (1) Ibid., pp. 254–8. (2) *Daikaishi Dai Yonhyakusanjugo-go Bessatsu* 大海指第四百三十五號別冊 (Imperial General Headquarters Navy Directive No. 435, Annex) 26 Jul 44.
70 The First Air Fleet had under its command the 15th Air Regiment (Army), which was attached on 16 May 1944 to assist in long-range reconnaissance. (1) *Daikaishi Dai Sambyakushichijuku-go* 大海指第三百七十九號 (Imperial General Headquarters Navy Directive No. 379) 16 May 44. (2) *Teraoka Nikki* 寺岡日記 (Diary of Vice Adm. Kimpei Teraoka) First Air Fleet Commander.

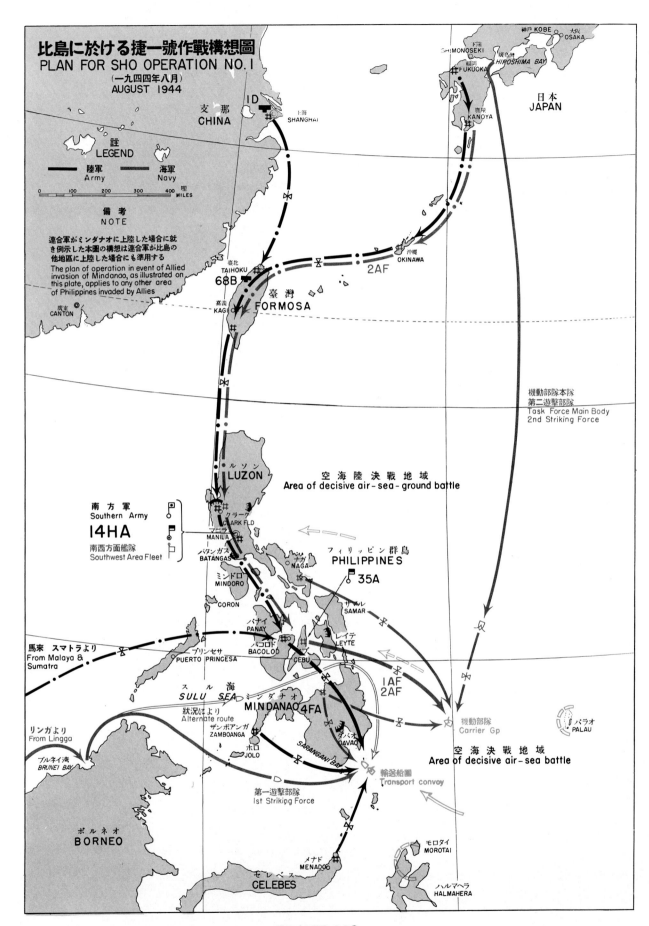

PLATE NO. 79
Plans for *Sho* Operation No. 1, August 1944

Following the issuance of the Army's basic *Sho-Go* Operations order on 24 July, the Navy Section of Imperial General Headquarters on 26 July issued a new directive fitting its previous outline of naval policy for urgent operations into the framework of the *Sho-Go* Operations plan.[79] To implement both these directives, the Combined Fleet on 1 August issued Combined Fleet Top Secret Operations Order No. 83, which specified the following general missions of the naval forces in the *Sho-Go* Operations:[80]

1. *Operational Policy:*

   a. *The Combined Fleet will cooperate with the Army according to the operational procedures specified by Imperial General Headquarters for the Sho-Go Operations in order to intercept and destroy the invading enemy in decisive battle at sea and to maintain an impregnable strategical position.*

2. *Outline of Operations:*

   a. *Preparations:*

   *(1) Air bases will be prepared as rapidly as possible in the Philippines to permit deployment of the entire air strength of the First and Second Air Fleets. Air bases in the Clark Field and Bacolod areas will be organized rapidly in accordance with the Army-Navy Central Agreement.*

   b. *Operations:*

   *(1) Enemy aircraft carriers will be destroyed first by concentrated attacks of the base air forces.*

   *(2) Transport convoys will be destroyed jointly by the surface and air forces. If the enemy succeeds in landing, transports carrying reinforcements and the troops already on land will be the principal targets so as to annihilate them at the beachhead.*

   *(3) Surface forces will sortie against the enemy landing point within two days after the enemy begins landing. All-out air attacks will be launched two days prior to the attack by the surface forces.*

Combined Fleet Top Secret Operations Order No. 84, also issued on 1 August, fixed the new tactical grouping of naval forces for the *Sho-Go* Operations. Almost the entire surface combat strength of the Fleet was included in a Task Force placed under the overall command of the First Mobile Fleet Commander, Vice-Adm. Jisaburo Ozawa. This force was broken down into three tactical groups: (1) the Task Force Main Body, directly commanded by Vice-Adm. Ozawa and consisting of most of the Third Fleet (carrier forces): (2) the First Striking Force, commanded by Vice Adm. Takeo Kurita and made up of the Second Fleet with part of the 10th Destroyer Squadron attached: (3) the Second Striking Force, commanded by Vice Adm. Kiyohide Shima and composed of the Fifth Fleet plus two destroyer divisions and the battleships *Fuso* and *Yamashiro*.[81]

The manner in which these tactical forces were to be employed in the planned decisive battle operations was set forth in more detail in an "outline of operations" annexed to Combined Fleet Top Secret Operations Order No.

---

79 *Daikaishi Dai Yonhyakusanjugo-go* 大海指第四百三十五號 (Imperial General Headquarters Navy Directive No. 435) 26 Jul 44.

80 These missions were set forth in a separate annex to the order. The original text of this annex is not available, but the essential portions paraphrased in this volume were reconstructed from the following sources: (1) Combined Fleet Top Secret Operations Order No. 84, 1 Aug 44; Task Force Top Secret Operations Order No. 76, 10 Aug 44; Second Striking Force Top Secret Operations Order No. 1, 10 Aug 44. ATIS Limited Distribution Translation No. 39, Part VIII, 4 Jun 45, pp. 226-33; Part I, 22 Apr 45, pp. 3-8; Part V, 28 May 45, pp. 5-11. (2) Philippine Naval Operations Part I, op. cit., pp. 14-19. (3) Statement by Comdr. Sakuo Mikami, Staff Officer (Operations), Imperial General Headquarters, Navy Section.

81 ATIS Limited Distribution Translaation No. 39, op. cit. Part VIII, pp. 227-33 and Part V, pp. 5-11.

85, issued on 4 August. The gist of this outline applying to surface force operations was as follows:[82] (Plate No. 79)

*1. Disposition of forces: The First Striking Force will be stationed at Lingga Anchorage, while the Task Force Main Body and the Second Striking Force will be stationed in the western part of the Inland Sea. However, if an enemy attack becomes expected, the First Striking Force will advance from Lingga Anchorage to Brunei, Coron or Guimaras; the Task Force Main Body and the Second Striking Force will remain in the Inland Sea and prepare to attack the north flank of the enemy task force.*

*2. Combat operations: If the enemy attack reaches the stage of landing operations, the First Striking Force, in conjunction with the base Air forces, will attack the enemy in the landing area.*

*a. If the enemy attack occurs before the end of August, the Second Striking Force, plus the 4th Carrier Division and part of the 3d Carrier Division, will facilitate the operations of the First Striking Force by launching effective attacks against the enemy and diverting his task forces to the northeast.*

*b. If the enemy attack occurs after the end of August, the Second Striking Force will be incorporated under the command of the Task Force Main Body as a vanguard force. The Main Body will then assume the mission of diverting the enemy task forces to the northeast in order to facilitate the attack of the First Striking Force, and will also carry out an attack against the flank of the enemy task forces.*[83]

During August, the Navy Section of Imperial General Headquarters also took action to give the Combined Fleet more unified operational control of naval forces in order to facilitate the execution of the *Sho-Go* plans. On 9 August the General Escort Command and units assigned to naval stations were placed under operational command of the Combined Fleet, and on 21 August the China Area Fleet was similarly placed under Combined Fleet command.[84]

## Preparations for Battle

In line with the broad plans handed down by Imperial General Headquarters, Army and Navy preparations for decisive battle in the Philippines area were pushed ahead on a first priority basis during August.[85] Particular urgency was attached to the early completion of preparations by the Air forces, which were to play the key role in the initial phases of an enemy invasion.

Shipping to the Philippines continued to be severely limited, but air reinforcements and supplies arrived steadily. Meanwhile, ground units made every effort to speed the airfield construction program. By the end of September, over 60 fields considered good enough for

---

82 Original text of the operational outline annexed to Combined Fleet Top Secret Operations Order No. 85 is not available. The substance of this outline relating to surface force operations is given here on the same sources as listed in n. 80.

83 The outline further contained a paragraph regarding the employment of surface special attack forces. This paragraph provided that the use of such forces would be subject to direct control by Imperial General Headquarters, and that the latter would issue a special order for their employment in combat. Philippine Naval Operations, Part I, op. cit., p. 21.

84 (1) *Daikairei Dai Sanjusan-go* 大海令第三十三號 (Imperial General Headquarters Navy Order No. 33) 9 Aug 44. (2) *Daikairei Dai Sanjugo-go* 大海令第三十五號 (Imperial General Headquarters Navy Order No. 35) 21 Aug 44.

85 Concurrently with these military and naval preparations, steps were taken to assure coordinated action on the political and diplomatic fronts. An Imperial conference held on 19 August called for a thorough mobilization of national strength by the end of the current year and for decisive action to improve the Empire's diplomatic position. (Statement by Col. Hattori, previously cited.)

all-weather use were in operational condition. (Plate No. 80)

Completion of the movement of the 2d and 4th Air Divisions from Manchuria brought the total strength of the Fourth Air Army in the Philippines up to approximately 420 aircraft of all types by the latter part of August.[86] The 2d Air Division, which contained all the combat flying units, commanded five air brigades, one air regiment, and other small elements.[87] These were deployed principally at the Clark Field and Bacolod bases, where they had been undergoing intensive training since their arrival from Manchuria.

On 7 September, the 2d Air Division Commander with part of the headquarters staff moved to Menado in the Celebes.[88] Prior to his departure, the 2d Air Division Commander ordered battle preparations for three of the devision's air regiments in order to bolster the air strength of the 7th Air Division,[89] which was deployed in the Menado area for support of Second Area Army operations. Minor elements of the 2d Air Division were also dispatched to bases on North Borneo.

Parallel with the strengthening of the Fourth Air Army, the reorganization and replenishment of the naval land-based air forces also proceeded according to plan. By the end of July, the combat flying elements of the 23d Air Flotilla in the Celebes and the 26th and 61st Air Flotillas at Davao had been reconstituted as the 153d, 201st and 761st Air Groups, respectively.[90] These three combat air groups, under an Imperial General Headquarters Navy order of 10 July, were to be detached from their respective flotillas and operate under direct command of the First Air Fleet.[91]

The headquarters of the First Air Fleet, which had been virtually wiped out in the Marianas operations, was under reorganization in Japan. On 7 August the reorganization was completed, and the newly appointed Air Fleet commander, Vice Adm. Kimpei Teraoka, left shortly thereafter with his staff to set up the headquarters at Davao.[92] On 10 August, to unify the command of naval forces in the Philippines, Imperial General Headquarters transferred the First Air Fleet from direct Combined Fleet command to that of the South-

---

86 The 2d Air Division had 400 planes, and the 4th Air Division (mostly base maintenance personnel) about 20. The 7th Air Division operating in the Second Area Army zone had about 70 planes. The 6th Air Division had remained inoperational since losing its last strength at Hollandia and was finally deactivated on 19 August. (1) Philippine Air Operations Record, Phase Two, op. cit., pp. 25–6. (2) *Dairikumei Dai Senhyakugo-go* 大陸命第千百五號 (Imperial General Headquarters Army Order No. 1105) 19 Aug 44.

87 Principal units assigned to the 2d Air Division at the end of August were: 6th Air Brigade (65th and 66th Fighter-Bomber Regiments); 7th Air Brigade (12th and 62d Heavy Bomber Regiments); 10th Air Brigade (27th and 45th Fighter-Bomber Regiments); 13th Air Brigade (30th and 31st Fighter Regiments); 22d Air Brigade (17th and 19th Fighter Regiments); 2d Air Regiment; one squadron, 28th Air Regiment and 31st Independent Reconnaissance Squadron. Philippine Air Operations Record, Phase Two, op. cit., pp. 22–3.

88 One reconnaissance company also made the move to Menado. *Washi Sakumei Ko Dai Rokuju-go* 鷲作命甲第六十號 (2d Air Division Order A, No. 60) 22 Ang 44.

89 Principal units operating under the 7th Air Division at this time were: 3d Air Brigade (13th Fighter Regiment and 75th Light Bomber Regiment); 9th Air Brigade (24th Fighter Regiment and 61st Heavy Bomber Regiment); two reconnaissance squadrons. (1) Ibid. (2) Philippine Air Operation Record, Phase Two, op, cit., p. 23.

90 *Nihon Kaigun Hensei Suii oyobi Heiryoku Soshitsu Hyo* 日本海軍編成推移及兵力喪失表 (Tables Showing Organizational Changes and Losses of Japanese Naval Forces) 2d Demobilization Bureau, Oct 49, pp. L-31-2.

91 (1) Ibid., pp. L-31-3. (2) *Teikoku Kaigun Senji Hensei* 帝國海軍戰時編成 (Wartime Organization of the Imperial Navy) Navy General Staff, Vol. II, 15 Aug 44, p. 32. (3) At the same time the service units of the 26th Air Flotilla had been organized as the Philippines Airfield Unit.

92 Philippine Naval Operations, Part I, op. cit., p. 30.

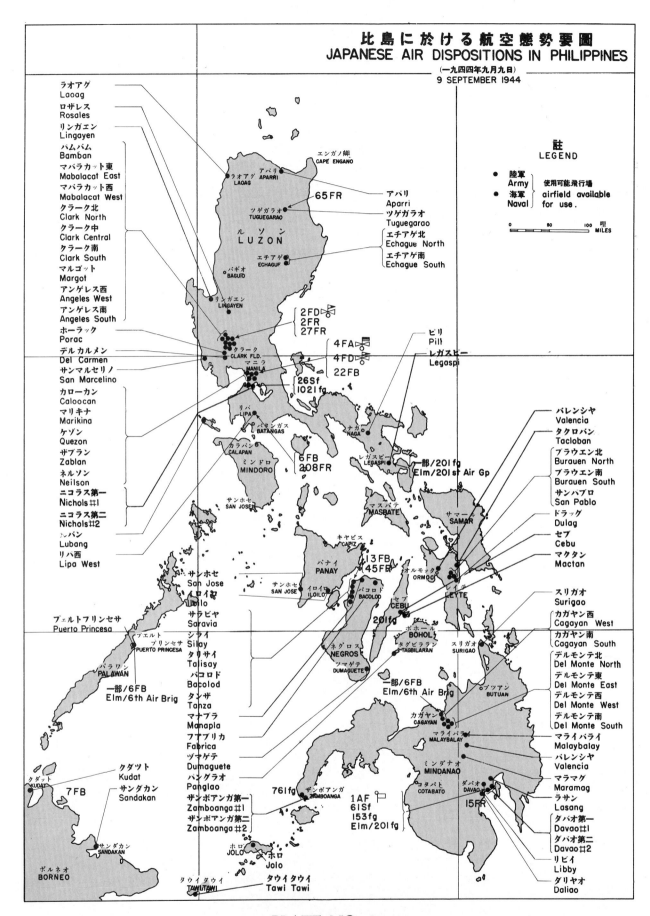

PLATE NO. 80
Japanese Air Dispositions in the Philippines, 9 September 1944

west Area Fleet, which already controlled all naval ground and surface elements in the Philippines area.[93]

With the establishment of First Air Fleet headquarters at Davao, the flying elements began an intensive training program to raise combat efficiency. By the early part of September most of these elements were deployed at bases on Mindanao and Cebu in readiness to carry out the missions assigned to the Navy Air forces. The combat air groups, however, were still short of both trained flying personnel and aircraft.[94] Despite the attachment of the Army's 15th Air Regiment equipped with long-range reconnaissance planes, the First Air Fleet was unable to perform its preliminary mission of patrolling the waters east of the Philippines with complete adequacy.

Back at Homeland bases, the Second Air Fleet, which had been activated on 15 June, was conducting a program of specialized training in preparation for its scheduled deployment to Formosa and Ryukyu Island bases in September.[95] Under the Army-Navy Central Agreement on air operations (cf. p. 294), the Second Air Fleet was to advance from these bases to the Philippines and reinforce the First Air Fleet upon the activation of Sho Operation No. 1.

To the Second Air Fleet was assigned the particular mission of attacking enemy carriers. Its flying units were therefore specially trained and equipped for this purpose, and a special force designated as the "T" Attack Force was organized to carry out surprise attacks at night or under adverse weather conditions.[96] The bulk of the best pilots in the Navy Air forces were assigned to this unit.[97] In addition, the 7th and 98th Air Regiments (Army), equipped with new Type IV twin-engine bombers, were attached to the Second Air Fleet on 25 July.[98] The bombers were modified for carrying torpedoes, and training instituted in executing attacks on carriers.

Detailed plans for air operations under the *Sho* No. 1 plan were meanwhile under joint study by the Army and Navy High Commands in Manila. By early September, the main lines of these plans had been worked out as follows:[99]

*1. Prior to the start of decisive battle operations, surprise hit-and-run attacks will be directed against enemy land-based air forces in order to gradually reduce their strength. Enemy air attacks against our bases will be intercepted in planned localized actions so as to minimize the dissipation of our combat strength.*

*2. In the event of enemy task force raids, designated air units will execute attacks on the enemy force at night or in poor weather. Under favorable circumstances, daylight attacks may also be carried out.*

*3. In the event of enemy invasion operations, the invading forces will be drawn as close as possible before the full weight of our Air forces is thrown into*

---

93 Wartime Organization of the Imperial Navy, op. cit. Vol. II, p. 32.

94 The 153d, 201st and 761st Air Groups had a total strength of about 400 aircraft, only about half of which were in operational condition. Philippine Naval Operations, Part I, op. cit., pp. 32-3.

95 Statement by Capt. Ohmae, previously cited.

96 The "T" Attack Force was so designated because of its ability to fight even in adverse weather. The letter "T" stood for *taifu*, the Japanese word for "typhoon".

97 Philippine Naval Operations, Part II, op. cit., p. 3.

98 (1) *Daikairei Dai Sanjuichi-go* 大海令第三十一號 (Imperial General Headquarters Navy Order No. 31) 24 Jul 44. (2) *Dairikumei Dai Senhachiju-go* 大陸命第千八十號 (Imperial General Headquarters Army Order No. 1080) 22 Jul 44.

99 (1) Philippine Air Operations Record, Phase Two, op. cit., pp. 5-8, 14-15, 33-9. (2) *Dai Roku Kichi Koku Butai Meirei Saku Dai Roku-go Bessatsu* 第六基地航空部隊命令作第六號別冊 (Supplement to Sixth Base Air Force Operations Order No. 6) 5 Sep 44. (3) *Hito Homen Koku Sakusen ni kansuru Riku-Kaigun Genchi Kyotei* 比島方面航空作戦に關する陸海軍現地協定 (Army-Navy Local Agreement Concerning Philippines Air Operation) 1 Sept 44.

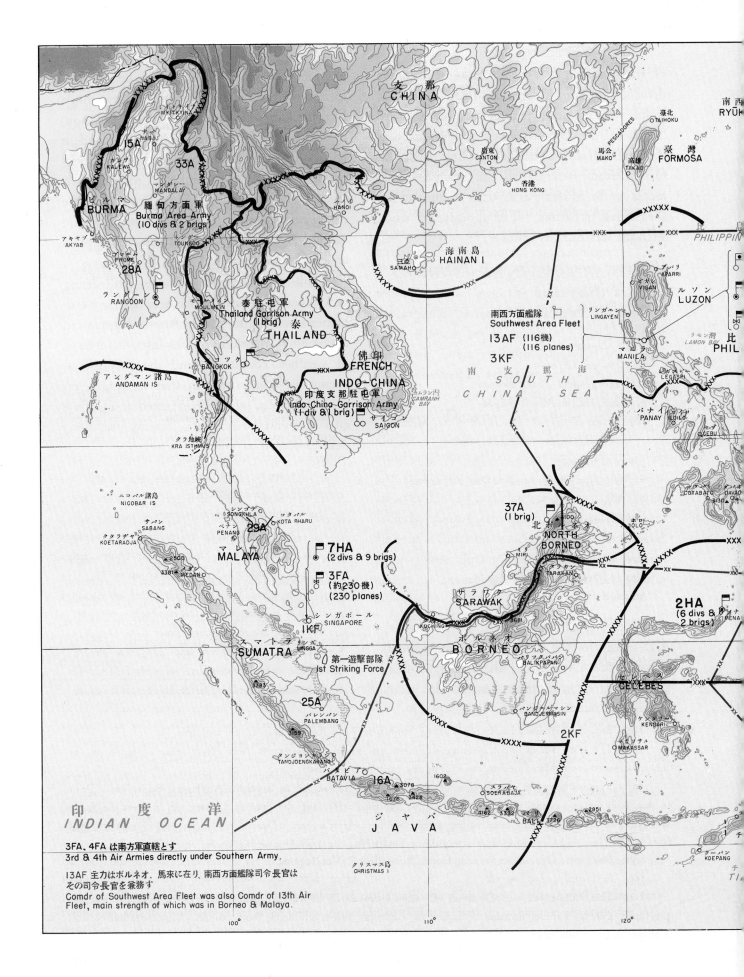

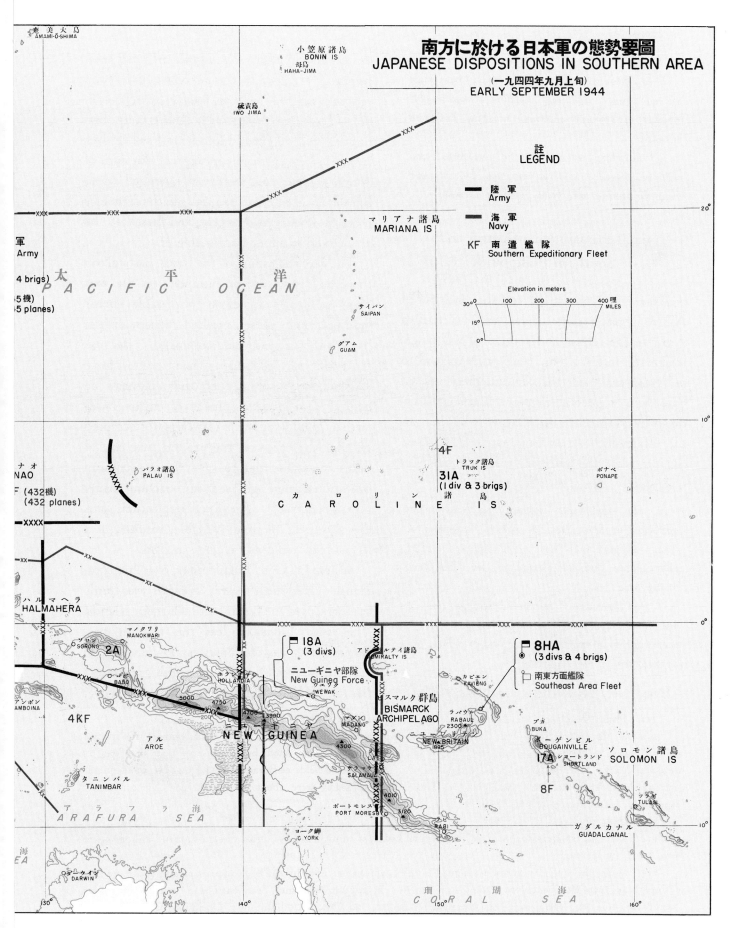

PLATE NO. 81
Japanese Dispositions in Southern Area, September 1944

the islands of the Visayan Sea. The 30th Division was on the northern tip of Mindanao around Surigao. The 100th Division occupied other key points on Mindanao from Zamboanga on the west to Dansalan on the north and Davao on the southeast. A force of only about one battalion was stationed in the Davao area. The 54th Independent Mixed Brigade was in the vicinity of Cebu.

Lt. Gen. Suzuki summoned his subordinate commanders to Cebu on 18 August for a conference on the operational plans to be employed in the event of an Allied landing in the Army area.[116] The substance of these plans was as follows:[117]

*1. Operational objectives: The Army will secure the central and southern Philippines, particularly the air bases near Davao and on Leyte, and will destroy enemy landing forces in coordination with the decisive operations of the sea and air forces.*

*2. Outline of Operations:*

*a. The Army will maintain a tight defense in the Davao sector and the Leyte Gulf area with the 100th Division and the 16th Division, respectively. The main body of the 30th Division and elements of the 102d Division will constitute a mobile reserve to be committed to any key area which the enemy may attack.*

*b. Suzu Operation No. 1: If the principal effort of the enemy invasion is directed at the Davao sector, the main body of the 30th Division, three reinforced infantry battalions of the 102d Division, and other forces will be committed to the area.*

*c. Suzu Operation No. 2: If the principal effort of the enemy landing is directed at the Leyte Gulf area, the main body of the 30th Division, two reinforced infantry battalions of the 102d Division, and other forces will be landed at Ormoc to reinforce the 16th Division.*

*d. If the enemy lands powerful forces at both Davao and Leyte Gulf, it is tentatively planned to commit the main body of the 30th Division to Davao and elements of the 102d Division to Leyte.*

In accordance with these plans, the 100th Division was immediately ordered to concentrate its main strength in the Davao area, while the 54th Independent Mixed Brigade was dispatched from Cebu to take over the mission of defending western Mindanao and Jolo Island. The main strength of the 30th Division, consisting of the division headquarters and two reinforced infantry regiments, was directed to move from Surigao to the vicinity of Malaybalay and Cagayan, a centralized location more suited to the division's mission as mobile reserve.[118] The 16th Division on Leyte and 102d Division in the Visayan area were not affected by this regrouping.

Naval ground forces in the central and southern Philippines were also being reinforced and regrouped. During August and September, nine naval construction units with a total strength of about 9,000 arrived in the Philippines, a portion of this strength being allocated to the central and southern islands.[119] The 33d Special Base Force was activated early in August at Cebu. The 36th Naval Guard unit at Guimaras Anchorage was ordered to move to Leyte in early October to expedite defense preparations in the vicinity of the naval airfield at Tacloban. The 32d Special Base Force still remained responsible for the defense of naval and harbor installations at Davao. These

---

116 At the time he assumed command, Lt. Gen. Suzuki estimated that the most probable target of the enemy's initial assault would be Davao, with the beaches along Leyte the next most likely landing spot. Philippine Operations Record, Phase Three, op. cit. Vol. II Supplement, pp. 7, 21–2.

117 Ibid., pp. 23–5.

118 Ibid., pp. 18–19.

119 Philippine Naval Operations, Part I, op. cit., pp. 38–9.

various units were primarily concerned with the construction of such fortifications as were necessary for the direct protection for naval installations.

The primary mission of the 16th Division on Leyte was to secure the vital air bases at Tacloban, Dulag and Burauen. Until midsummer, however, the division was so occupied in the construction of new airstrips and in antiguerrilla operations that organization of ground defenses had not proceeded beyond the construction of coastal positions facing Leyte Gulf.[120] The construction of inland positions did not get under way until July, when the main strength of the 9th Infantry Regiment was moved back from Samar to Leyte to speed defense preparations.

Concerned by the 16th Division's overconcentration on beach defenses, Thirty-fifth Army in August directed Lt. Gen. Makino to place greater emphasis on the preparation of defenses in depth and suggested that strong positions for the main body of the division be organized along an axis running through Dagami and Burauen. In compliance with these instructions, work on inland positions was accelerated in September, although seriously hampered by the difficult terrain and guerrilla activity.

Concurrently with these preparations, the Army began building up reserves of ammunition and rations with a view to the possible interruption of supplies from the rear during an enemy attack. Each division stocked sufficient food to be self-sustained for a period of one month and from 1,050 to 1,500 tons of ammunition.[121] In addition, a reserve supply of one month's rations and 750 tons of ammunition was stored on Cebu.

In late August Southern Army headquarters at Manila decided that the defenses of the Thirty-fifth Army area were inadequate and ordered Fourteenth Area Army to reinforce the troop strength in specified sectors up to prescribed minimum levels. These levels, in terms of nuclear infantry strength only, were as follows:[122]

| | |
|---|---|
| *Davao sector:* | At least one division |
| *Leyte Gulf sector:* | One division |
| *Sarangani:* | Three battalions |
| *Zamboanga:* | Three battalions |
| *Jolo Island:* | Two battalions |
| *Surigao:* | Strong elements |

To provide Thirty-Fifth Army with the necessary additional strength to implement these orders, Fourteenth Area Army released to it the 33d Infantry Regiment/16th Division and the 55th Independent Mixed Brigade, which had previously been designated as Area Army reserve. At the same time Lt. Gen. Suzuki was ordered to effect a further regrouping of his forces to meet the prescribed troop levels fixed by Southern Army.

The 30th Division, the main body of which had not yet completed its movement to the Malaybalay–Cagayan area, was now ordered to dispatch one regiment to Sarangani to release the 100th Division elements stationed there. Upon being relieved, these elements were to move to the Davao area, rejoining the main

---

120  These coastal defenses consisted of a series of strongpoints built at strategic points along the coast between Palo and Abuyog. Lt. Gen. Makino, 16th Division commander, ordered key emplacements to be constructed strongly enough to resist 15-cm howitzer shells.

121  This overall tonnage was broken down as follows: 300 rounds per rifle; 20,000 rounds per machine gun; 10,000 hand-grenades per division; 300 rounds per "knee" mortar; 1,500 rounds per 7.5 cm artillery piece. (Statement by Col. Ryoichiro Aoshima, Staff Officer (Line of Communications), Fourteenth Area Army.)

122  (1) Philippine Operations Record, Phase Three, op. cit. Vol. II Supplement, pp. 25–6. (2) Fourteeth Area Army Operations Orders. Published in XXIV Corps ADVATIS Translation XXIV CAET No. 7, 12 Nov 44. (3) Thirty-fifth Army Operations Orders. XI Corps ADVATIS Translations No. 38, 14 Jan 45.

body of the division.[123] The 54th Independent Mixed Brigade, previously allocated to garrison both Zamboanga and Jolo, had not yet completed its movement from Cebu and was now relieved of responsibility for the defense of Jolo. This latter mission was assigned to the 55th Independent Mixed Brigade.[124]

While these new shifts in troop assignments caused a certain delay in ground defense preparations on Mindanao, the forces in the central Philippines were able to continue making ready for impending Allied attack. Heavy rains during this period, however, impeded troop movements over the inadequate road nets and also generally retarded the construction of defensive fortifications.[125]

As the first half of September wore on, indications mounted that the Allied forces in Western New Guinea and the Marianas were about to launch new offensive operations. Guerilla activities in the Philippines increased sharply. More significant, intelligence gathered from enemy intercepts and a sudden increase in the scale of air and submarine activity appeared to foreshadow an imminent move against either the Halmaheras or the western Carolines.[126]

These were the final barriers which stood in the path of the Allied advance upon the Philippines.

---

[123] The 166th Independent Infantry Battalion of the 100th Division, stationed around Cotabato, was not pulled back to Davao, but was transferred to 30th Division command and remained in the vicinity of Cotabato. (Statement by Col. Muneichi Hattori, Chief of Staff, 100th Division.)

[124] The 55th Independent Mixed Brigade moved first from Luzon to Cebu, re-embarking there for Jolo. The last elements of the brigade reached Jolo on 5 October. (Statement by Maj. Tokichi Temmyo, Commander, 365th Battalion, 55th Independent Mixed Brigade.)

[125] Statement by Col. M. Hattori, previously cited.

[126] *Dai Niji Sekai Taisen Ryakureki*, (*Otsu*) 第二次世界大戰略曆(乙) (Abridged Chronicle of World War II, (B) 2d Demobilization Bureau, Mar 46, Part III, p. 17.

# CHAPTER XII

# PRELUDE TO DECISIVE BATTLE

## Initial Air Strikes

While the Japanese forces in the Philippines hastened to complete preparations against anticipated Allied invasion, enemy carrier-borne aircraft served sudden warning on 9 September 1944 that the date of this invasion was fast drawing near.[1] In the first large-scale air operation by the Allies against the Philippines, an estimated 400 carrier planes staged a devastating ten-hour offensive against southern Mindanao, concentrating their attacks on Davao, Sarangani, Cagayan and Digos.

Since Japanese air patrols had failed to discover the enemy task force,[2] the attacks achieved complete surprise and inflicted widespread and severe damage to ground installations, airfields, anchorages, and lines of communication. Reconnaissance units of the First Air Fleet immediately flew off search missions, which revealed that the attacks originated from three enemy naval task groups boldly maneuvering in the waters southeast of Mindanao. Two of these groups were reported to have nuclei of two aircraft carriers each; the composition of the third was not ascertained.

The First Air Fleet's 153d Air Group was the only combat flying unit actually based at fields in the Davao area at the time of the strike.[3] Despite damage to some of its fighter aircraft which were caught on the ground, this unit, as well as the 761st Air Group's torpedo bombers based at Zamboanga, were in a position to attack the enemy carrier groups had Vice Adm. Teraoka, First Air Fleet Commander, ordered such action. However, the *Sho-Go* Operation plans covering employment of the air forces rested on the basic tactical principle of not committing those forces against pre-invasion raids by enemy task forces, but conserving their strength for all-out attacks when the enemy was about to launch actual landing operations. The First Air Fleet therefore withheld retaliatory action pending further developments.

Ground and naval units in the Davao area were nevertheless ordered on the alert to meet the possible contingency that an invasion attempt would follow the air strikes, and the Japanese armed forces throughout southern Mindanao became tense with expectancy. A

---

1 This chapter was originally prepared in Japanese by Maj. Toshiro Magari, Imperial Japanese Army. For duty assignments of this officer, cf. n. 1, Chapter XI. All source materials cited in this chapter are located in G-2 Historical Section Files, GHQ FEC.

2 The effectiveness of Japanese air patrols was reduced by the fact that radar equipment was still in the developmental stage. Also the shortage of planes made it impossible for the First Air Fleet to cover all sectors in its air search and patrol operations. *Hito Homen Kaigun Sakusen Sono Ichi* 比島方面海軍作戦其一 (Philippine Area Naval Operations, Part I) 2d Demobilization Bureau, Aug 47, pp. 43, 46.

3 The 201st Air Group, containing the bulk of First Air Fleet fighter strength, had displaced from Davao to Cebu between 3 and 6 September owing to the increasing frequency of raids on Davao bases by enemy land-based bombers operations from Sansapor. These raids had resulted in the destruction of a considerable number of aircraft. The 761st Air Group had moved earlier to bases at Zamboanga, on western Mindanao, and on Jolo Island, in the Sulu Archipelago. No Army air units of any importance were stationed anywhere on Mindanao at this time. (1) Ibid., p. 45. (2) *Teraoka Nikki* 寺岡日記 (Diary of Vice Adm. Kimpei Teraoka) First Air Fleet Commander.

feeling of nervousness gripped the weak local forces at Davao[4] and rapidly spread to the large Japanese civilian colony. A wave of wild rumors swept the city. On 10 September, a second series of heavy enemy raids aggravated this state of alarm. The city and harbor were reduced to a shambles and communications paralyzed. Panic and civil disorder broke out.

In the midst of the alarm and confusion caused by the air strikes, a 32d Naval Base Force lookout post on Davao Gulf suddenly sent in a report at 0930 on 10 September that enemy landing craft were approaching the shore.[5] The Base Force headquarters hastily transmitted the report to the First Air Fleet, which in turn radioed all navy commands affected. Not until mid-afternoon, several hours after the report had been broadcast, was it established by air reconnaissance over the gulf that there were actually no enemy ships present. The First Air Fleet thereupon radioed at 1630 that the previous report was erroneous.[6]

In the interim, however, higher army and navy headquarters had reacted swiftly. Admiral Soemu Toyoda, Commander-in-Chief of the Combined Fleet, had ordered all naval forces alerted for the execution of *Sho* Operation No. 1. Thirty-fifth Army headquarters at Cebu had simultaneously issued an alert for *Suzu* No. 1 Operation, applicable to an enemy invasion of the Davao area.[7] The 30th Division main strength in the Cagayan area was ordered to prepare immediately to move to Davao to reinforce the 100th Division, and the 102d Division in the Visayas was directed to release two infantry battalions for dispatch to Mindanao. The Fourth Air Army meanwhile issued orders directing the 2d Air Division elements which had just advanced to Menado for the purpose of reinforcing the 7th Air Division[8] to return immediately to Bacolod.[9]

Following receipt of the First Air Fleet's retraction of the earlier invasion report, the Combined Fleet and Thirty-fifth Army cancelled the *Sho* No. 1 and *Suzu* No. 1 alerts late on 10 September. The whole incident, however, had a vital influence on later developments. The acute embarrassment caused by the false landing scare made military and naval commanders excessively chary of accepting later invasion reports at face value.[10]

Less than 48 hours after the termination of the raids on Mindanao, the enemy struck again, this time in the central Philippines. On the morning of 12 September, a navy radar picket station on Suluan Island, in Leyte Gulf,

---

[4] "The 100th Division was hurriedly concentrating near Davao, but the military strength immediately available in that area in the middle of September consisted of not more than two infantry battalions. Moreover, no defense installations of any kind had been built in the vicinity." (Statement by Col. Muneichi Hattori, Chief of Staff, 100th Division.)

[5] This erroneous report was evidently due to the fact that the lookout post observers, nervously expecting an enemy invasion, mistook some unusual wave contours on the horizon level at the entrance of Davao Gulf for ships and promptly reported that enemy landing craft were approaching. Philippine Naval Operations, Part I, op. cit., pp. 48-9.

[6] Diary of Vice Adm. Kimpei Teraoka, op. cit.

[7] Cf. Chapter XI, p. 314. *Hito Sakusen Kiroku Dai Sanki Dai Nikan Furoku: Reite Sakusen Kiroku* 比島作戰記録第三期第二卷附録レイテ作戰記録 (Philippine Operations Record, Phase Three, Vol. II Supplement: Leyte Operations) 1st Demobilization Bureau, Oct 46, pp. 28-9.

[8] Cf. Chapter XI, p. 307.

[9] In accordance with these orders which reached Menado on the night of 10 Sep, Lt. Gen. Masao Yamase, 2d Air Division commander, took his headquarters back to Bacolod. *Kimitsu Sakusen Nisshi* 機密作戰日誌 (Top Secret Operations Log) Aug-Sep 44, Fourth Air Army Staff Files: 2d Air Division Detailed Action Report, 16 Sep 44.

[10] Cf. Chapter XIII, p. 338.

broadcast over the general air-warning net that a vast formation of enemy carrier planes was heading westward toward the Visayas. Since the Suluan Island lookout was only about twenty minutes' flying time from Cebu, the air forces there could not be alerted quickly enough to put up an effective defense. By 0920 the enemy planes were already swarming over the Cebu airfields, where the main fighter strength of the First Air Fleet was based following its transfer from Davao. Although the attacks extended over the entire Visayan area and later took in Tawitawi, in the Sulu Archipelago, the Cebu fields appeared to be the principal objective.

In the three days over which this air offensive continued, the First Air Fleet suffered damage to 50 Zero fighters on Cebu alone, while in other areas 30 additional aircraft of all types were rendered non-operational. Flight personnel suffered numerous casualties, and training was disrupted.[11] Heavy damage was also sustained by Army air units. The 13th Air Brigade, made up of Type I fighters, was so hard hit that it had to be ordered back to Japan for regrouping, while the 45th Fighter-bomber Regiment was reduced to half strength. In addition, 11 transports totalling 27,000 gross tons and 13 naval combat vessels were sent to the bottom of Cebu harbor.[12]

The carrier raids on Mindanao and the Visayas at once strengthened the conviction of the Southern Army command that the Allies were preparing for an early invasion of the Philippines. At the same time, they had shown all too clearly that the tactical policy of not committing available air strength against raiding enemy task forces was open to serious question as a means of conserving that strength for subsequent decisive battle. After carefully studying the over-all situation, Field Marshal Terauchi and his staff therefore prepared recommendations to Imperial General Headquarters substantially as follows:[13]

*1. It is recommended that Imperial General Headquarters activate Sho Operation No. 1 as soon as possible and accelerate the planned reinforcement of the Philippines.*

*Justification: It is estimated that the Philippines will be the next target of enemy invasion, and that the attack will come very soon. Our intelligence cannot be relied upon to provide us with timely warning, and much time is still needed for the assembly of forces, particularly air units. If the activation is delayed, these units may be obliged to enter the theater with insufficient time to prepare for battle and familiarize themselves with the terrain.*

*2. It is further recommended that the Fourth Air Army be given immediate authorization to employ its main strength against enemy carrier task forces.*

*Justification: The policy previously fixed by Imperial General Headquarters and providing for conserving our air forces in order to strike with full force at the moment of an attempted enemy landing is impossible to implement on the local level since our airdrome defenses do not appear capable of protecting the air strength which we are trying to conserve. If, on the other hand, we at once attack and destroy the enemy's carrier task forces, we will gain time and freedom of action to complete further operational preparations.*

---

11 Training in skip-bombing had been under way since late August for fighter units. Basic training was scheduled to be completed in mid-September, and the flying personnel were gradually developing confidence in the new technique. However, the Cebu raids caused training to be broken off, and all units became so preoccupied with combat operations and maintenance that the program was never completed. Philippine Naval Operations, Part I, op. cit., pp. 35–6.

12 Naval vessels sunk were: 8 PT boats, 3 converted gun-boats, 1 converted minesweeper, and 1 submarine chaser. Three torpedo boats were heavily damaged, and six others received lesser damages. Ibid., pp. 51–2.

13 *Nampo Gun Sakusen Kiroku* 南方軍作戰記錄 (Southern Army Operations Record) 1st Demobilization Bureau, Jul 46, pp. 143–5.

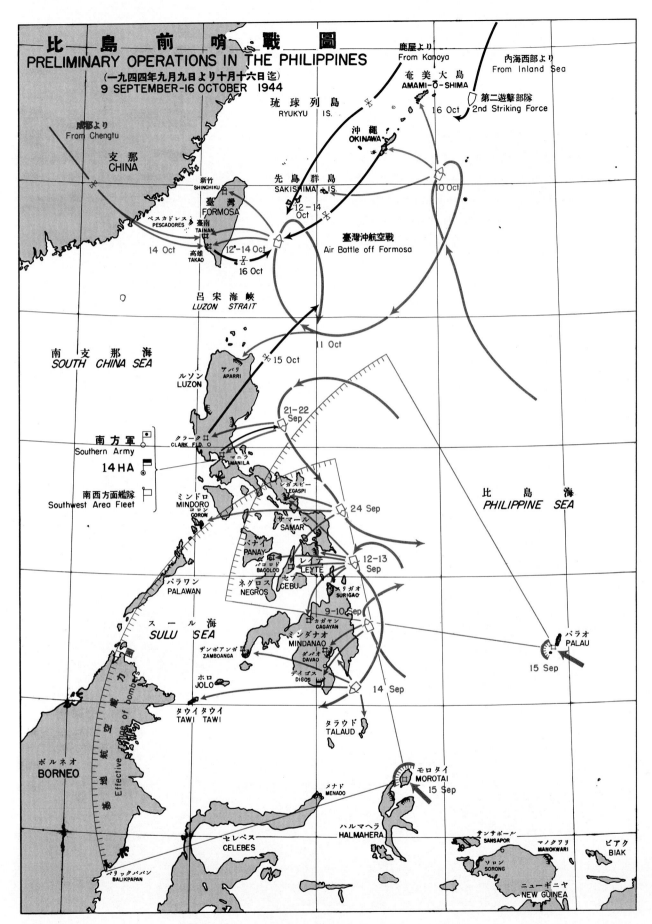

PLATE NO. 83
Preliminary Operations in the Philippines
9 September—16 October 1944

Although these recommendations were put into final form prior to 15 September, Marshal Terauchi desired to back them up with a simultaneous and full report on the damage done by the Allied carrier air strikes of 9 and 12 September. He therefore delayed forwarding them pending receipt at Manila of reports from all sectors which had been attacked. Col. Yozo Miyama, senior operations officer of Southern Army, was ordered to proceed to Tokyo by air to place the recommendations and report before Imperial General Headquarters, finally leaving Manila on 18 September.[14]

Meanwhile, it had already become apparent that the enemy's carrier strikes against Mindanao and the Visayas were not the prelude to a direct invasion of the Philippines themselves, but a cover for the launching of preliminary amphibious assaults on two vital defensive outposts—the Palau Islands in the western Carolines and Morotai in the northern Moluccas.

## Invasion of Palau

Indications that the enemy contemplated an imminent invasion of the Palau group, strategic eastern gateway to the Philippines, had been mounting for some time. Following a three-day carrier air strike against the islands late in July, enemy planes had continued small-scale attacks and reconnaissance activity throughout August. With the beginning of September, powerful carrier-borne forces launched a new offensive of full pre-invasion intensity, carrying out daily attacks which continued almost without interruption through 14 September. By the latter date, these attacks had done severe damage to antiaircraft installations, gun emplacements, beach defenses, and vital supply dumps.[15]

While the enemy's carrier aircraft pounded targets throughout the Palau group, strong surface elements also subjected the southernmost islands of Peleliu and Angaur to a series of heavy naval gunfire bombardments directed against shore defense positions. These bombardments reached greatest intensity on 12 September, when the island of Peleliu received a concentration of 2,200 rounds of gunfire, knocking out important defense installations and communications facilities.[16]

At 0730 on 15 September, following a final sharp naval gunfire and air preparation, the enemy began landing on Peleliu with an estimated strength of one infantry division and more than 150 tanks. The landing caught the bulk of the Japanese defense forces concentrated on Babelthuap, the main island of the Palau group, with only minor 14th Division and other elements present on Peleliu to contest the invasion.[17] The heavily outnumbered garrison fought tenaciously, but the enemy suc-

---

14 Col. Miyama reached Tokyo late on 18 September and laid the Southern Army recommendations before the Army Section of Imperial General Headquarters the following day. (Statement by Col. Yozo Miyama, Chief, Operations Section, Southern Army.

15 Between 6 and 14 September inclusive, enemy aircraft flew a total of 1,647 sorties against Peleliu, Koror, and Babelthuap Islands. The attacks were heaviest on 7 September, when a total of 583 sorties was recorded. *Pereriu Angauru-to Sakusen no Kyokun* ペリリューアンガウル島作戦の教訓 (Lessons of the Peleliu and Angaur Operations) Imperial General Headquarters, Feb 44, pp. 4–6.

16 Naval surface bombardments were carried out on 7, 12, 13 and 14 September. Ibid., p. 5.

17 The main strength of the 14th Division, 53d Independent Mixed Brigade, and 30th Special Naval Base Force was disposed on Babelthuap and the adjacent island of Koror. Units garrisoning Peleliu were: 2d Infantry Regiment, 14th Division; one battalion, 15th Infantry Regiment, 14th Division; one infantry battalion, 53d Independent Mixed Brigade; 14th Division Tank Unit; 33d, 35th, and 38th Provisional Machine Cannon Units; elements, 14th Division Special Troops. Naval Units were: Headquarters, West Carolines Airfield Unit; elements, 45th Naval Garrison Unit. (1) Ibid. Attached Charts II, III, and V. (2) *Chubu Taiheiyo Homen Sakusen Kiroku* 中部太平洋方面作戦記録 (Central Pacific Operations Record) 1st Demobilization Bureau, Vol. II, pp. 2–3, 98–9.

ceeded in expanding the initial beachhead so rapidly that, by 19 September, the fighting had moved into the central highlands.

Troop reinforcements were subsequently ferried in to bolster the defense, and naval seaplanes operating from secret bases on Babelthuap carried out night attacks on the American forces. Nevertheless, Peleliu airfield remained securely in the enemy's hands. American fighter aircraft began using the airdrome operationally from 27 September.[18]

Meanwhile, the enemy had already moved to expand his foothold in the Palau group by invading the small island of Angaur, southwest of Peleliu. At 0900 on 17 September, a strong force, supported by the usual air and naval gunfire preparation, began landing in the face of scattered resistance by the small Japanese garrison of one infantry battalion, an artillery battery, and a handful of miscellaneous troops.[19] The island was quickly overrun, and the fate of the defenders was never known.

Babelthuap still remained in Japanese possession, but the enemy had apparently achieved his objectives with the capture of Peleliu and Angaur and made no attempt to invade the main island. From rapidly developed bases on Peleliu and Angaur, enemy air power not only could keep the forces on Babelthuap helplessly pinned down in their hill positions,[20] but could effectively deprive the entire western Carolines of any further value to the Japanese as a defensive outpost guarding the eastern sea approaches to the Philippines.[21]

## Defense of Morotai

Concurrently with the enemy advance to Palau on the Central Pacific front, General MacArthur's forces in Western New Guinea had also taken an essential preliminary step toward the final reinvasion of the Philippines by landing on the strategically situated island of Morotai, off the northeast coast of Halmahera.

Ever since the seizure of Sansapor by MacArthur's forces in July, the Second Area Army command at Menado had anticipated an early enemy invasion of the Moluccas, estimating that the main island of Halmahera would be the most probable target of attack. Throughout August and the first part of September, Allied air raids on Halmahera steadily increased in both weight and frequency. When a Japanese reconnaissance aircraft, on 11 September, reported a heavy concentration of enemy invasion shipping in Humboldt Bay, Hollandia,[22] it appeared likely that the anticipated drive was about to get under way.

---

18   General narrative on the Peleliu fighting is based on Lessons of the Peleliu and Angaur Operations, op. cit.

19   Japanese units on Angaur at the time of the enemy landing were: 1st Bn. 59th Infantry Regiment, 14th Division; one battery, 59th Infantry Regimental Artillery Battalion; elements, 14th Division Special Troops. (1) Ibid., Attached Charts II and IV. (2) Central Pacific Operations Record, op. cit. Vol. II, pp. 101-3.

20   The troops on Babelthuap were powerless even to prevent Allied use of Kossol anchorage, situated just north of the main island. The Japanese expected that this anchorage would be made the main advance base of the enemy fleet for subsequent operations against the Philippines. However, a submarine reconnaissance on 7 October revealed that Ulithi Atoll, in the northwestern Carolines, had been occupied by the enemy and was being used instead of Kossol as the main advance fleet base. The Japanese had no forces on Ulithi and were unaware until this discovery that the enemy had captured the atoll. (Statement by Capt. Toshikazu Ohmae, Staff Officer (Operations), First Mobile Fleet.)

21   Until the enemy invasion, seaplanes of the 30th Base Force and small elements of the First Air Fleet used Palau as a reconnaissance base. Stoppage of this activity meant that the Japanese now became virtually blind to enemy fleet movements in the western Carolines and Philippine Sea areas.

22   This concentration was reported to include two aircraft carriers, three battleships, eight cruisers or destroyers, and 110 transports. *Gohoku Sakusen Kiroku Furoku Dai Ni: Dai Sanjuni Shidan Morotai To Sento Gaishi* 濠北作戰記錄附錄第二第三十二師團モロタイ島戰鬪概史 (North of Australia Operations Record, Supplement 2: 32d Division Operations on Morotai) 1st Demobilization Bureau, Jul 46, p. 4.

Inadequate troop as well as air strength had seriously impeded Second Area Army efforts to bolster the defenses of the Halmahera-Morotai area. The 32d Division under Lt. Gen. Yoshio Ishii, which was the principal combat force charged with the defense of the area, was understrength due to heavy losses suffered en route from China in May.[23] Lt. Gen. Ishii initially assigned two battalions of the 211th Infantry Regiment to garrison Morotai, but in mid-July, as General MacArthur's offensive neared the western tip of New Guinea, this force was withdrawn to bolster the thinly-spread Japanese troops on Halmahera itself.[24]

Upon the withdrawal of the 211th Infantry elements, Lt. Gen. Ishii assigned the mission of securing Morotai to a small, provisionally-organized force designated as the 2d Provisional Raiding Unit.[25] The advance echelon of this force arrived on Morotai on 12 July, but its meager strength led the 32d Division, on 30 July, to order the construction of dummy positions and encampments, the lighting of campfires throughout the jungle, and other measures of deception to lead the enemy to believe that the island was strongly held.[26]

By 19 August the remaining strength of the 2d Provisional Raiding Unit had arrived, followed on 13 September by elements of the 36th Division Sea Transport Unit. Troop strength still remained dangerously low, however, and had to be so thinly disposed that it was completely impossible to plan an effective defense.[27] Maj. Takenobu Kawashima, 2d

---

23 These losses, suffered as a result of submarine attacks on the *Take* convoy, reduced the 32d Division to only five infantry battalions and one and a half artillery battalions. Cf. Chapter X, p. 252.

24 It appeared probable at this time that Second Area Army would receive neither air nor ground reinforcements for the defense of the vital northern Moluccas. The Area Army expected that the enemy's attack would be directed at Halmahera and therefore considered it necessary to concentrate the bulk of its meager troop strength on that island. (Interrogation of Lt. Gen. Takazo Numata, Chief of Staff, Second Area Army.)

25 The raiding unit (*yugekitai* 遊撃隊) was a relatively new permanent-type organization established by Imperial General Headquarters as a result of the successes achieved in eastern New Guinea by provisionally-organized volunteer groups using infiltration and guerrilla tactics. Cadres for the new permanent units were trained at a special school in Tokyo under the direction of the Director of Military Intelligence, Army Section, Imperial General Headquarters. The training course covered infiltration tactics, demolition, and use of special weapons and equipment. Due to the late date at which the organization and training of such units began, their deployment to active fronts was delayed. On 15 January 1944 the 1st Raiding Unit, planned to consist of ten raiding companies, was added to the order of battle of Second Area Army, and in April and May two companies and the headquarters were ordered to Western New Guinea. Four other companies were still in process of organization in Japan, and four were to be activated by Second Area Army in the field. Although the headquarters reached Western New Guinea and was attached to Second Army, shipping difficulties held up the movement of subordinate units to such an extent that none had yet arrived by the time the enemy capture of Sansapor virtually terminated the New Guinea campaign. One company which had reached Luzon was subsequently assigned to Fourth Air Army for conversion to an airborne raiding unit, and elements on Halmahera were reorganized by Second Area Army in July as the 2d Provisional Raiding Unit, attached to the 32d Division. (1) Statements by Maj. Takenobu Kawashima, Commander, 2d Provisional Raiding Unit, and Lt. Col. Kotaro Katogawa, Staff Officer (Operations), Second Area Army. (2) Second Army Operations Order No. A-142, 9 Jul 44. ATIS Bulletin No. 1457, 20 Sep 44.

26 32d Division Operations Order No. A-491, 20 Jul 44. ATIS Bulletin No. 1570, 9 Nov 44, p. 1.

27 Japanese units present on Morotai at the time of the enemy landing were:
    2d Provisional Raiding Unit Headquarters
        4 provisional raiding companies
        1 plat., 11th Co., 211th Infantry Regt.
        1 plat., 1st Co., 32d Engineer Regt.
        Elms, 8th Field MP Unit
        Elms, 36th Div. Sea Transport Unit
        Elms, 26th Special Naval Base Force

(1) North of Australia Operations Record, Suppl. 2, op. cit., p. 2. (2) Statement by Maj. Kawashima, previously cited. (3) Miscellaneous field orders, official files, letters, notebooks and diaries published in the following ATIS Bulletins: No. 1542, 29 Oct 44, p. 1; No. 1583, 14 Nov 44, p. 7; No. 1632, 22 Dec 44, p. 5; ADVATIS Bulletin No. 161, 18 Jan 45, p. 2.

Provisional Raiding Unit commander, deployed his small combat force chiefly in the southwest sector of the island, while the remaining miscellaneous elements were scattered in lookout posts and security detachments around the island perimeter.

This was the situation when, at 0600 on 15 September, an enemy amphibious task force of about 80 ships appeared off Cape Gila and began shelling the entire southwest corner of Morotai. Following this gunfire preparation, reinforced by attacks from the air, the enemy put ashore a force estimated at one division. The 2d Provisional Raiding Unit, unable to offer effective resistance to the overwhelming enemy force, retired in good order, and by early morning of the 16th, the beachhead had been expanded to the Tjao River.[28] (Plate No. 84)

While Maj. Kawashima endeavored to assemble sufficient strength for a small-scale counterattack, 7th Air Division planes, operating from bases on Ceram and the Celebes, launched a series of nightly hit-and-run raids with small numbers of aircraft, aiming principally at enemy shipping.[29] These attacks had little more than a harassing effect, and the enemy, having reached the Tjao River, paused to consolidate his gains, at the same time hastening construction on the airfield at Doroeba.

On 18 September the main body of the 2d Provisional Raiding Unit, which had moved into position along the upper Tjao, launched a strong night infiltration attack with the objective of disrupting the enemy's rear area in the vicinity of Doroeba and Gotalalmo. Although deep penetration of the enemy lines was achieved and considerable casualties inflicted, the attack failed to reduce the beachhead or to interfere with the enemy's rapid preparation of Doroeba airfield. On 20 September enemy fighters began using the strip.

The 32d Division command on Halmahera had realized from the very beginning that successful development of an enemy base of air operations anywhere in the Moluccas would seriously compromise the future defense of the Philippines. Lt. Gen. Ishii therefore took immediate steps to reinforce Maj. Kawashima's forces, ordering the 211th and 212th Infantry Regiments and the 10th Expeditionary Unit to organize temporary raiding detachments for immediate dispatch to Morotai. The 210th Infantry Regiment was also ordered to prepare one battalion as a follow-up force. On 25 September the three raiding detachments were ordered to proceed to Morotai as follows:

1. *The 1st Raiding Detachment (from the 212th Infantry) will embark at Bololo on the night of 26 September and will land at Cape Posiposi at dawn of the 27th.*

2. *The 2d Raiding Detachment (from the 211th Infantry) will embark at Cape Djere on the night of 26 September and will land in the area between Wadjaboela and Tilai on the morning of the 27th.*

3. *The 3d Raiding Detachment (from the 10th Expeditionary Unit) will embark on the night of 27 September at Nupu. The landing point will be near Tilai.*

Although the detachments successfully car-

---

28  General narrative of the Morotai campaign is based on the following sources: (1) *Butai Ryakureki Dai Ni Yugekitai* 部隊略暦第二遊撃隊 (Unit History, 2d Provisional Raiding Unit). (2) North of Australia Operations Record, op. cit, Suppl. 2, pp. 4–6. (3) Statement by Maj. Kawashima, previously cited.

29  During the period 15–19 September, the 7th Air Division flew a total of 33 sorties against the enemy at Morotai. Reported results were: 30–40 enemy landing craft sunk or damaged; one cruiser and one large transport heavily damaged. Five aircraft failed to return. The 7th Air Division continued similar small-scale night raids during the next three months, at least two of these attacks (on 22 and 30 November) doing considerable damage to enemy aircraft and installations on Doroeba airdrome. North of Australia Operations Record, op. cit., pp. 24–6.

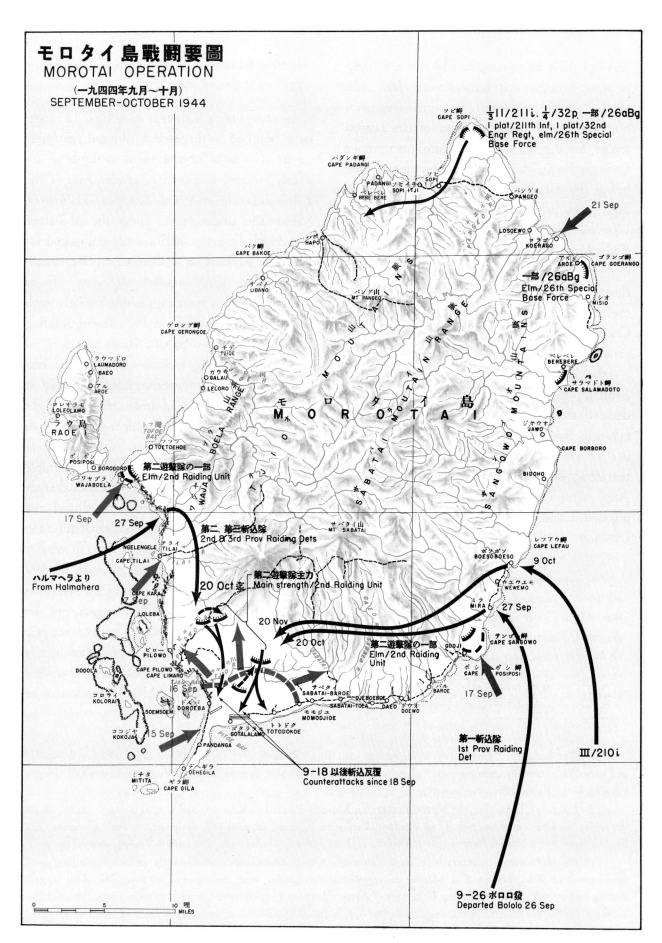

PLATE NO. 84
Morotai Operation, September–October 1944

ters Army order to the same effect.³⁸ This order read:³⁹

> *1. Imperial General Headquarters tentatively designates the Philippine Islands as the area of decisive battle and estimates that the time of this battle will be sometime during or after the last ten days of October.*
>
> *2. The Commanders-in-Chief of the Southern Army and the China Expeditionary Army and the Commander of the Formosa Army will generally complete operational preparations by the last part of October for the accomplishment of their respective missions.*

Further implementing this decision, the Army Section of Imperial General Headquarters ordered the 1st Division, hitherto scheduled under the *Sho-Go* plans to be held at Shanghai as strategic reserve until the activation of actual decisive battle operations in one of the *Sho* areas, to move immediately to the Philippines. Plans were also made to assign ten surface Raiding Regiments to the Philippine area.⁴⁰

With respect to Southern Army's request for authorization to employ the main strength of the Fourth Air Army against raiding enemy carrier forces, Imperial General Headquarters demurred on the ground that such action would probably entail losses of aircraft and pilots incommensurate with the amount of damage which could be inflicted on the enemy. Authorization was granted, however, to carry out hit-and-run attacks with small elements whenever the situation appeared especially favorable for such operations.⁴¹

On 22 September Imperial General Headquarters also acted to implement plans for the reinforcement of the Fourth Air Army. The 16th Air Brigade (51st and 52d Fighter Regiments) was ordered to proceed to the Philippines at once, and the 12th Air Brigade (1st, 11th and 22d Fighter Regiments) was directed to prepare for subsequent movement upon the activation of *Sho* No. 1. In addition, three more fighter regiments, one light bomber regiment, three heavy bomber regiments and one reconnaissance regiment were allocated to Fourth Air Army, to advance to the Philippines upon the activation of *Sho* No. 1.⁴² On 11 October, a further order activated the 30th

---

38  Imperial General Headquarters Army and Navy Section orders required Imperial signature, whereas directives were issued in the name of the Army and Navy Chiefs of General Staff. Since the Army Section on this occasion issued an order, its issuance was delayed one day by the necessity of obtaining the Imperial signature.

39  *Dairikumei Dai Senhyakusanjugo-go* 大陸命第千百三十五號 (Imperial General Headquarters Army Order No. 1135) 22 Sep 44.

40  *Dairikumei Dai Senhyakusanjuroku-go oyobi Daisenhyakusanjuhachi-go* 大陸命第千百三十六號及第千百三十八號 (Imperial General Headquarters Army Orders No. 1136 and 1138) 22 Sep 44.

41  Under the *Sho-Go* plans, enemy task forces conducting raids prior to invasion were to be attacked only by designated air units. These units, with the exception of some army air units equipped with Type IV bombers and undergoing special training in Japan Proper, were entirely navy. The Fourth Air Army at this time had no units equipped with Type IV bombers or trained in attacking carriers. Under the more flexible policy established by Imperial General Headquarters, Fourth Air Army units executed a number of hit-and-run raids on enemy carrier groups during the latter part of September, but no appreciable results were obtained.

42  This brought the number of stand-by reinforcement regiments to be sent to the Philippines to 11, three more than provided for in the original *Sho-Go* plans. In addition, the 67th Fighter-Bomber Regiment was ordered to the Philippines on 22 September for anti-submarine patrol work. (1) *Dairikumei Dai Senhyakuyonju-go* 大陸命第千百四十號 (1) Imperial General Headquarters Army Order No. 1140) 22 Sep 44, (2) *Dairikushi Dai Nisenhyakushichijugo, Dai Nisenhyakushichijuroku, Dai Nisenhyakushijushichi-go,* 陸指第二千百七十五、第二千百七十六、第二千百七十七號 (Imperial General Headquarters Army Directives No. 2175, 2176, 2177) 22 Sep 44.

PLATE NO. 85
Air Force Day: Propaganda Poster

not accompanied by an invasion convoy of transports,[50] the situation did not for the moment appear to be one which called for the immediate and full activation by Imperial General Headquarters of the *Sho-Go* plans. However, the Navy considered that decisive action by its own air forces was imperative.

The opening of the attack on the Ryukyus found Admiral Soemu Toyoda, Commander-in-Chief of the Combined Fleet, at Shinchiku, northern Formosa, on his way back to Tokyo from a command inspection trip to the Philippines. This had a vital effect on subsequent events, for it meant not only that Admiral Toyoda's decisions were psychologically influenced by his presence virtually on the front line of battle, but that full, direct consultation with the Navy High Command was rendered impossible. During the ensuing action, Admiral Toyoda, while delegating the power to make minor decisions to his Chief of Staff in Tokyo, actually directed operations from Formosa, issuing some orders direct and others through Combined Fleet headquarters in Tokyo.

To Admiral Toyoda, it seemed that the enemy, by sending his carrier forces into the northern Philippine Sea within striking range of the major concentrations of Japanese land-based air strength, had presented an opportunity that might never arise again, to deal the enemy fleet a crippling blow and disrupt the entire Allied invasion timetable. He therefore decided to gamble all available naval air strength in a determined effort to destroy the enemy carrier forces.

This meant a sharp divergence from the tactical concepts which formed the basis of the original *Sho-Go* plans. The central idea of those plans was to husband air, sea and ground strength until a major enemy invasion attempt against any of the areas constituting Japan's inner defense line, and then to commit all forces in decisive battle. Accordingly, while the use of minor elements of naval air strength against raiding enemy task forces prior to an invasion was authorized, commitment of the main strength of both Army and Navy air forces was to await Imperial General Headquarters decision activating one of the *Sho* operations.[51]

Actually, experience in the earlier Philippine strikes had shown that passive tactics against enemy task force raids were of doubtful effectiveness in conserving air strength. Moreover, discussions between the Navy's top operational commanders and the Naval General Staff had emphasized the impossibility under all circumstances of rigidly adhering to the *Sho-Go* plans with regard to air action against enemy task forces, and had resulted in agreement that a large measure of discretion must be left to the Combined Fleet command[52] to determine the opportune moment for committing the naval air strength. Now, that moment appeared to be at hand.

Admiral Toyoda promptly decided to remain

---

50  Naval search planes established at 1540 on 10 October that two task groups were operating to the east-southeast of Okinawa, at distances of about 100 and 140 miles, respectively, from Naha. One group was reported to have a nucleus of three carriers, and the other to consist of two carriers and about ten cruisers and destroyers. Philippine Naval Operations, Part II, op. cit., p. 5.

51  Cf. Chapter XI, p. 296.

52  The *Sho-Go* plan provisions regarding air operations were discussed at a Combined Fleet operational conference at Kure on 22 August and again at a conference in Tokyo on 8 September, attended by Admiral Toyoda, Second Air Fleet Commander Vice Adm. Fukudome, and Vice Chief of Navy General Staff Vice Adm. Seiichi Ito. These conferences resulted in a decision that the time for initiating general attack by the naval air forces against enemy task forces must be left flexible, to be determined by the Combined Fleet on the basis of circumstances as they arose. (Diary Notes of Capt. Bunzo Shibata, Staff Officer (Operation), Second Air Fleet.

PLATE NO. 86
Transoceanic Air Raid During Typhoon

on Formosa and assume personal direction of battle operations. At 0925 on 10 October, approximately three hours after the start of the enemy air assault on the Ryukyus, Combined Fleet headquarters in Tokyo, acting at Admiral Toyoda's direction, alerted all naval land-based air forces for *Sho* Operation No. 2. At 1214 the same day Admiral Toyoda, by order from Shinchiku, extended the alert to include *Sho* No. 1 as well.[53]

On 11 October the enemy carrier groups turned south to effect small-scale reconnaissance raids over the Aparri area, on northern Luzon. The following day, however, the air offensive was resumed in full force, this time against Formosa and adjacent islands. Admiral Toyoda now decided that it was time to strike. Again acting through Combined Fleet headquarters in Tokyo, he ordered the naval base air forces, at 1030 on 12 October, to execute *Sho* Operations Nos. 1 and 2, with the objective of destroying the enemy carrier forces in the northern Philippine Sea.[54]

Vice Adm. Shigeru Fukudome's Second Air Fleet, main strength of which was still deployed at bases in southern Kyushu,[55] immediately prepared to attack. Meanwhile, to throw as many aircraft as possible into the battle, Combined Fleet headquarters in Tokyo ordered Vice Adm. Ozawa, First Mobile Fleet Commander, to release the newly-reconstituted flying groups of the 3d and 4th Carrier Divisions, which had not yet completed their combat training in the Inland Sea, to temporary command of the Second Air Fleet. These groups were immediately ordered to bases in southern Kyushu and the Nansei Islands to operate with the land-based air forces.[56]

The first attack on the enemy carrier groups, three of which were now reported operating off the east coast of Formosa, was carried out between 1900 and 2020 on 12 October. Taking off from Kanoya air base, 56 planes of the "T" Attack Force struck at the enemy within the perimeter of a sudden typhoon and then put down on Formosan bases. The pilots reported four enemy carriers sunk, and ten other major units set afire. Meanwhile, a separate force of 45 torpedo planes and Army Type-IV torpedo-bombers[57] sortied from bases on Okinawa and carried out an attack, in which two unidentified fleet units were reported set aflame.[58]

Despite these reported successes, the enemy carrier forces renewed their assault on Formosa on 13 October, sending over a total of about 600 aircraft during the day. Damage in these raids was light, and the "T" Attack Force sortied from Kanoya late in the afternoon to strike back. Locating two enemy carrier groups southwest of Ishigaki Island, in the southern Ryukyus, the attack formation of 32 planes struck at dusk, reporting four ships sunk, of which two were carriers, and a third carrier left in flames.[59]

---

53 Philippine Naval Operations, Part II, op. cit., p. 5.
54 Ibid., pp. 5–6.
55 Original plans called for the transfer of the Second Air Fleet main strength to Formosan bases in September. Because of incomplete training, however, only a portion of this strength had moved to Formosa by 12 October.
56 These air groups were then located at Oita and Kagoshima in Kyushu, at Kure and Iwakuni in western Honshu, and at Tokushima on Shikoku. The air groups of the 4th Carrier Division completed concentration at southern Kyushu bases by the evening of 13 October, while those of the 3d Carrier Division sent off their first echelon for Okinawa early on the 15th. (1) Philippine Naval Operations, Part II, Op. Cit., p. 6. (2) Statement by Capt. Ohmae, previously cited.
57 These Army air units were assigned to the Second Air Fleet. Cf. Chapter XI, p. 308.
58 Philippine Naval Operations, Part II, op. cit., pp. 6, 9.
59 Ibid., pp. 6–7, 9.

On the basis of the reported results of the attacks thus far, it appeared that at least one segment of the enemy task forces had been decisively crushed. This estimate was seemingly confirmed by the fact that carrier-plane raids on Formosa were resumed on a sharply reduced scale at 0700 on 14 October and ceased completely at 0930. It appeared that the enemy forces had initiated a retirement to the southeast.

Complete victory now appeared almost within grasp. To accomplish the total destruction of the damaged and withdrawing enemy, the Second Air Fleet ordered its entire strength of 450 planes to sortie from southern Kyushu.

Admiral Toyoda ordered the Second Striking Force under Vice Adm. Kiyohide Shima to sail from the Inland Sea and sweep the waters east of Formosa to mop up remnants of the reportedly crippled enemy task forces.[60]

On the afternoon of 14 October, 100 B-29 bombers evidently operating from China struck at Formosa in what was believed to be a covering operation for the retirement of the enemy fleet. Meanwhile, at 1525, the first wave of Second Air Fleet planes (124 aircraft) attacked an enemy group southwest of Ishigaki Island, claiming hits on one carrier and three cruisers. A second attack wave of 225 planes sortied but was unsuccessful in finding the enemy. The third, striking after nightfall with 70 aircraft, including Army torpedo bombers, claimed two carriers, one battleship and one heavy cruiser sunk, and one small carrier, one battleship and one light cruiser set afire.[61]

Events on 15 October caused optimism to remain at a high pitch. Second Air Fleet search planes reported that one aircraft carrier and two battleships, all trailing oil slicks and without steerage way, were spotted off the coast of Formosa under guard of 11 destroyers.[62] Admiral Toyoda ordered naval air units to continue the attack despite heavy plane losses. Meanwhile, the Second Striking Force was already racing south from the Inland Sea at high speed to assist the air forces in cleaning up the enemy remnants.

Farther south, an enemy task group, with four carriers still intact, appeared off the east coast of Luzon and at 1000 on 15 October sent off a force of 80 planes to attack Manila. In interception operations, Japanese fighters claimed 32 enemy aircraft shot down or damaged, while two separate attacks on the enemy task group by a total of 115 Army and Navy planes from Philippine bases were reported to have sunk one of the carriers and set afire the flight decks of two others.[63]

On 16 October regular morning search missions over the western Philippine Sea brought in disquieting reports that did not seem to tally with earlier claims of damage to the enemy forces. Three separate task groups with a total of 13 carriers were reported navigating in the area.[64] Forces aggregating 247 naval aircraft immediately sortied from Okinawa, Formosa and Luzon to search for the enemy groups. These units swept wide areas of the Philippine Sea but only a small number of the planes found the carriers.

---

60  This was a special operation conceived by Admiral Toyoda outside the framework of the *Sho-Go* plans covering surface forces. As constituted for this special mission, the Second Striking Force consisted of two heavy cruisers, one light cruiser, and seven destroyers. After executing its mission, the force was to return to the Inland Sea and hold itself in readiness to execute the planned *Sho-Go* Operation for the surface forces. (Statement by Capt. Ohmae, previously cited.)

61  Philippine Naval Operations, Part II, op. cit., pp. 7, 9.

62  *Gunreibu Socho no Sojosho* 軍令部總長の奏上書 (Report to the Throne by the Chief of Navy General Staff) 16 Oct 44.

63  Philippine Naval Operations, Part II, op. cit., p. 8.

64  Ibid., pp. 9–10.

Despite the conflicting reports, Admiral Toyoda and the Navy High Command were still inclined to believe that the enemy was attempting to cover the retirement of badly damaged and disorganized carrier task forces. If it were true, however, that enemy strength in the area was still as large as indicated by the reconnaissance reports of 16 October, the Second Striking Force, then passing east of the Ryukyus, was sailing directly into an engagement in which it would be heavily outweighed. The Chief of Staff of Combined Fleet therefore radioed a suggestion to Vice Adm. Shima that he change course to the west, pass through the northern Nansei Islands, and run south through the East China Sea in order to stay out of range of Allied carrier planes. This was followed by an order from Admiral Toyoda directing Vice Adm Shima to prepare to sortie again into the Pacific and fight a night engagement if an enemy force of appropriate size presented itself. If no such opportunity arose, the Second Striking Force was to proceed to the Pescadores and await further orders.[65]

Final reconnaissance reports on 17 October confirmed that considerable enemy strength remained present in the waters east of Formosa and Luzon, but also indicated that substantial damage had been inflicted. Of four separate task groups spotted, one of about 20 ships, including three carriers and three battleships, was reported withdrawing eastward at a reduced speed of ten knots with one of the battleships under tow. This strongly suggested that the group was composed of damaged ships retiring from action. Orders to attack were immediately issued, but contact was subsequently lost and the attacks could not be carried out.[66]

The Formosa air battle had now ended, and the Navy High Command undertook to assess the damage done to the enemy's carrier fleet. The necessity of avoiding any exaggeration of enemy losses was clearly recognized because of the importance to future operational planning. Combined Fleet staff officers thoroughly studied and sifted the action reports of the combat flying units. Although these reports were considered of dubious reliability, Second Air Fleet strongly insisted upon their accuracy, and in the absence of adequate post-attack reconnaissance, the Navy Section of Imperial General Headquarters had no choice but to base its assessment on the reports at hand. Enemy losses were finally listed as follows:[67]

Sunk:     *11 carriers, 2 battleships, 3 cruisers, 1 destroyer (or light cruiser).*

Damaged:   *8 carriers, 2 battleships, 4 cruisers, 1 destroyer (or light cruiser), 13 unidentified ships. In addition, at least 12 other ships set afire.*

These results, officially accepted and announced, added up to the most phenomenal success achieved by the Japanese Navy since the attack on Pearl Harbor. The nation was swept by a sudden wave of exhilaration which dispelled overnight the growing pessimism over the unfavorable trend of the war. Mass celebrations were held in many cities throughout the country, and government spokesmen proclaimed that "victory is within our grasp!" All Army and Navy units concerned were honored by the issuance of an Imperial Rescript.

---

65 On the afternoon of 16 October, Combined Fleet also alerted Vice Adm. Kurita's First Striking Force to be ready to sortie from Lingga anchorage. His plan was to throw this force against the remnants of the enemy carrier groups after further damage had been inflicted in continued attack operations by the naval air forces. Philippine Naval Operations, Part II, op. cit., pp. 38–9.

66 Ibid., p. 10.

67 Imperial General Headquarters Communique, 19 Oct 44. *Asahi Shimbun* 朝日新聞 (Tokyo Asahi Newspaper) Tokyo, 20 Oct 44.

However, while the nation thrilled to a victory which events soon proved to be a bitter illusion, the situation brought about by the Formosa air battle was actually fraught with potential disaster. The battle had cost the air forces 312 planes of all types, a level of losses which they could ill afford to sustain. The Second Air Fleet, comprising the main strength of the Navy's base air forces, had lost 50 per cent of its strength and was reduced to 230 operational aircraft.[68] The First Air Fleet and Fourth Air Army in the Philippines were left with a combined operational strength of only a little over 100 aircraft.[69] Of 143 carrier planes used to reinforce the Second Air Fleet, about one-third, with their flight crews, had been lost.[70]

The losses in carrier aircraft and flying personnel meant further delay in remanning the 3d and 4th Carrier Divisions, which Admiral Toyoda had hoped to send south to join the First Striking Force, thus providing it with desperately needed air striking elements.[71] Moreover, the Second Striking Force, scheduled under the *Sho-Go* plans to operate as a vanguard to Vice Adm. Ozawa's Task Force Main Body, was now far from its base and had consumed tons of precious fuel in a fruitless operation. It was hoped that the damage inflicted on the enemy's carrier forces would slow up his invasion schedule long enough to permit the replenishment of aircraft losses and the redeployment of the surface forces. However, this hope was to prove vain.

The credence temporarily placed in the Navy's claims regarding the Formosa air battle also paved the way for a momentous change in plans regarding decisive ground operations in the Philippines. On the basis of the results officially claimed by the Navy for the Formosa Air Battle, it appeared likely that an enemy invasion of the Philippines would be delayed, or if undertaken soon, would be unsupported by strong carrier forces. Consequently, the Army Section of Imperial General Headquarters now became more favorably inclined toward modifying the *Sho-Go* plans along the lines of Southern Army thinking. However, there seemed to be ample time to study the matter in detail before reaching a final decision.[72]

That decision was still pending on 17 October, when the American invasion of the Philippines began in earnest.

---

68 Philippine Naval Operations, Part II, op. cit., p. 10.
69 Ibid.
70 Statement by Capt. Ohmae, previously cited.
71 It had been tentatively decided at the end of September that Vice Adm. Ozawa, First Mobile Fleet commander, would go south with the carrier forces as soon as the refitting of the ships and the training of the air groups were completed. These forces were to join the First Striking Force, which henceforth would operate under Vice Adm. Ozawa's direct command. This was a long-range plan, and it was not believed that it could be carried into effect until November. For that reason, no orders were issued, and the task organization of the fleet under the *Sho-Go* plans remained unchanged. (1) Statement by Capt. Ohmae, previously cited. (2) United States Strategic Bombing Survey (Pacific), Naval Analysis Division, *Interrogations of Japanese Officials*, Vol. I, pp. 219–20. (Interrogation of Vice Adm. Jisaburo Ozawa.)
72 Statement by Col. Takushiro Hattori, previously cited.

# REPORTS OF GENERAL MacARTHUR

VOL I:  The Campaigns of MacArthur in the Pacific
VOL I:  Supplement: MacArthur in Japan: The Occupation, Military Phase
VOL II: Japanese Operations in the Southwest Pacific Area

". . . This report has been prepared by the General Staff to serve as a background for, and introduction to the detailed operational histories of the various tactical commands involved.

The pressure of other duties having prevented my personal participation in its preparation, it has been entrusted by me to that magnificent staff group which actually conducted the staff work during the progress of the campaigns. They speak with that sincere and accurate knowledge which is possessed only by those who have personally participated in the operations which they record . . ."

Preface by General Douglas MacArthur.

### Senior Commanders: Southwest Pacific Areas

Gen W. Krueger: Sixth Army; Lt Gen R. L. Eichelberger: Eighth Army; Gen Sir Thomas Blamey: Aust. Imp. Forces; Lt Gen G. C. Kenny: AAF; Adm T. H. Kinkaid, USN: Seventh Fleet

### The General Staff: GHQ: Southwest Pacific Area

Lt Gen. R. K. Sutherland, CofS; Maj Gen R. J. Marshall, D CofS; Maj Gen C. P. Stivers, G–1: Maj Gen C. A. Willoughby, G–2; Maj Gen S. J. Chamberlin, G–3; Maj Gen L. J. Whitlock, G–4; Maj Gen S. B. Akin, CSO; Maj Gen W. F. Marquat, AAO; Maj Gen H. J. Casey, CE; Brig Gen B. M. Fitch, AG; Brig Gen L. A. Diller, PRO.

### Editor in Chief

Maj Gen Charles A. Willoughby, G–2

### Senior Editors

Col E. H. F. Svensson, G–2; Gordon W. Prange PhD; Mr. Stewart Thorn

### Associate and Contributing Editors

Brig Gen H. E. Eastwood, G–4; Col F. H. Wilson, G–2; Col R. L. Ring, G–2; Col W. J. Niederpruem, G–3; Lt Col M. K. Schiffman, G–2; Maj J. M. Roberts, G–3; Capt J. C. Bateman, G–2; Capt Mary Guyette, G–2; Capt John L. Moore, G–2; Lt Stanley Falk, G–2; Mr. Jerome Forrest; Mr. Kenneth W. Myers; Miss Joan Corrigan.

### Translation–Interrogation–Production

Lt Col W. H. Brown, G–2; Louis W. Doll, PhD; Capt E. B. Ryckaert, G–2; Capt K. J. Knapp, Jr., G–2; Lt Y. G. Kanegai, G–2; Mr. James J. Wickel; Mr. John Shelton, ATIS; Mr. Norman Sparnon, ATIS; SFC H. Y. Uno, G–2; Mr. K. Takeuchi; Mr. S. Wada.